AF517898

Mushrooms: Edible and Medicinal

Cultivation, Conservation, Strain Improvement with their Marketing

The Authors

Dr. Bhola Nath Verma was born on 8th June 1943 and was superannuated from the active service as University Professor and Head, University Department of Botany, B.R.A. Bihar University, Muzaffarpur, Bihar (India). He is the pioneer Phycologist who initiated and propagated Phycological Research in Bihar. During thirty three years of active service, he successfully conducted a number of major research projects sponsored by UGC, CSIR, DST, Ministry of Environment and Forests under various ecological as well as Man and Biosphere Programme and Ministry of Non-conventional source of Energy, Govt of India, New Delhi. At least twenty research students were awarded Ph.D. degree under his guidance. He supervised CSIR sponsored post-doctoral research and made a basic contribution on the phenolics of fresh water algae in collaboration with Indian Toxicological Research Institute, Lucknow. He conducted a maiden exploration of edible mushrooms in Arunachal Pradesh. He has published more than hundred research papers including establishment of many new taxa of Green Algae in scholarly journals of National and International repute and authored three reference books. In addition to teaching, research and writings, Prof. Verma actively participated in academic deliberations and chaired a number of technical sessions of National as well as International levels Symposia/Conferences besides delivering invited lectures. He has been associated and Life Member of seven learned Societies, and as executive of several of them. He is Fellow of four Leaned Societies. He was conferred Scientists of the Year 2001 award by National Environmental Science Academy of India and was felicited by Tribhuban University, Kathmandu, Nepal at its Birganj Campus in 2002 on the recognition of his outstanding contributions in the field of Plant Science. He was honored by Indian Botanical Society in its 26th Annual Conference held in the year 2003 at Zamia Hamdard, New Delhi and by Phycological Society of India in a National Symposium held in the year 2004 at University of Madras, Chennai.

Dr. K.K. Sahu is Head and Reader in Botany, B.R.B.College, Samastipur under L.N. Mithila University, Darbhanga, Bihar (India) with 26 years of undergraduate teaching and research experience.

He availed UGC Teacher Fellowship for his scholarly research work under the supervision of Professor B.N. Verma. During this three and half years of research work on Cyanobacterial physiology and biochemistry, he turned into technocrat as he simultaneously developed many software for statistical analysis for his own research work under the very stress and undeveloped world of computer science in early 80's.

He has published many research papers in National Journal of repute and presented more than 20 papers in different National/International seminars and symposia. He has been Associate Member of Computer Society of India. Dr. Sahu has regularly been invited as a Resource Person in Orientation and Refresher course under the aegis of UGC's Academic Staff College. Currently, Dr. Sahu as a Principal Investigator, under UGC Research Project, is preparing data-set for Cyanobacterial taxonomical analysis using DELTA (DEscriptive Language for TAxonomy) software.

Dr. Prem Kumar Prasad is well known for his consistent high academic performance as a student, as a researcher and as a teacher. He secured first position in first class securing high percentage of marks at honours level and awarded CSIR fellowship for doctoral research. He was awarded Ph. D. degree in the year 1989 from B.R.A. Bihar University. He has contributed a maiden research on the relationship of Plant Parasitic Nematodes and Cyanobacteria *in vitro* and *in vivo*. He made commendable contribution on the soil algae of North Bihar and adjoining area of Nepal after establishing a self designed methodology to quantify the soil algae. He is leading a team of devoted research students in different fields of phycology and many of them have been awarded Ph. D. degree in his guidance. He has acquired 26 years of graduate and post graduate teaching experience besides proving his administrative potency as the Principal of a constituent unit of Lalit Narayan Mithila University, Darbhanga, Bihar. He is associated with about a dozen learned societies of India and abroad in different capacities. He has actively participated in a number of National and International conferences/symposia besides organizing successfully conferences of national level to his sole credit. He has established his worth in publication of more than 20 research articles in the journals of International repute, and two reference books on lower as well as higher plants.

Mushrooms: Edible and Medicinal

Cultivation, Conservation, Strain Improvement with their Marketing

B.N. Verma
Retd. Professor and Head
University Department of Botany,
B.R.A. Bihar University, Muzaffarpur

Prem Kumar Prasad
Principal
B.M. College, Rahika
Madhubani, Bihar

K.K. Sahu
Reader
Department of Botany
B.R.B. College
Samastipur, Bihar

2013

Daya Publishing House®
A Division of
Astral International Pvt. Ltd.
New Delhi – 110 002

ISBN 978-81-7035-802-2

Cover Page Photograph by
Taylor F. Lock Wood
300 Pinecrest
Dr. Mount Dora
S. L. 32757 (U.S.A.)

Published by : **Daya Publishing House®**
A Division of
Astral International Pvt. Ltd.
81, Darya Ganj, Near Hindi Park,
Delhi Medical Association Road,
New Delhi - 110 002
Phone: 011-4354 9197, 2327 8134
Fax: +91-11-2324 3060
E-mail: info@astralint.com
Website: www.astralint.com

Laser Typesetting : **Classic Computer Services**
Delhi - 110 035

Printed at : **Salasar Imaging Systems**
Delhi - 110 035

PRINTED IN INDIA

Foreword

The present scenario in India provides a crucial juncture, as far as mushroom fungi are concerned. On one hand, Indian mushroom resources are under severe threat due to increasing pressure on our forest wealth on account of infrastructural development, industrialization, urbanization and climate change; while on the other hand, increasing awareness about mushroom as a health-food as well as an eco-fiiendly cash-crop and enhancement of people's purchasing power have led to an appreciable increase in the demand and supply of cultivated mushrooms across the country. Wild mushrooms are the basic genetic resources at our command and their proper identification, assessment, conservation and sustained utilization can only provide the required support to a thriving mushroom-farming enterprise in our country. At the same time, the cultivation technology need to be regularly updated, modernized and made competitive and cost-effective for a variety of mushroom species as per market-demand and the same need to be disseminated fast to the growers located far and wide. All these are easily told than done.

Books are the most reliable messengers to quickly transfer the message of changes and advancements .Most of the books in the field of mushrooms or say the Fleshy fungi published in the past or even in the recent past have been lop-sided emphasizing only on one aspect, either Basics, confining to taxonomic description & systematic identification of mushroom fungi, or Applied side only, describing the methods of cultivation, processing, marketing etc. ignoring the fact that the two aspects are closely inter-linked and must be dealt together.

"Mushrooms: Edible and Medicinal", encompasses both the aspects referred to above and is an attempt in the right direction. It provides an updated status of mushroom fungi so far reported from different parts of India along with their diagnostic characteristics and photographs, in a consolidated format, which is likely to serve as

an updated Catalogue of fungal resources of our country. Some supplementary information given in the book, *viz.* on biology, taxonomy and method of collection and preservation of mushrooms will encourage the new entrants in the field.

On the applied side, the authors have briefly described the methods of cultivation of some mushrooms popular in India, their pests and diseases, post-harvest management, economics of mushroom-farming etc. for the benefit of those who may be interested in this enterprise.

A few other important aspects dealt in the book are Conservation of mushroom-biodiversity, Economic importance of mushrooms and Genetic improvement of mushrooms, which add to its usefulness.

I hope that the book will find good readership amongst the students, researchers and entrepreneurs in the fields of both basic and applied aspects of mushrooms and their cultivation and will serve its purpose well.

"Ashirvad", Rabindra-Nagar,
RANCHI-834008

R. N. Verma
Former Director and Project Coordinator (AICMIP)
National Research Centre for Mushroom,
Solan – 173 213, H.P.

Contents

From Author's Desk

Today, when manuscript of this book is ready in my hand, I am reminded of co-operations and co-ordinations received from various persons to whom I feel indebeted, and wish to share with my readers as a sense of complements and acknowledgements.

A name Dr. Rishikes Mishra, associated with me right from Intermediate stage, flashes first whose begotten indulgence made a phycologist like me to supervise his doctoral research on "Edible mushrooms of Arunachal Pradesh". It is this which eventually inspired me even after superannuation from active services to bring out a comprehensive account of Indian Edible Mushrooms which are scattered in the pages of periodicals accumulated after publication of the monumental compilation by Purkayastha and Chandra (1985). Unfortunately, Dr. Mishra did not join me in this venture due to his own limitations. However, two students from my Phycological Laboratory Dr. Prem Kumar Prasad and Dr. K. K. Sahu came forward to strengthen me and made this venture a reality.

It took more than seven years to shape this endeavour during which many ups and downs came and sometimes the whole efforts appeared collapsing but Dr. (Mrs.) Krishna Verma, my life partner catalysed all of us with befitting impetus, and also coordinated with Mr. Sumangal Pandey who bore all the inherent difficulties of typing a manuscript amidst his busy schedule not as a typist but as a sincere grandson to conclude this project. In spite of being a social scientist, Mrs. Verma inspired and helped me in many ways not less than a natural scientist in shaping the manuscript. The help of Adarsh Krishna, my son, in hauling the website to pick out coloured photographs of desired mushroom taxa amidst his tiring corporate engagements can not be underestimated. I have no word to express my feelings.

I would be remiss if I did not acknowledge my debt and heartfelt appreciation to Prof. N. S. Atri, Department of Botany, Punjabi University, Patiala, Punjab from whom

I got technical assistance time to time. I must acknowledge my gratitude to Dr. R. N. Verma, Ex Director and Dr. R. C. Upadhyay, National Research Centre for Mushroom, Solan, Himanchal Pradesh and Prof. A. S. Ahluwalia, Department of Botany, Punjab University, Chandigarh, Punjab, for their cooperations.

Any publication of Natural Science is incomplete unless it is not supplemented with live citations. This could have been possible with generous permission granted by Dr. Robert Thomas, Doothy, B. Orr, Albert P. Baker, Elsbernd Karren, Danielle Castronovo, Dr. Nick Kurrenko and Jo-Ann Ordens California Academy of Science. Ron Wolf, Jon W. Wall, Mark Radz, Ted Swiecki, C. A. USA, R. P. Olowin and Alfred Burousseau, St. Mary's College, California; Taylor F. Lockwood, 300 Pinecrest; Dr. Mount Dora, S.L. 32757 (USA); Lorny Evans; Amdej Troczy, Dr. Nick Kurzenk, Russian Academy of Sciences, Valdivostok and Michelle Cloud, Hughes, San Diego State University, U.S.A. There is no word to acknowledge their co-operations with a high sense of gratitude.

Space will not permit me to list all the assistance and help received at various levels from the authority of Library of different Universities and Institutes, so I simply say thank to all of them.

Mushroom is heterogenous assemblage of distantly related fungal taxa belonging to different classes of fungi. It displays self designed fruit bodies characteristic of its own. Mushroom taxa are thus described after clubing them either in respect of their systematic position or on the basis of their fruit type. In either case, it becomes very difficult for a common man to locate out the desired mushroom taxa within the text of any book. Thus, the mushroom taxa in the present book have been arranged in ascending series of generic alphabet ignoring technical base and approach.

Edible mushrooms are not only highly nutritive and tasty food but also good for therapeutic treatment increasing immunity and adding flavour. They may be either exclusively edible or medicinal or both. In the present text, all the taxa falling under these three categories have been described in the text and rest all have been ingnored. Distribution of the taxa has been described only with reference ot Indian habitats.

Edible mushrooms described in this text have been screened from the original literature of the mushroom taxa reported with their edibility as well as from the exhaustive global list of wild edible fungi compiled by E. R. Boa (2004). While describing the mushroom taxa, overlapping of one over the other was noticed mainly due to Synonyms. Since this overleaping was unilateral and morphological identity was distinct it has been ignored without entering into technicality.

We hope that this book will contribute to the recognition of edible mushrooms as an important nutritive and tasty food. We welcome the comments and criticism of readers of the book.

B.N. Verma

Chapter 1
Introduction

No leaves, no buds, no flowers yet fruits, this is the miracle played only by mushroom. This unique fruit is basically a gift of nature to the poor as evident from its appearance on thatched house and rotten woods just after first shower. But its taste and flavour soon introduced its delicious dish in the kitchen of aristocrats and elites of the society. Increasing knowledge opened more and more dimensions of its utility provoking extensive cultivation of mushroom world wide and its popularisation in every sphere of life as well as in every sects of the society. Wider knowledge not only widened the utility of mushrooms but also gave wider coverage to the term.

Mushroom is the fleshy, spore bearing body of a fungus, typically produced above ground on soil or on its food source. The standard for the name "Mushroom" is the cultivated white button mushroom, *Agaricus bisporus*, hence the word mushroom is most often applied to those fungi (Basidiomyçetes, Agaricales) that have a stem (stipe), a cap (pileus) and gills (lamellae *sing.* lamella) on the underside of the cap, just as do store–bought white mushrooms.

The word mushroom can also be used for wide variety of gilled fungi, with and without stems, and the term is used even more generously to describe both the fleshy fruiting bodies of some Ascomycetes and the woody and leathery fruiting bodies of some other Basidiomycetes depending on the context of the word.

Forms deviating from the standard usually have some more specific names, such as "Puffball", "Stinkhorn", "Morel" and gilled mushrooms themselves are often called "agarics" in reference to their similarity in the order Agaricales. By extension "mushroom" can also designate the entire fungus when in culture or the thallus (called a mycelium) of species forming the fruiting bodies.

The term "mushroom" and "toadstool" go back centuries and were never precisely defined nor was there consensus on application.

The term "toadstool" was often but not exclusively applied to poisonous mushroom. The German *todesstuhl* translates as "death's stool" or to those that have the classic umbrella-like cap-and-stem form.

The term "mushroom" and its variation may have been derived from the French word *mousserom* in reference to moss (mausse). The toadstool's connection to toads may be direct in reference to some species of poisonous toad or may just be a case of phono-semantic matching from the German word (Miles and Chang, 2004). However, there is no clear cut distinction between edible and poisonous fungi. Term "toadstool" now a days used in storytelling when referring to poisonous or suspect mushrooms.

Many spceies of mushroom seemingly appear overnight, growing or expanding rapidly. This phenomenon is the source of several common expressions in English language in "to mushroom" or "mushrooming" (expanding rapidly in size or scope) and "to pop up like a mushroom" (to appear unexpectedly and quickly). In actually all species of mushrooms take several days to form primordial mushroom fruit bodies, though they do expand rapidly by the absorption of fluids. There are some ephemeral mushrooms, like *Parasola plicantilis* (formerly *Coprinus plicatilis*) that literally appear overnight and may disappear by late afternoon on a hot day after rainfall. The primordia form at ground level in lawns in humid spaces under the thatch and after heavy rain fall or in dewy conditions balloon to full size in a few hours, release spores and then collapse. However, not all mushrooms expand overnight, some grow very slowly and add tissue to their fruit bodies by growing from the edges of the colony or by inserting hyphae.

Mushrooms have been found in fossilized wood that are estimated to be 300 million years old (Editorial, 1997) and almost certainly prehistoric man has used mushrooms collected in the wild as food but has no authentic record. The earliest record of living mushroom can be traced from *Vedic* period in Hindu religion. More recently the description of the desert truffle mushroom, *Terfezia amenaria* can be found as "bread from heaven" in Bible and also "Manna of the Israelites" (Pegler, 2002). The people in India although were conscious about the utility of mushroom since ancient time, its scientific studies were initiated very late in 18th century with identification of *Podaxis pistillaris* (L. ex Fr.) Morse by Linnaeus out of collection sent to him by Koening from Tamil Nadu. Fries, Berkley, Cooke, Massea, Lloyad Murrill, Butler and Bisby and Rea were among other foreign workers who subsequently made significant contributions towards the Indian Mushrooms. The initial collections of mushrooms were only from Himalayas (Berkeley, 1851, 1852 and 1854). Bose (1918) was the poineer to record some mushrooms from plains of West Bengal. Much later in 1970's, the agarics of South and South West India started receiving greater attention and it was actively followed by the workers spread over almost every corner of the country in the 20th century. State wise contributions made by the workers on the mushroom flora of India may be summarized as follows.

Real floristic work was initiated in Maharashtra as early as 1943 by Kamat and it was followed by Sathe and his associates (Sathe and Rahalkar, 1976; Sathe and

Sasangan, 1977, 1978; Sathe and Deshpande, 1976; Sathe and Daniel, 1980 and Sathe and Kulkarni, 1980). Sathe and Rahalkar (1975) reviewed the status of Indian Agaricales taking 1825 as the base. Additions of 40 different members of Agaricales were subsequently made by Trivedi (1972), Thite and Kulkarni (1975), Thite *et al.* (1976) and Chavan and Barge (1978) from Nagpur, Kolhapur and Satara areas, respectively. Watling and Gregory (1980) presented a comprehensive list of 119 taxa from Jammu and Kashmir. Abraham (1991) published a list of agarics with ecological notes from Kashmir Himalayas reporting 250 species. From a survey of North-West Himalayas, Lakhanpal and his associates (Lakhanpal, 1988, 1992, 1995, 1996, 1997; Lakhanpal and Shad, 1986a, 1986b, 1999; Lakhanpal *et al.*, 1985, 1986, 1988; Bhatt and Lakhanpal, 1988 a-d; Sharma and Lakhanpal, 1981; Lakhanpal and Sagar, 1989 and Lakhanpal and Sharma, 1988) made around 700 collections belonging to 223 species, 59 genera and 15 families of Agaricales. This survey provides an inventory of the species occurring in the Himalayas, a list of species of mushroom which enter into mycorhizal relationship with forest trees; and a list as well as description of non-conventional and edible species discovered during the surveys. While reviewing the work on Himalayan Agarics, Lakhanpal (1993) concluded that the taxa recorded are least commensurate with the vastness and diversity of mushrooms in the lofty mountainous range is endowed with. Saini and Atri (1995) reviewed the exploratory works on mushroom from Punjab and listed 94 taxa spread over 24 genera. Out of these, 73 species spread over 16 genera were Agarics while rest were mushrooms from other groups. Atri *et al.* (2000a, b; 2001) and Kaur (2000) also made significant contributions to the mushroom flora of Punjab. While critically analyzing the mushroom genetic resource of Punjab, Atri (2001) reported 27 genera spread over 152 species and 8 families. While working on the taxonomy of the mushroom from North-West India, Gupta (1994) made 261 collections falling in 66 taxa which include 4 new species, 10 new varieties and 40 new records from India. From South-Indian region, excluding Kerala, Natarajan (1995) reported 457 species of Agarics spread over 76 genera. Work on mushroom from Kerala has been reviewed by Bhawani Devi (1995). Patil *et al.* (1995) listed 95 additional species of mushrooms from North-Eastern hills. Mishra (1999) after extensive survey of Arunachal Pradesh, reported the occurrence of 38 taxa falling under 23 genera. Manohracharya and Vijay Gopal (1991) gave an account of mycofloristics of Agaricales from Andhra Pradesh. Those who made good contributions to the Agaric flora of West Bengal are Purkayastha and Chandra (1974a and b, 1975, 1976, 1985), Kar and Dewan (1975), Chakravorty and Purkayastha (1976), Chakravorty and Sarkar (1982) and Ray and Samajpati (1979). Purkayastha and Chandra (1976) presented a consolidated account of 283 mushroom taxa collected from Indian habitats. Moses (1948) collected and identified many edible species of mushrooms from Gujarat (Baroda) including *Volvariella diplasia, Pleurotus ostreatus, Boletus crocatus, Agaricus arvensis* and some puffballs. Khalita *et al.* (1997) recorded occurrence of 19 species of edible mushrooms from Assam. Rai (1997) made collection of six wild species of edible mushrooms from Seoni Tribal areas of Madhya Pradesh. Kumar *et al.* (1991) and Kumar and Shukla (1995) have also made good contribution on the mushrooms of Central India. M.P. Sinha and Padhi (1978) reported 3 new species of *Tricholoma* from Kapilash and Dobjharan forests of Orissa. Ghosh and Pathak (1965) described few edible species of *Macrolepiota* and

Ghosh *et al.* (1967) some edible species belonging to the order Agaricales from U.P. The workers who made significant contributions on the mushroom flora of Uttarakhand are Thind and Sethi (1957), Joshi *et al.* (1982) and Sharda *et al.* (1997). The large tract of Indogangetic plains of U.P., Bihar, Bihar Plateau (now Jharkhand) and tribal hilly areas of Orissa, although are known to abound in wild fleshy fungi, have remained almost unattended. The extreme North-Eastern hills bordering China, Burma, Bangladesh and marked by high rainfall have also yielded rich mushroom flora as a result of extensive survey made by Verma *et al.* (1985, 1987, 1989, 1995). Exploratory work in different agro-ecological zones obviously witness diverse taxa from desert ecosystem of Rajasthan (Doshi and Sharma, 1990; Nag *et al.*, 1991; Sharma *et al.*, 1992). Doshi and Sharma (1997a and b) reported more than 60 species of wild mushroom from Rajasthan. Sharma and Doshi (1996) made collection of wild mushroom *Phellorinia inquinans* from Rajasthan and also analyzed its nutritional values.

If we look into the global scenario, 14000 mushroom species are known to exist. Of these about 7000 species are considered to possess varying degree of edibility and almost 3000 species from 31 genera are regarded as prime edible mushroom. To date only 200 of them are experimentally grown, 100 of them are economically cultivated, approximately 60 are commercially cultivated and about 10 have reached to industrial scale production in many countries (Tewari, 2005). While estimating the list of Agaricales, *Tramallales* and *Auriculariales* recorded in India, Verma and Upadhyay (1998) mentioned a total of 912 species belonging to 121 genera from 16 families as compared to 538 species belonging to 115 genera of 20 families reported by Manjula (1983). The number of mushroom taxa recorded so far, although has not been updated, appears far less than expected keeping in view of the vast potential of Himalayas, Arawali, Bindh and other ranges of mountain coupled with diversity of climatic conditions prevalent in the country. It is feared that this rich diversity of mushroom may be lost without our having known them and their potential, therefore, it is highly desirable to prepare exhaustive inventory of mushroom existing in the country. Further that systematic and intensive exploration of undescribed mushroom taxa be made and their conservation in National Gene Bank being established at the National Research Center for Mushroom to overcome the threat of fastly receding natural habitats.

Human being inhabiting this planet lived in hunter–gatherer society for most of their evolutionary history. Only 10000 years back they settled in agricultural societies and only about 200 years ago entered the industrial age with the population of one billion and biological diversity probably at an all time high. Biological resource was freely used for development and this coupled with the colonial policy, immensely impoverished the biosphere. Human have, of late, been rather complacent about conservation. The greatest realization of its importance came when pictures sent from moon by Neil Armstrong in late 1960s showed mother earth as a small fragile ball and it was recognized as a spaceship floating in the sea of nothingness, the life support systems of which are finite resources. Conservation of biological diversity has been the subject of intense debate all over the world for over three decades or so. Environmental conference in 1972 at Stockholm, Sweden can be said to be the starting point of international debate and is considered critical to the health and stability of the biosphere. In spite of this initiative, efforts on *in situ* conservation of fungi in general is still in its infancy world over. The reliable sources on geography and ecology are dependent on a sound taxonomic and nomenclatural basis. Unfortunately traditional floristic and monographic work on macrofungi is itself endangered in majority of countries. A plea has been made by some workers for international cooperation in monitoring programme and compilation of supranational Red Lists. In India conservation work on fungi in general and mushroom in particular has not been attempted properly so far. The "Red Data Books" on endangered flowering plants have been published but without reference of fungi. It is, therefore, high time that preparation of Red Lists of macrofungi is taken in hand and comprehensive database of mushroom flora be prepared.

The humanity is passing today through difficult competitive area of its civilization. For ever increasing population, the demands are multiplying. In the last two decades, there is almost stabilization and/or stagnation in agricultural production. It is high time to divert resources and energy to develop alternative source of employment and quality food production. Undoubtedly, mushroom farming looks not only promising but also enterprising and adaptable because of their high nutritive quality, land independent cultivation by utilizing worthless agrowastes besides earning foreign currency and generating self employment. Perhaps it is the reason that biotechnology of mushroom production though has late origin compared to its edibility, progressed fastly during the previous decade.

India is basically an agriculture based country. Variety of agro-industrial wastes are produced every year in plenty, disposal of which could have been problem. Mushrooms have been successfully tamed to utilize substrates differing widely in the physical and chemical property from cereal grains and their brans to hard wood.

Composting is an important event of mushroom cultivation and still it rules the entire technology. There are many fundamental questions which chase the mind of the researcher such as what goes inside the compost piles in the bunkers and bulk chambers to make it selective and productive and what triggers and regulates the fruiting and flushing of mushrooms have been deciphered. Due attention has, therefore, been paid for faster and better preparation methods.

Traditionally in India compost is produced by long and short methods of composting that involve larger space area, costly machinery if produced on commercial scale, production of obnoxious gases or compounds in the reduced states *viz.* Hydrogen sulphide (rotten eggs), Acetic acid (vinegar), Valaic acid (vomit) and alcohols and phenolic compounds (putrefying smell). All these gases and compound pollutes the atmosphere and create nuisence to the people. Need is therefore felt to operate the composting process such a way so as to minimize the pollution level obtained in traditional processes. A new method of composting called express composting or total indoor composting has been developed in which compost is produced in total/partial enclosed structure under fully aerobic conditions (both phase I and phase II) exploiting the use of thermophilic organism. Emission of harmful gases polluting the atmosphere is almost negligible in such compost. Due to stringent environmental legislations, indoor composting has made rapid strides in Australia,

Austria, Belgium, Holland, Italy, Switzerland and in France. In India too, many farms are switching over to indoor composting with favourable results.

The full yield potential of a crop species is often difficult to realize due to losses caused by biotic and abiotic stresses. In mushrooms also, a number of extremely harmful pests and diseases cause losses both in quality and quantity of the produce. Mushrooms themselves being fungi pose special problems in adopting chemical control measures, particularly against the diseases. The problem of pesticide residue is rather more alarming in mushroom as the waiting time is very small. Hence strains with genetic resistance or tolerance to the biotic and abiotic stresses should be the preferred strategy. In future, therefore, breeding strategy must include to ensure yield stability through improvement of strains for resistance to insects, diseases, moulds/ weeds and abiotic stresses like low oxygen, high temperature, low humidity, excess watering and tolerance to pollutants and fumes of machines.

A perfect breeding strategy must not confine to yield potential and yield stability only if the ultimate aim is to obtain higher production per unit time/cost. Many other attributes like nutrient use efficiency and crop duration parameter carry a lot of meaning in a crop like mushroom. Similarly, for getting higher profit strains capable to suffer least post-harvest losses and possessing high nutritional and culiniary qualities will be essential. Hence, future strategies for breeding mushroom strains will have to include breeding for short duration, for nutrient-use efficiency, for more nutritious and tasty strains and for increased shelf-life. During the last two decades, highly refined biotechnological tools and techniques have been deploped not only to understand the mushroom biodiversity, but also for the genetic improvement with respect to the yield quality and disease resistance of the commercial mushrooms. Hybridization seems to offer the best prospect for real, especially multiple gene transfer mediated by protoplast fusion. With the advanced technique of protoplast fusion, there is great potential to make both inter and intraspecies protoplast fusion and to create new hybrid mushrooms which is otherwise not possible through classical breeding techniques. With molecular tools available today and likely addition of more and more biotechnological techniques in future, it would be very much possible to incorporate genes governing the required traits from wild germplasm of the same species, different species or from a different genus, and such a possibility may not be far off.

Mushroom mycelia and spores are often microscopic usually filamentous organisms that have a few phenotypic markers that can be used to differentiate between individuals in a population for breeding programme. This limitation of taxonomic identification of mushroom strains can be overcome by the use of DNA markers like RFLP (Restriction Fragment Length Polymorphism) and PCR (Polymerase Chain Reaction) based RAPDs (Random Amplified Polymorphic DNA), since DNA markers detect variations directly at DNA level and are not influenced by environment. The PCR procedure is based on repeatation of a set of three steps all conducted in a succession under somewhat different controlled temperature condition. These three steps are denaturation, annealing of extension primers and primer extension (amplification step), (Williams *et al.*, 1990).

Mushrooms are highly perishable and get spoiled due to wilting, veil-opening, browning, liquefaction, loss of texture, aroma, flavour etc. making it unusable (Azad *et al.*, 1987). Most of the mushrooms, being high in moisture and delicate in texture cannot be stored for more than 24 hours at the ambient conditions prevailing in the tropical country like India. Researchers who are interested in spoilage of fresh mushrooms earlier believed the primary cause to be the enzymatic reactions in the living tissue. Later it was suggested that mushroom spoilage might be caused by the action of bacteria on the mushroom tissue and browning of mushroom was due to a combination of autoenzymatic and microbial action on the tissue. Sound post-harvest practices have since been developed to extend the shelf-life of the fresh mushroom by developing suitable strains besides improving post-harvest processing technique and marketing. In India not much attention has been paid to strain improvement work and in fact the strain introduced about 35-40 years ago, is still popular. However, this strain does not have very good shelf-life and therefore there is need for strains having better keeping quality and higher yield. India is a tropical country but it is unfortunate that still temperate species are cultivated. Mushroom cultivation in India started as early as in 1943. Even after 68 years of mushroom cultivation history, the Indian mushroom industry yet has only three species being commercially grown. Diversification of Indian mushroom industry through domestication of wild tropical species is required to be explored.

There is intensive industrial interest in a novel class of compounds extractable from mushroom. Between 80-85 per cent of all medicinal mushroom products are derived from the fruiting bodies, which have been either commercially farmed or collected from the wild. Rest of the 15 per cent of all the products are based on extracts from mycelia or consist of the actual powdered mycelia plus the substrates used for growing.

The cultivation of many mushrooms of medicinal importance has not been successful till now. Secondly the cultivation of mushroom for fruiting body production is a long term process taking one to several months depending upon species and substrates. In contrast, the production of mushroom mycelium in submerged culture would allow acceleration in growth and to obtain high yield of biomass with constant composition round the year. Therefore popularization of mycelial culture on commercial scale might be the trend of future. The compounds extracted from mycelium are called mushroom "nutricenticals", exhibit either medicinal or toxic qualities and have immense potential as dietary supplements for use in the prevention and treatment of various human diseases.

From the foregoing discussion it is apparent that every hope rests with mushroom to fight out with the future crisis emerging from population explosion, receding land and ever increasing demands. It is the reason that mushroom has witnessed unhindered contribution of the environmentalists on the conversion of waste material to food concomitantly with the avoidance of pollution commonly associated with the disposal of wastes, microbiologists interested in thermophillic organisms involved in composting process, geneticists concerned with strain improvement, horticulturists interested in the development of efficient cultivation practices, nutritionists involved in the assay and evaluation of mushroom nutrients and pathologists studying mushroom diseases. The contributions made by them have been reviewed from time to time but they are scattered in the pages of Journals and in the proceeding of the symposia/conferences. In the present endeavour attempt has been made to consolidate the attainments made so far on different aspects of mushroom for ready reference of the students and workers interested in this field.

Chapter 2

Biology of Mushrooms and its Fruit Types

All the cultivated mushrooms today (except truffles of Ascomycetes) belong to the class Basidiomycetes under orders Agaricales and Tremellales yet they exhibit a lot of variations, not only in their morphology, taste and texture, but also in their biology, sexuality and genetics. The general pattern of the life cycle of a Basidiomycetous fungus, as understood today is depicted in Figure 1.

Life Cycle

The sequence of events in the life cycle of Basidiomycetes in general are:

1. Germination of basidiospore
2. Formation of haploid monokaryotic mycelium
3. Plasmogamy
4. Formation of fertile mycelium *i.e.* dikaryon
5. Development of fruit body
6. Formation of hymenium within the fruit body which consists of terminal spore bearing cells called basidia,
7. Fusion (karyogamy) of dikaryotic nuclei to form a diploid nucleus inside the basidium.
8. Meiosis resulting into the production of four haploid nuclei
9. Migration of each post meiotic nucleus into the basidipspores born externally on the basidium and ultimately discharged for the repetition of life cycle.

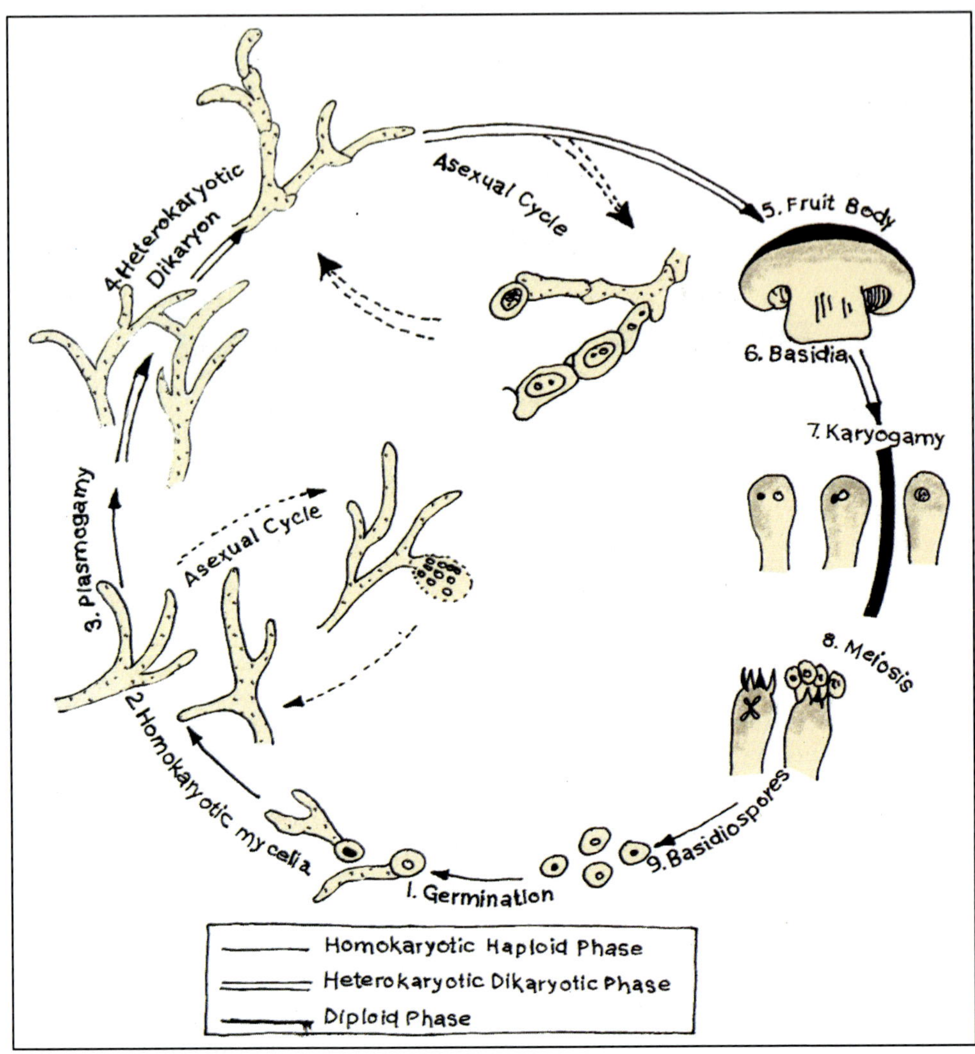

Figure 1: Life Cycle of a Typical Basidomycetes–Steps 1-9

1. Basidiospore-germination; 2. Development of homo-karyotic mycelium; 3. Plasmogamy through mating of two compatible homokaryons; 4. Development of hetero-karyotic dikaryon; 5. Initiation of basidiocarp (fruit body); 6. Development of basidia in the fruit bodies; 7. Karyogamy *i.e.* fusion of the two parental nuclei; 8. Fusion nucleus undergoes meiosis. 9. Formation of basidiospores each receiving one of the 4 daughter nuclei.

In general the life cycle of Basidiomycetes is the feature of alternation of an indefinite haploid phase with one celled diploid phase, usually with a heterokaryotic phase of dirkaryotic structure. Under appropriate environmental conditions, this heterokaryon is induced to produce fruit bodies which is the sexual reproductive phase of mushrooms consisting of three cardinal events of plasmogamy, karyogamy and meiosis. Plasmogamy is the fusion of two protoplasts, bringing together to two different nuclei into the same cell. The karyogamy phase is a short lived phase and takes place in the basidium which is immediately followed by meiosis.

Entire phase of life cycle except fruit formation are completed under humus, the latter on above it. Basidiospores under congenial circumstances germinate into white string like strands called rhizomorph. Rhizomorphs are composed of many interwoven slender filaments hyphae, collectively called mycelium.

As the mycelium grows through the humus, it absorbs food, some of which is stored in that mycelium. With suitable temperatures and adequate rains, knob like developments appear on the strands-the young mushrooms, popularly called "buttons". With sufficient moisture these buttons within a few hours, develop into mature mushrooms. Thus, all mushrooms, in the course of their development, exhibit a button stage.

Sporophores grow from a mat of thin mycelium, or spawn, which is located below the soil surface. Each mycelium grow new sporophores during the annual fruiting season. Mycelial life spans range from several months to centuries, depending on nutrient and moisture sources and suitable temperature. A honey mushroom mycelium in Michigen once spread across 40 acres over fifteen hundred years.

The typical full grown mushroom or fruiting body (Figure 2), is composed of a stalk which in most instances, extends upright and bears an umbrella–shaped cap (pileus). In turn, the cap bears, on its lower surface, radiating, vertically–arranged plates–the lamellae, or gills. On the faces of these gills the reproductive cells, called spores, develop in fabulous numbers–millions on one fruiting body. At maturity, the spores are ejected free of the gill and float slowly to the ground. In still air, they fall directly to the humus below, and in mass reveal the colour characteristic of the species. Spores in mass, as seen in a spore deposit, may be white, cream, yellow, clay, brown, pink, lilac, purplish, or black or various shades and tint, depending on the species.

Spore colour can be readily viewed from spore deposit or spore print preparation of which is discussed in identification section. If a spore deposit is not available, spore color may usually be inferred from the gill colour.

Some of the normally discharged spores ultimately fall on a favorable substratum where they germinate (sprout) to form new hyphae and finally a mycelium. When these hyphae develop ropelike strands (rhizomorphs), the buttons may begin to form.

If the young mushroom is cut longitudinally through the stalk and cap, there is revealed, in some mushrooms but not all, a webby or membranous curtain called the inner veil, which connects the edge of the cap with the stalk. As the cap expands, this curtain breaks, according to the species, in one of three ways; (1) the veil-remnants may remain around the stalk as a ring annulus; (2) the veil may break clear of the stalk but remain, for a short time, as a hanging fringe on the cap-margin; (3) the veil may be so sparse or rudimentary as to leave no trace of itself. In many mushrooms, the veil is entirely absent.

In some mushrooms, such as species of *Amanita* another important structure, the outer, or universal, veil, is observed when the young mushroom is cut vertically

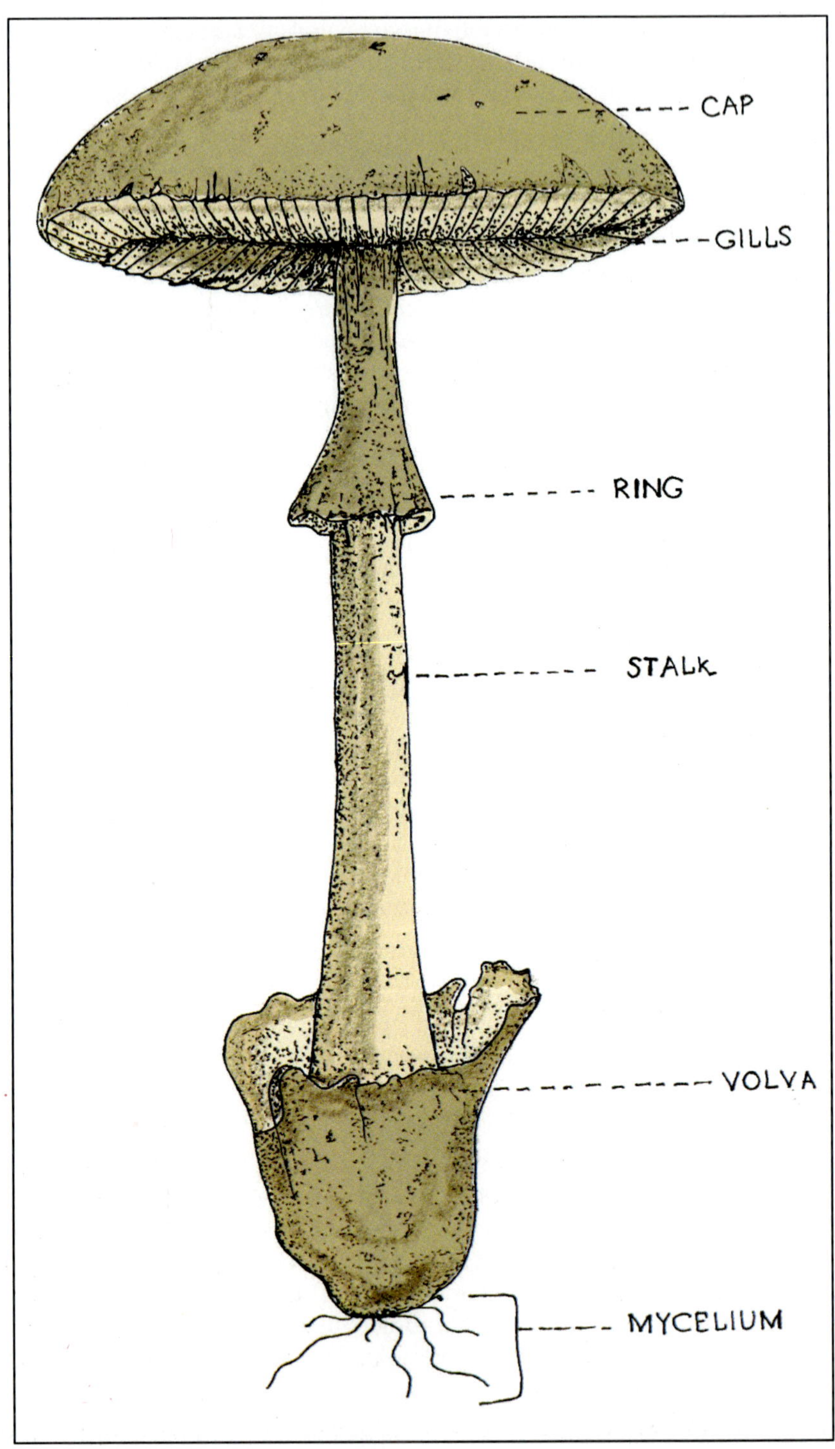

Figure 2: A Typical Full Grown Fruiting Body of Basidiomycetous Mushroom

through the cap and stalk. In the young stage, this veil envelops the whole button. As the cap expands and the stalk elongates, this enveloping veil breaks in different ways, according to the species.

Many non-gilled mushrooms donot follow this general pattern of a pileus and stipe with hymenophore on the lower surface of the pileus. Diversity in Hymenomycete fruiting body has been depicted in Figure 3. In may polypores, the fruiting body is a self like pileus with a tubular hymenophore on the lower surface (Figure 3i); in some hedgehog mushrooms the hymenophore consists of spines which hang down from a fleshy mass of tissue or an elaborately branched framework in the coral fungi (Figure 3a, b); The fruiting body may be much branched with upright to ascending branches which are covered, at least on the sides, by the hymenium.

The basidiospores at maturity fall free of the fruiting body and are carried away by air current in Hymenomycetes. In contrast the basidiospores in Gastromycetes are not exactly discharged from the basidia at maturity and they depart from the fruit body in a variety of ways (Figure 4). In some puffballs (*Lycoperdon*), the basidiospores are in a sack with an apical opening. When rain strikes the sack spores are forced out. In *Clavatia* the mass of spores is exposed at maturity. Wind and man may disperse the spore. In some cases (*Bovista pila*), the fruiting body is blown by the wind and spores sift out of it. The insects and rodents may eat fruiting bodies and disperse the spores in their droppings.

The hymenium of Gastromycetes is partly to completely enclosed within the fruiting body, at least until the spores are nearly mature. The spores producing tissue in these fungi is gleba (Figure 4) and some would say the hymenophore is globulose. The gleba in puffballs may include sterile threads, called capillitial threads (Figure 4k) which together with the spores form the capillitium. In some puffballs, the capillitial threads radiate it out from a more or less distinct region at the base of the gleba called the pseudocolumella (Figure 4c, f). The gleba is enclosed by a wall called peridium (*pl.* peridia) which may be one or more layered.

The consistency of the gleba varies from powdery in the puffballs (order Lycoperdales, Sclerodermatales and Tulostomatales) to pill like in the birds-nest fungi (Nidulariales), to slimy in the stinkhorns (Phallales). In some Gasteromycetes, the peridium does not have a specialized manner of opening, in others such as the earth stars, it opens in an elabroate manner. A stalk may be absent, or if present, take one of a variety of forms (Figure 4). Sometimes there is an internally chambered to cottonly sterile base (so called pseudostem) (Figure 4a, b). Sometimes a true stipe of more or less vertically arranged hyphae is present (Figure 4d). In the stinkhorn, the spongy structure that elevates the gleba is termed a receptaculum (Figure 4g).

In Gastroid Agaricales (connecting link between Gastromycetes and Hymenomycetes), fruiting body consists of pileus bearing glebulose hymenophore is typically convoluted and charmbered but resemble gills or tubes. The stipe is short penetrating gleba and is connected with pileus is the columella. The whole structure is called stipe-columella (Figure 4h). In Tuber-like Basidiomycetes, the pileus is replaced by a peridium which may or may not arise from the sterile base (Figure 4i). The divisions between the chambers of gleba are called tramal plates.

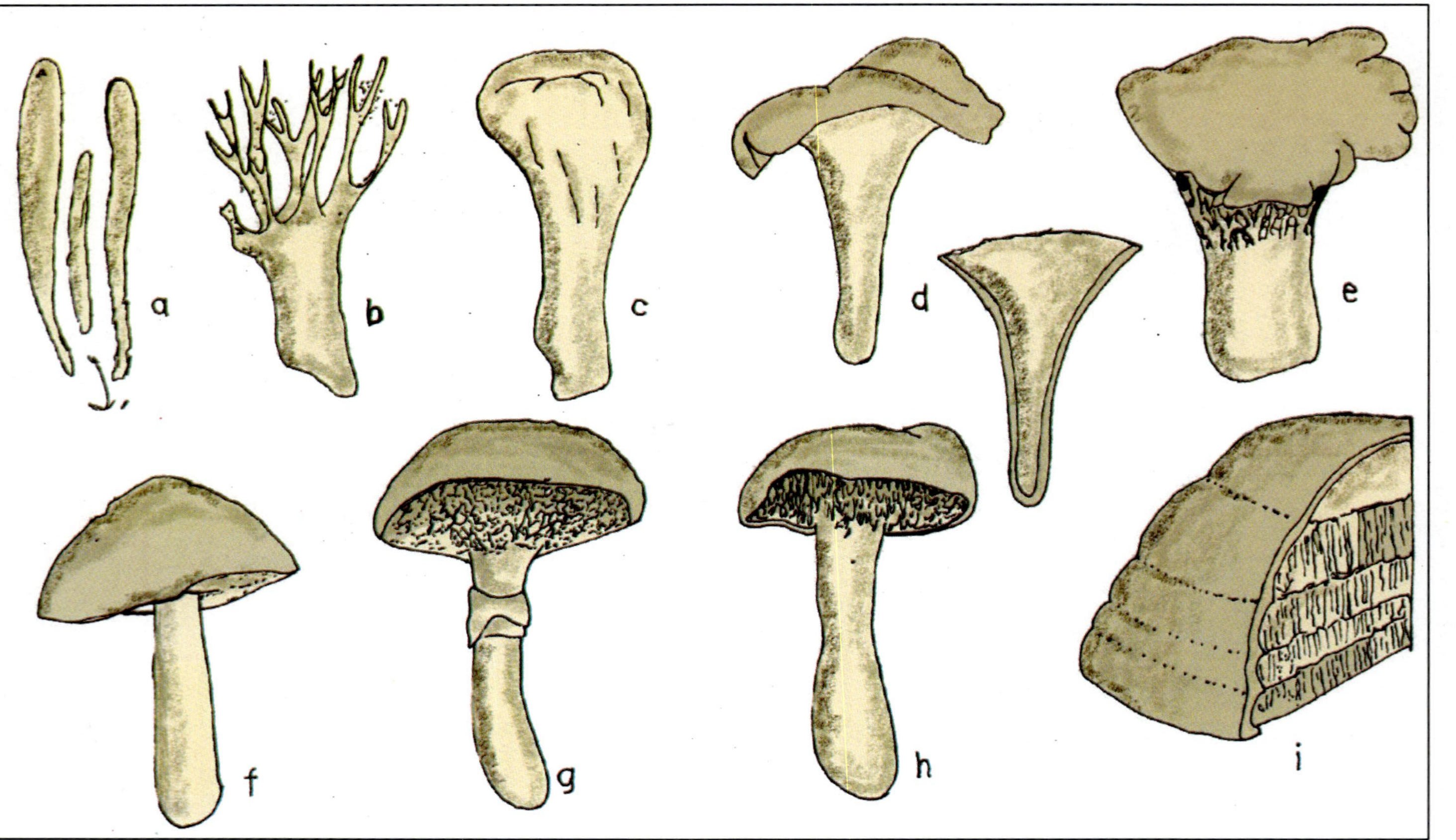

Figure 3: Representative Type of Hymenomycete Fruiting Body

a: Clavate; b: Coral-like; c: Clavate-truncate; d: Vasent; e: Cantherelloid; f: Gilled mushroom; g: Bolete (Tubular hymenophore); h: Hedgehog mushroom (spinose) hymenophore; i: Shelf fungus (tubular hymenophore).

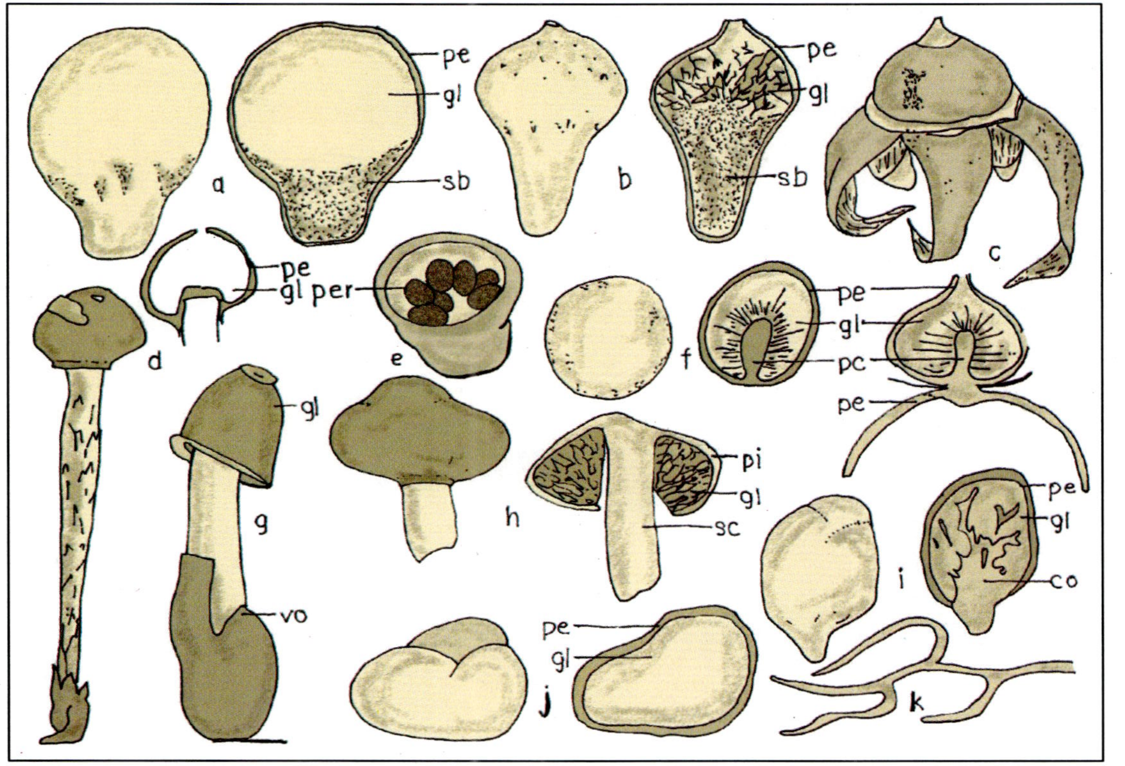

Figure 4: Representative Type of Gasteromycete Fruiting Body

a, b: Puffballs: surface view and in longitudinal section (a: *Calvatia*, b: *Lycoperdon*); c: Earth star (*Geasturm*), surface view and in longitudinal section; e: Bird's nest fungus (*Crucibulum*); f: *Radiigera* surface view and in longitudinal section; g: Stinkhorn (*Phallus*); h: Gastroid agaric (*Thaxterogaster*), surface view and in longitudinal section; i, j: Tuber-like Basidiomycete (i: *Truncocolumella*, j: *Rhizopogon*) surface view and in longitudinal section; k: Capillitial thread from a puffball, (pe: Peridium; gl: Gleba; sb: Sterile base; pc: Pseudocolumella; per: Peridiole; pi: Pileus; sc: Stipe-columella; co: Columella).

The spores (ballistospores) in Hymenomycetes fall free of the fruiting body and asymmetrically attached by a basal projection called the apiculus, to the sterigma of the basidium (Figure 5d). In the majority of Gastromycetes and many Tuber-like

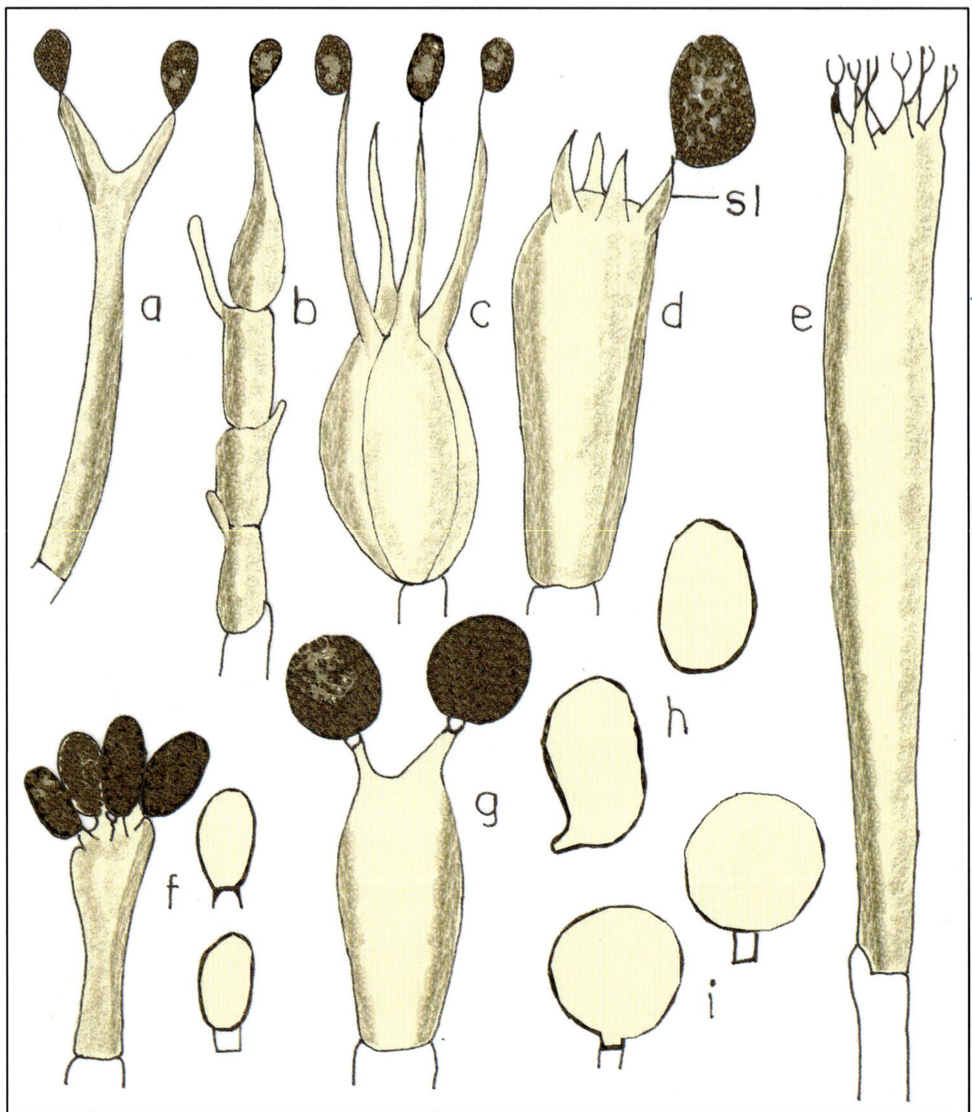

Figure 5: Representative Types of Basidia

a: Tuning fork (Dacrymycetales); b: Septate (Auriculariales); c: Septate (Tremellales); d: Clavate (st: Sterigma; many Agaricales and Apphyllophorales); e: Narrow and with young spores (Canthartelloid Apphyllophorales); f, g: Gastroid with statismospores (f: *Rhizopogon*; g: Zlleromyces); h: Ballistospores, face view (upper), profile view (lower); i: Statismospores, surface view (upper), section showing relation of spore and pedicel (lower); a-e may discharge spores.

Basidiomycetes, the spores are not forcibly discharged from the basidia (statismospores). They are radially symmetric and are symmetrically placed on the sterigma; in place of apiculus they may have a pedicel which is a combination of part of the sterigma and part of the spore itself (Figure 5f,i). In Gastorid Agaricales and some Tuber-like Basidiomycetes, the spores have morphology of ballistospores but are not discharged so they function as statismospores.

The Ascomycetes have more varied fruiting bodies than the Basidiomycetes, but have fewer fleshy representatives. The most common type of fleshy fruiting body is a cup or modified cup (an apothecium technically) with the hymenium borne on the upper or outer surface. Ascomycetes with this type of fruiting body form the class Discomycetes, often referred as cup fungi. The asci shoot the spores up into the air current instead of letting them free of the fruiting body as in Hymenomycetes. In some Discomycetes, the area of hymenophore is increased by reflexing the cup and by lobing, wrinkling or pitting the resulting structure (Figure 8 g-j) as in the morels and lorchels. Such spore producing region is not a pileus because the hymenium covers the upper, not the lower surface of the fertile portion of the fruiting body.

Depending on the structure and manner of dehiscence of asci Discomycetes are divided into two orders: the Pezizales with asci opening by lid (operculate Discomyctes) and Helotiales in which the spores exit through canal or pore that penetrates the thickened ascus apex (Figure 6 c-e) Members of this order are commonly called Inoperculate Discomycetes.

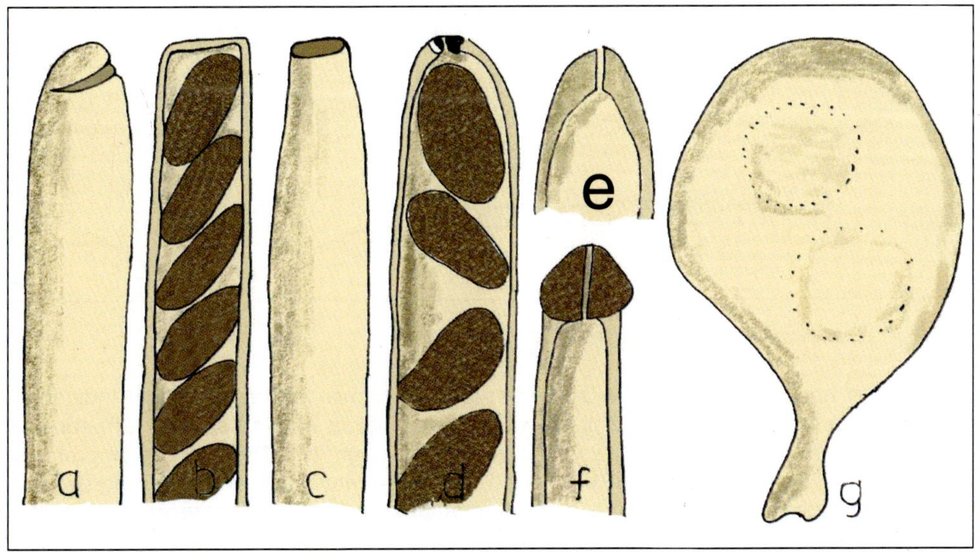

Figure 6: Representative Asci (Apex only in a-f)

a: Mature and empty operculate ascus; b: Immature operculate ascus; c: Mature and empty inoperculate ascus; d: Inoperculate ascus with spores; e, f: Additional types of inoperculate asci; g: Indehiscent ascus of a Tuber.

The ascospores produced by the fruiting body in Ascomycetes do vary in regard to its organisation, content, surface and shape as shown in the Figures 7 a–h.

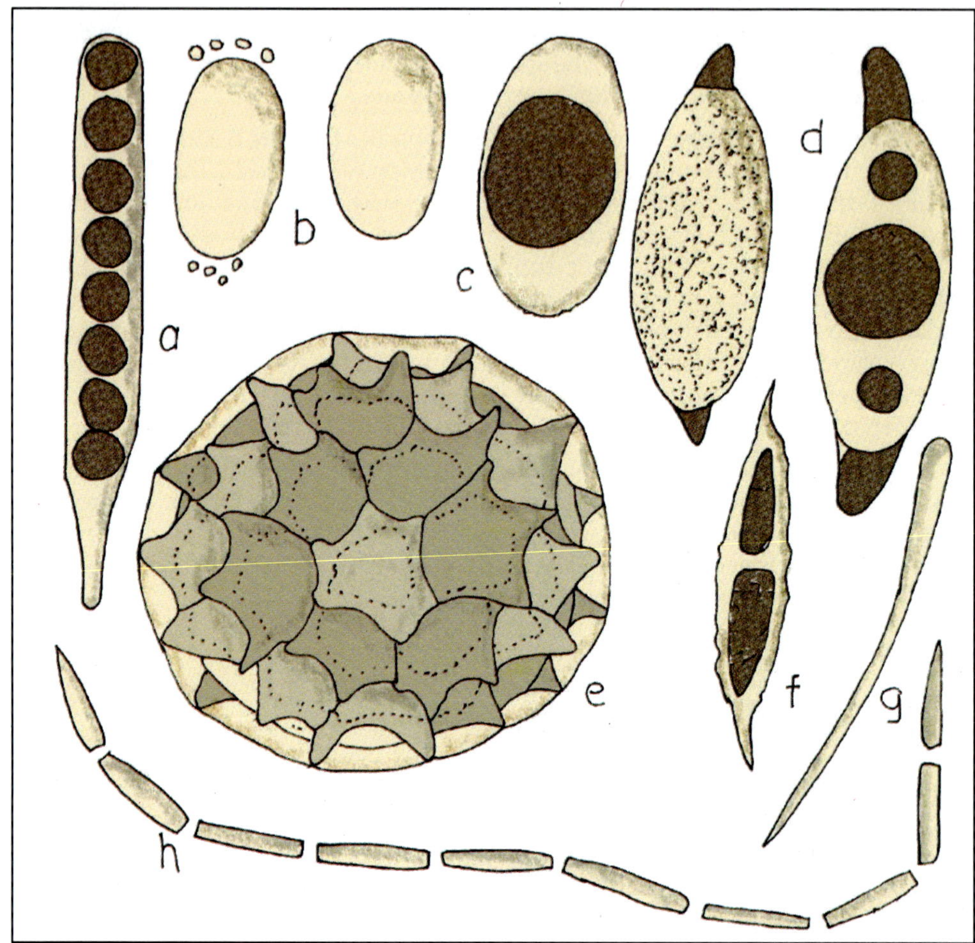

Figure 7: Representative Ascospores

a: Spores in an ascus; b: *Morchella* spores, left with crown of droplets, right indicating homogeneous content; c: *Helcella* spore with oil globule; d: *Discina* spores, left; surface view as seen in Melizer's reagent, optical section as seen in KOH showing oil droplets (note prominent apiculi at both ends of each spore); e: Reticulate-alveolate spores of a Tuber; f: Apiculate, 2-celled spore of a Hypomyces; g: Slender spore of a *Cudonia*; h: *Cordyceps* spores breaking into segments.

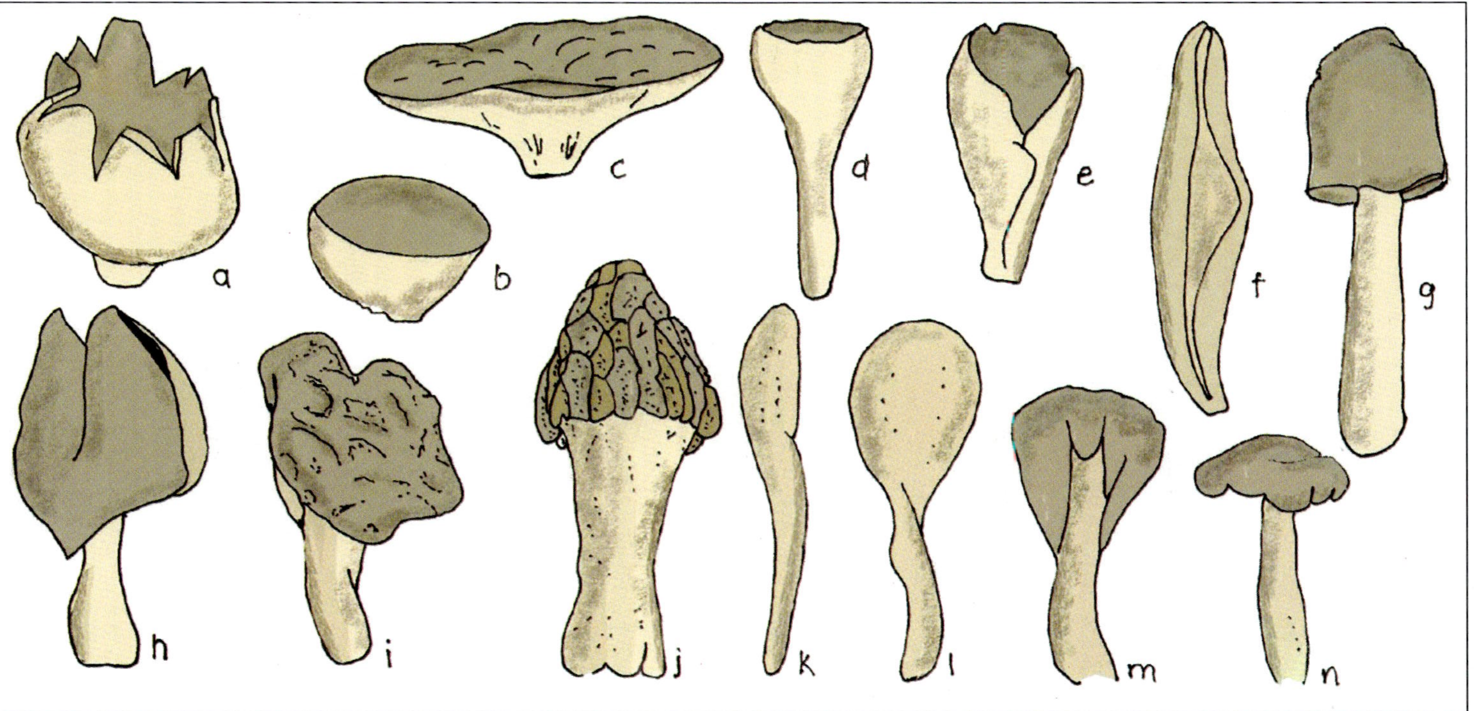

Figure 8: Representative Types of Discomycete Fruiting Body

a, b: Cup-shaped; c: Saucer-shaped; d: Urn-shaped; e: Truncate and split down one side; f: Rabbit-ear-like; g: Campanulate-stipitate; h: Saddle-shaped and stipitate; i: Brain-like (gyrose) and stipitate; j: Pitted and stipitate; k: Clavate; l: Flattened elavate; m: Spathulate; n: "mushroom-like."

Chapter 3

Collection, Preservation and Identification of Mushrooms

Collection and identification of wild mushrooms are important for the study of mushroom biodiversity and their ecological role. The discovery of new mushroom species will lead to their exploitation by an expanding mushroom industry. The collection and identification of wild mushrooms of species already known will provide the genetic variability required for breeding better mushrooms with higher yield and also provide the phenotypic traits which may be desirable by the mushroom industry or useful for research purposes because the mushrooms themselves are the only source of this genetic material. Extinction of a single strain or species would mean the potential loss of many thousands of unique genes that might be used for breeding desirable new strains.

The main limiting factor encountered in the collection and identification of mushroom is the lack of proper uniform procedure for collection and recording of the morphological characters of the collected specimens. It is often a difficult proposition to record the morphological details of the collected specimens in its fresh form due to evanescent nature of most of the species. If a standard procedure for collection will be adopted *i.e.*, enumeration of morphological character, place of collection, it will be easy to identify these fungi.

Mushroom collection requires only the simplest of equipment: a flat bottomed basket or box, a roll of waxed paper, a digging tool and a pencil and paper for notes.

Be sure to collect the entire mushroom, including the base only fresh, young specimens that are free of insect damage is taken. Each type of mushroom should be wrapped separately in waxed paper (not plastic wrap, which hasten decay) alongwith any notes you might want to take about the habitat and appearance of the mushroom.

Its a good idea to note where the mushroom is growing (on wood, soil, moss), whether it is single or in clusters, the colours of the caps, gills and stem; and any other distinctive features. The more you observe about the mushroom in the field, the easier it will be to identify at home.

Some basic precautions should be taken to keep the collected materials in proper condition. Most fresh mushrooms are relatively fragile, and they should be protected from vibration and impact by careful packing. When gathering collection for identification, care must be taken to obtain entire, intact specimen. If specimens are available at different developmental stages, all of them should be collected whenever possible. Mushroom grows on the ground may have structures at the soil surface or even below it. Collectors should check and record observations. A permanent marker pen can be used to record such as date, time, location, smell, substrate (host) colour, exudate (if present), habitat and any thing at all unusual about the specimen. Some important characteristics for identification disappear rapidly as the mushroom matures. These characteristics have to be recorded accurately at the time of collection, All measurements and illustrations should be made with aid of *Camera lucida* drawings.

Some mushroom species have a symbiotic relationship with vegetation particularly trees. Therefore the substratum (host) should be carefully recorded as this can be an important feature in identification and in classification *i.e.* whether the mushroom grows on dung, wood, bark living trees, litter, soil etc. is important information. The circular structures and hymenophoral trama are to be studied by cutting sections 15mm thick with the help of freeze Microtome (Spensor). Microscopic observations are made with 10 per cent KOH solution. The amyloid and dextroid reaction of spores and hyphae are determined by mounting materials in Molzer's reagent.

Preservation of Collected Mushrooms

Preservation of fresh collected wild delicate species of mushrooms for ready reference is done in FAA (Formalin–Acetic Acid–Alcohol). Preservation of mushroom for museum specimen is done in distilled water, alcohol and formalin (70:25:5) for the preparation of mushroom herbarium it is suggested that fleshy mushroom should be dried in air oven at moderate temperature or over a radiator, ventilation must be provided during process of drying. Small and delicate specimens may be sun dried or be dried at room temperature. Drying in an air-conditioned room, or with a vegetable dehydrator, or with an air dryer is the standard method of preparing permanent collections for a mushroom herbarium. It should be noted that fleshy mushrooms should not be dried at temperature above 35 to 40°C, because the hyphae and other microscopic structure become too strongly distorted, making later microscopic studies difficult.

Dried specimens are usually packed in waxed paper or brown paper packets and stored in boxes after dusting with a recommended insecticide or fungicide to check secondary infection.

Identification of Mushroom

Mushrooms have features of their own making them distinct from other group of fungi. Such distinction is due to their self styled fruit bodies varying from species to species. Fruit formation passes through different developmental stages leaving a number of morphological attributes over it which are also of immense taxonomic importance and help in identifying the species. Important taxonomic characters which play a dominant role in identification of the mushrooms are enumerated as below:

Veil

The veil in the widest sense, *i.e.* the involucrum in Persoon's terminology has been considered as being one of the utmost importance in taxonomy as far as Fries and his school were concerned and an unproportionate over emphasis was put on it in such classification as those produced by Kaster, Schroter and especially Earle. Several basically different organs have been confused in the term "veil".

The Volva

This general enveloping layer in the "egg stage" of the carpophores and subjects the primordium to a certain centripetal pressure. A volva has been observed in the Agaricaceae in *Termitomyces, Amanita, Volvariella, Coprinus* sp., *Macrometrula* and in more reduced, indistinct or fungaceous from other related genera and in some species of *Thocybe* and *Cortinarius*. The "annuliform" *i.e.* ring–like portion of the volva may be referred to as an 'annular volva' or the 'volval' portion of the annulus". The counter part of the agaric 'volva' in Gasteromycetes is the volva or so called peridium of the Phallineae, the genera *Montagnea, Gyrophragmium, Braumilla* and *Torrendia.*

The Pellicular Veil and the Cortina

These are remnants of a layer or all the layers of cortical tissue of primordium and are later ruptured by the expansion of the pileus whereby they are extended and thinned. If the cortical layer is gelatinized and the stipe absent, this kind of veil is called pellicular veil by Lohwag; if it is dry and arachnoid and the carpophores are stipitate, it is known as cortina. The thinness of this veil is an important feature because the layer taking part in its formation is thin layer from the start, and is reduced by the tension from the margin when the pileus expands.

The Marginal Veil

This type of veil formation is due to the incurving margin whereby the covering layer of the apex of the stipe is brought in intimate contact with the tissue of the, margin of the pileus. Later when the pileus is expanded, the final separation takes place near the surface of the stipe, the marginal veil then hangs down from, or adhere to the margin of the pileus; if however, the separation takes place farther outside, an annulus will be formed that remains on the apex of the stipe or slides down it. The marginal veil can be best studied in *Boletinus cavipes, Macrolepiota procera* and in *Chlorophyllum molybdites* etc. In the species of *Boletinus* the development of the hymenophore in the young carpophores is pseudoangiocarpous; in *Macrolepiota* and *Chlorophyllum* it is hemiangiocarpous.

The Annulus Superus

This organ has been distinguished in the Phallales (Gasteromycetes) where it has no velar character. However, the homologies of the *Phallus* and *Clathrus, Dictyophora* and other Phalloids can be seen with those of such agarics as *Amanita*.

The Spore Print

The only macroscopical character available that concerns the basidiospores in the Agaricales is the formation of a spore print and its colour.

Individual spores are too small to be seen with the naked eye but one can make a spore print that will show the colour of the spores in mass. This colour is important in identifying characterstic for any mushroom, especially the gilled fungi.

To make spore print, cut the stem off the mushrooms, and place the cap gill side or pore side down on a piece of white paper. To best see the spore colour, use one sheet of black paper and one of white, taped together side by side. Cover with a bowl or jar. If the mushroom is at the right stage not too young, too old or detoriated–the spores will slowly collect on the paper. A spore print will be visible in one to twelve hours.

On the basis of the colour of the spores following six groups of the mushrooms can be identified:

(*i*) The Green-Spored Group

This group of agarics, considered as a taxonomic unit in some artificial classification, belongs to various families and genera in the Agaricales, *e.g.* the Agaricaceae, Amanitaceae, Tricholomataceae and Boletaceae. The green colour spores has no place in Friesian classification because Fries, misinterpreted either the spore colour (in, *Phylloporus*), or the species (all the tropical green-spored agarics), it has also no place in the modern classification because it contains elements from four different spheres of affinities.

(*ii*) Pink-Spored Group

The largest constituents are *Rhodophyllus*, and the Plutaceae. The former group belongs to the family Plutaceae, neither of them related to each other. These are present in genera as *Rhodocybe, Phyllotopsis* and several common species of *Collybia, Gomphalina, Lepista* and *Termitomyces* species as well as some Amanitaceae and Agaricaceae.

(*iii*) The Yellow-Spored Group

There are several genera of "white spored agarics" with creamcoloured to ochraceous or citirinous spores but they have passed as white-spored because of errors of observation. Examples are pale (even pallid) in the Cortinariaceae (*Inocybe*, the genera *Habelomina* and *Leucocortinarius* as a whole, *Galerina*). *Phaeomarasmius* (Strophariaceae) *Pholiotina* (Bolbitiaceae): *Paxillus* (Paxillaceae)

(*iv*) The Black-Spored (Melanosporus) Group

This group integrades with the brown and purple-spored groups at certain levels, as has been recognised by Britizelmayr and other earlier but may also present some olive tints ex. Coprinaceae with single exception of the genus *Gomphidius*, which belongs to the Gomphidiaceae, near the boletes.

(*v*) The Brown-Spored (Ochrosporus) Group

This group is likewise not a homogenous group as was anticipated by Fries ex *Inocybe* and *Habeloma*.

(*vi*) The Gray-Spored Group

It is much neglected and at the same time taxonomically unimportant type. Species of *Rhodocybe* pink-spored in general, are sometime found to have distinctly sordid gray spore prints and thin deposit of spores of *Gomphidius* species may appear gray.

The above examples show that the spore print colours are not as such indicative of an affinity between groups according to general classes of colours (white, pinkish purple, black or brown etc.). The colour observed in fresh spore print are apt to change in the Herbarium as a consequence of dehydration and in some cases they loose the olive hue, so characteristic for the spore print of several genera of the Boletaceae, in other cases they black to almost white after having been a distinct vinacoeus pink in some species of *Tylopilus* (Boletaceae) while in *Russula, Melasleuca, Leucortinarius, Pseudoclitocybe* and other "White-spored" agarics, the pale coloured fresh spore print eventually darkens to decidedly cream colour or ochraceous, especially if prepared with some fixative.

The Mycelium

It has not been used for taxonomic purposes on a large scale. It is obvious, that differences of colour, zonation, consistency and manner of growth in standard cultures, should also be given importance as diagonstic value in the Agaricales. The Agaricales mycelium possessess the characters of luminescent mycelia while some species have characteristic odour. This character can be used for the determination of ectotrophic mycorrhiza. The mycorrhiza of *Russula punctata* and *R. dadmuni* has characteristic odour of idoform which can be obtained in test tube culture.

Some mycelia form sclerotia, rhizomorphs, oidia, conidia, chlamydospores, oliviferous hyphae and even mycelial basidia and cystidia. The latter have been named allocysts by Kuhner. These often resemble like the cystidia or cheilocystidia of the hymenophore of the same species or of allied species. Many non-mycorrhizal fungi can be grown on malt agar and on Lutz synthetic medium.

Miller used a cell type scheme for the mycelia which is borrowed from the Discomycetes.

Short-Celled Tissue

1. *Textura globosa*, characterized by round to ovoid cells, thin cells walls with intercellular spaces, often pigmented. *e.g. Armillaniella tabescens*.
2. *Textura angularis*, characterized by polyhedral cells, thin or, thick walls, without intercellular spaces, often pigmented *e.g. Exromphilina brunneola*.

Long-Celled Tissue

3. *Textura intricata,* characterized by interwoven hyphae with spaces between cells, usually thin, hyaline walls *e.g. Cheimonophyllum canadissimum.*
4. *Textura oblita,* Characterized by interwoven or parallel, thick walled; usually pigmented cells *e.g. Armillariella mellea.*

The word Cystidia in the broadest sense-replacing the earlier term "Micheleanbodies"- designates any sterile body that is interspersed in the hymenium or replaces the basidia in any part of the hymenophore or occur on one of the usually sterile surfaces of the carpophore but resemble the hymenophoral cystidia which are apparrently homologous with them.

Spores

In the Agaricales, uni-nucleate and bi-nucleate oidia, conidia and chlamydospores are comparatively less common than in some other orders of fungi although in culture under definite culture definitions on a definite media arthospores (oidia) are observed relatively frequently. The chlamydospores of *Aaterophora* and *Squamanita* have taxonomic importance as a generic character whereas the presence of conidia in the other groups has ordinarily not more than the value of an auxiliary specific character in as much as the conditions under which conidia are formed in nature, their constance, and even their existence in many species are unknown. In shape, the basidiospores vary from almost perfectly globose to strongly elongated from round to nodose, stellate, cross-shaped, or angular in circumference and from terete to laterally compressed or angular (polyedric) when seen from one end. The walls of the spores are either smooth and homogenous or ornamented. Three types of ornamentations can be distinguished *viz.* ornamentation primitive, secondary definite and perisporial. This classification cannot be applied exactly because it requires exact studies of fine structure of walls and their metachromatic prophases against a series of dyes and reagents.

The SEM is undoubtedly valuable for taxonomy of Agaricales. Pegler *et al.* (1993) studied the hilar appendages by using single stage carbon replica technique. They distinguished two types of hilum structure.

Nodulose Type

Hilum asymetrically borne on the hilar appendages pointing toward the inner (adaxial) side of the spore, consisting of an approximately circular area characterized by a number of small rounded protuberances-no spore. This type is common in hyaline-spored Aphyllophorales (*Schizophyllum, Cantharellus*) and Agaricales.

Open-Pore Type

Hilum similarly borne appearing as a depression, not noticeably often constituting a break in the wall continuity of the hilar appendage there is a conspicuous pore or tear in the wall. The two sides are frequently joined by a circum-apical slit which may widen to produce a single elongated tear. This type is common in Agaricaceae and perhaps exclusive in Coprinaceae, Bolbitaceae, Cortinaceae, Paxillaceae, Gomphidiaceae, Boletaceae and Strobilomycetae as well as in the Plutaceae and most Agaricaceae.

Chapter 4

Systematic Position of the Mushroom Taxa Described

Altogether five hundred sixty eight edible mushroom taxa belonging to one hundred fifty three genera reported from different habitats of India have been described in the text. Of these thirty five taxa belong to nineteen genera of Ascomycetes. Remaining five hundred thirty two taxa fall under one hundred thirty four genera of Basidiomycetes. Detailed account of the systematic position of the aforesaid taxa in the system of classification proposed by Ainsworth and Bisby (2001) may be read in tabular form as below:

Kingdom: Fungi

Phylum: Ascomycota

Class: Ascomycetes

Subclass: Sordariomycetidae

Order: Xylariales

Family: Xylariaceae

Xylaria

Order: Sordariales

Family: Lasiosphaeriaceae

Lasiosphaera

Order: Hypocreales

Family: Clavicipitaceae

Cordyceps

Subclass: Pezizomycetidae

Order: Pezizales

Family: Helvellaceae

Helvella, Elvella, Leptopodia, Microscyphus, Paxina

Family: Morchellaceae

Morchella, Verpa, Ptychoverpa

Family: Pezizaceae

Peziza, Daldinia

Family: Pyrenemataceae

Aleuria, Otidea, Sepultaria

Family: Sarcoscyphaceae

Cookenia, Sarcoscypha

Family: Tuberaceae

Tuber

Phylum: Basidiomycota

Class: Basidiomycetes

Subclass: Tremellomycetidae

Order: Auriculariales

Family: Auriculariaceae

Auricularia

Order: Dacrymycetales

Family: Dacrymycetaceae

Calocera, Dacryopinax

Order: Tremellales

Family: Tremellaceae

Tremella

Subclass: Agaricomycetidae

Order: Cantharellales

Family: Cantharellaceae

Cantharellus, Craterellus, Cratallarus

Family: Clavulinaceae

Clavulina

Family: Hydnaceae

Hydnum

Order: Phallales

Family: Geastraceae

Geastrum, Geaster

Family: Gomphaceae

Clavariadelphus, Gomphus

Family: Phallaceae

Phallus

Family: Ramariaceae

Ramaria

Order: Polyporales

Family: Albatrellaceae

Albatrellus

Family: Fomitopsidaceae

Daedalea, Piptoporus

Family: Ganodermataceae

Ganoderma

Family: Hapalopilaceae

Irciporus

Family: Meripilaceae

Rigidoporus

Family: Meruliaceae

Coriolus

Family: Podoscyphaceae

Podoscypha

Family: Polyporaceae

Amyloporia, Daedaliopsis, Favolus, Laetiporus, Lentinus, Lenzites, Microporus, Panus, Polyporus, Poria, Pycnoporus, Trametes, Tyromyces

Family: Sparassidaceae

Sparassis

Family: Steccherinaceae

Mycoleptodonoides, Steccherinum

Order: Thelephorales

Family: Bankeraceae

Sarcodon

Order: Russulales

Family: Auriscalpiaceae

Clavicorona, Lentinellus

Family: Bondarzewiaceae

Heterobasidiodon

Family: Hericiaceae

Hericium

Family: Peniophoraceae

Peniophora

Family: Russulaceae

Lactarius, Russula

Family: Stereaceae

Stereum

Order: Boletales

Family: Boletaceae

Austroboletus, Boletus, Leccinum, Strobilomyces

Family: Boletinellaceae

Phlebonus

Family: Gyroporaceae

Gyroporus

Family: Hygrophoropsidaceae

Hygrophoropsis

Family: Melanogastraceae

Melanogaster

Family: Paxillaceae

Paxillus

Family: Sclerodermataceae

Astraeus, Scleroderma

Family: Suillaceae

Suillus

Order: Agaricales

Family: Agaricaceae

Agaricus, Chlorolepiota, Chlorophyllum, Lepiota, Leucoagaricus, Leucocoprinus, Macrolepiota Micropsalliota, Podaxis

Family: Bolbitiaceae

Agrocybe, Bolbitius, Hebeloma Panaeolina, Panaeoluss

Family: Clavariaceae

Clavaria, Clavulinopsis

Family: Coprinaceae

Coprinus, Lacrymaria, Psathyrella

Family: Cortinariaceae

Cortinarius, Crepidotus, Galerina

Family: Entolomataceae

***Clitopilus, Entoloma, Rhodocybe, Rhodophyllus, Podabrella, Hirneola* (*Clitopilopsis*)**

Family: Fistulinaceae

Fistulina

Family: Hydnangiaceae

Laccaria

Family: Lycoperdaceae

Bovista, Calvatia, Langermannia, Lycoperdon, Vascellum

Family: Marasmiaceae

Armillaria, Flammulina, Lentinula, Marasmius, Oudesmansiella

Family: Mycenastraceae

Mycenastrum

Family: Nidulariaceae

Cyathus

Family: Phellorinaceae

Phellorinia

Family: Pleurotaceae

Hohenbuehelia, Pleurotus

Family: Pluteaceae

Amanita, Limacella, Pluteus, Volvariella

Family: Schizophyllaceae

Schizophyllum

Family: Strophariaceae

Pholiota, Stropharia, Hypholoma

Family: Tricholomataceae

Calocybe, Camerophyllus, Cantharellula, Clitocybe, Collybia, Cystoderma, Hygrocybe, Hygrophorus, Lepista, Leucopaxillus, Lyophyllum, Macrocybe, Melanoleuca, Mycena, Sinotermitomyces, Termitomyces, Tricholoma, Tricholomopsis, Trogia

Family: Tulostomataceae

Battarrea, Tulostoma

Chapter 5
Taxonomical Enumeration of Mushroom Taxa

1. *Agaricus abruptibulbus* Peck.

Synonym

Agaricus abruptibulbus (Pk.) Kauffm.

Common Name

Abruptly bulbous mushroom.

Habitat

It grows in scattered groups or in partial fairy rings, generally in hardwood forests, meadows and pastures.

Habit

Pileus 12-20 cm in diameter, thin, silky white (when scratched with fingernail or sharply bruised it turns yellow), hemispherical when young and convex to flat at maturity. Gills thin, narrow, and white but turns to pinkish to brown to blackish brown with age. Stipe tapering gracefully upwards from a base enlarged to a characteristic spherical bulb, annulate. Spores brown, elliptical and without ornamentation. Hyphae inamyloid and without clamp connections.

Distribution

Reported from Maharashtra (Sathe and Rahalkar, 1976; Patil *et al.*, 1995 and Sathe and Deshpande, 1979).

2. *Agaricus arvensis* Schaeff. Ex Sec.

Synonym

Psalliota arvensis (Schaeff. Ex Sec.) Kummer

Common Name

Horse mushroom.

Habitat

It grows in lawns, meadows, cultivated fields and in pastures etc., either solitary or scattered in groups of fairy rings.

Habit

Pileus 7-15 cm in diameter, white when young and turns yellow when touched or grow old, surface smooth, sub globose or flattened, sometimes covered with flat scales, margin usually decurved. Gills crowded, white but turns to grayish-pink to blackish brown with age. Stipe (5-13 cm×1-3 cm) central, usually cylindrical, hollow and bulbous, annulate and without volva. Spores purple-brown, broadly ellipsoid to ovoid.

Distribution

Reported from West Bengal (Bose and Bose, 1940 and Banerjee, 1947); Gujarat (Moses, 1948); Maharashtra (Trivedi, 1972; Sathe and Deshpande, 1979 and 1980); Uttar Pradesh (Pathak and Gupta, 1982); Madhya Pradesh (Rai, 1997); Arunanchal Pradesh (Mishra, 1999) Kerala (Bhavani Devi, 1995); and from Kolkata, Baroda and Nagpur (Nita Bahl, 1988).

3. *Agaricus augustus* Fries.

Synonym

Psalliota augustus (Fries) Quelet.

Common Name

The Prince; locally called "Mazakoon" in Kerala.

Habitat

It grows, solitary or scattered, in mixed forest mainly under the trees. *Quercus incanca* on the soil rich in humus in decaying leaf.

Habit

Pileus 6-20 cm in diameter, bright tawny orange to brown, convex then flattening, covered with concentric fibrillose scales, margin with tattered remnants of veil. Gills broad, almost white but soon turns brown. Stipe tall, white and covered below with floccose scales. Annulus large, pendent, skirt like and scaly. Spores (7-10 µm × 5-6 µm) ovate.

Distribution

Reported from Kerala (Bhavani Devi, 1995) and from Himachal Pradesh (Kumar and Lakhanpal, 1993).

4. *Agaricus basianulosus* Paracer and Chahal

Habitat

Sporophores scattered or in groups on hard soil in the field, underground at first but gradually comes out on the surface (1.8-2.8 cm high)

Habit

Pileus plobose but becomes planoconvex at maturity (5.6-9.0 cm in diameter), white but tan when bruised, texture velvety to fibrillose. Gills free, crowded, pink but dark brown or chocolate brown when old. Stipe (4.5-7.6×2.2-3.0 cm) with tapering basal part, solid within, brittle and easily separable from the pileus, annulus present towards the base of stipe, showing as incomplete rings of scale.

Distribution

Reported from Punjab (Paracer and Chahal, 1962).

5. *Agaricus bisporus* (Lange) Sing.

Synonym

Agaricus brunnescens.

Common Name

Cultivated mushroom, Champignon.

Habitat

Grows on soil, mannure heaps, in gardens, in the soil green houses meadows and pasture and along road sides on scattered horse manure.

Habit

Sporophores usually centrally stipitate. Pileus (3.5-10.0 cm in diameter), convex but turns flattened when old, whitish to stramineous or brown, surface smooth to finely scaly, margin thick and projecting, occasionally fibrilose-striate. Gills crowded, distinctly formed, free, at first whitish, then pink, finally becoming sepia coloured, narrow to moderately broad, edge fimbriate. Stipe central, equal or shightly attenuated at the top, sometimes slightly bulbous at the base (3.0-12×1.0-1.8 cm), whitish, sometimes upward fleshy coloured with basal white rhizomorph, annulus well developed without volva, white, soft, membranous.

Distribution

Reported from Solan, Himachal Pradesh (Sohi *et al.*, 1965; Seth, 1971); Seoni district of M. P. (Rai, 1997) and from Maharashtra (Sathe and Deshpande, 1979).

6. *Agaricus bitorquis* (Quel.) Sacc.

Synonym

Agaricus campestris subsp. *bitorquis* (Quel.) Konrad and Maubi. *Agaricus edulis* (Witt.) Moller and Schaeff. *Agaricus rodmanii* Peck.

Common Name

Town or Street mushroom or Pavement mushroom or Banded Agaricus

Habitat

Solitary to scattered in grassy places, meadows and pastures. Common on roadsides and woodland edges, often forces up paving stones.

Habit

Cup 6-15 cm, soon noticeably flattened with margin slightly inrolled, smooth or slightly cracking, white then discoloured pale brown. Gills very narrow and crowded, dull pink then deep reddish brown. Stem short and stout, thick, veil sheathing the base, rather like a volva. Often a thick narrow ring high on stem. Flesh thick, white, slowly becoming slightly brownish. Spores brown, subglobose, 5-7×4-5 μm.

Distribution

Reported from Punjab (Garcha, 1980; Gupta *vide* Saini and Atri, 1995); Karnataka (Sathe and Rahalker, 1978).

7. *Agaricus brunnescens* Peck

Synonym

Agaricus bisporus (Lange) Sing.

Common Name

Cultivated mushroom, Common mushroom, White button mushroom, European mushroom or Temperate mushroom.

Habitat

Sporophores solitary sometimes in groups on manure heaps, in garden or on soil rich in manure.

Habit

Pileus 5-10 cm, wide, convex but flattened when old, whitish to pale brown, soft, surface smooth to finally scaly, fibrillose, striate. Stipe central, uniform or slightly attenuated at the top, sometime with bulbous base (3-12×1-1.8 cm). Gills crowded and free, dull pink turning to dark brown.

Distribution

Reported from Arunanchal Pradesh (Mishra 1999); Himachal Pradesh (Sohi *et al.*, 1965; Seth, 1971); Jammu and Kashmir (Kaul, 1978); Maharashtra (Sathe and Desphande, 1980); Madhya Pradesh (Rai, 1997).

8. *Agaricus campastris* L. ex. Fr.

Synonym

Psalliota compestris (Fries) Quelet

Common Name

Field Mushroom or Meadow Mushroom.

Habitat

Grows scattered or in groups sometimes in fairy rings or partial rings on open grassy field, meadows and pastures, This is the widest known of all mushrooms.

Habit

Sporophores centrally stipitate, white when young and old. Pileus (4.0-8.0 cm in diam.) almost round but soon becomes convex to flattened, usually white or brownish, surface smooth sometimes scaly, non-striate. Gills crowded, unequal, free, separable. Initially white, then pink and finally purple brown to sepia at maturity. Stipe central, cylindrical tapering a little towards the base (4.0-8.0×0.8-1.5 cm). Thick, not hollow, without volva. Pileus and stipe not becoming yellow where bruised.

Distribution

Reported from Punjab (Baden Powell, 1863 in Bose and Bose, 1940; Atri *et al.*, 1991, 2001); Calcutta, West Bengal (Bose and Bose, 1940; Banerjee,1947; Nita Bahl, 1988); India (Chopra and Chopra, 1955); Bihar, North Western Himalayas (Butler and Bisbey, 1960; Nita Bahl, 1988); Nagpur (Trivedi, 1972; Nita Bahl, 1988); Kerala (Bhavani Devi, 1995); N.W. and N.E. Himalayas, Bihar and Bengal (Bose and Bose, 1940; Smith, 1949); Jammu and Kashmir (Nita Bahl, 1988; Watling and Gregory, 1980; Abraham, 1991; Pandotra, 1966), South West India (Sathe and Rahalkar, 1975; Sathe and Sasangen, 1977); Orissa (Dhancholia and Sinha, 1988); Andhra Pradesh (Manoharacharya and Vijaya Gopal, 1991) and from Maharashtra (Sathe and Deshpande, 1979).

9. *Agaricus campastris var. equestris* Moller.

Habitat

Grows gregariously below *Bougainvillea glabra*.

Habit

Carpophores upto 6 cm high. Pileus upto 4 cm broad convexo-flattened to applanate, yellowish white to greyish yellow, greyish brown on maturity, surface dry, fibrillose, margin irregular, splitting at maturity. Gills free, close, narrow, unequal, not in series. Stipe upto 5.7 cm long, obclavate, concolourous, hollow, fibrillose, annulated, pseudorrhiza present. Spores (5-7.5×4-5 µm) ellipsoidal, double walled, pigmented, apiculate.

Distribution

Reported from Punjab (Atri *et al.*, 1991)

10. *Agaricus campastris var. fusco-pilosella* Moller.

Habitat

Grows isolated in grassland, near angiospermic tree including *Dalbergia sissoo*, *Mangifera indica* etc.

Habit

Carpophores up to 7 cm high. Pileus up to 5.5 cm broad convex to umbonate then expanded, pale orange then brownish grey, margin almost regular, surface dry, scaly. Gills free, close, unequal in series of two. Stipe upto 5.5 cm long, broader towards lower half then attenuated, dull white, fibrillose, annulated. Spores (5.5-7.5×4-6.5 µm) ellipsoidal to oval, smooth, double walled, brown.

Distribution

Reported from Punjab (Atri *et al.*, 1991)

11. *Agaricus campastris var. indica* Atri

Habitat

Grows in small groups among *Cyanodon dactylon* along the road side.

Habit

Carpophores upto 5 cm high. Pileus upto 4.3 cm broad, conical to convex when young, applanate at maturity, greyish brown to dark brown, margin irregular, splitting, surface dry and scaly. Gills free, subdistant, unequal, not in series. Stipe upto 4.5 cm long, obclavate, annulus white, disappears at maturity. Spores (5-7.5×4.5-6 µm) broadly ellipsoid, double walled, pigmented, apiculate.

Distribution

Reported from Punjab (Atri *et al.*, 1991)

12. *Agaricus campastris var. isabellinus* Moller.

Habitat

Grows in humus soil under the hedge of *Clerodendron enorme.*

Habit

Carpophores upto 5.5 cm in height. Pileus upto 5 cm broad convex flattened, umbo broad, greyish orange in the centre, pinkish white towards the margin, feebly umbonate, surface dry and scaly, margin irregular. Gills free, crowded, unequal, not in series. Stipe upto 5 cm long, orange white above the annulus, white and fibrillose below. Spores (7-8×4-5 µm) broadly ellipsoidal, double walled, outer wall thick and dark, pigmented, apiculate and guttulate.

Distribution

Reported from Punjab (Atri *et al.*, 1991).

13. *Agaricus campastris var. singeri* Atri

Habitat

Grows in small groups below *Dalbergia lanceolasia.*

Habit

Carpophores upto 5.5 cm high. Pileus upto 5.5 cm broad, convex, flattened to applanate, umbo absent, greyish brown, surface moist, margin irregular. Gills free, close to crowded, broad, unequal, not in series. Stipe upto 5 cm long, white, solid, fibrillose below annulus. Annulus membranous, sheathed above. Spores (6-7.5×4-4.5 μm) ellipsoidal, double walled, pigmented, apiculate, guttulate.

Distribution

Reported from Punjab (Atri *et al.*, 1991)

14. *Agaricus comtulus* Fr. Epicr.

Habitat

Grows in small group below the *Syzygium cumini* (Linn.) Skeels.

Habit

Carpophore upto 4 cm in height, compestroid. Pileus upto 3.7 cm broad, yellowish white to light brown with dark greyish brown central disc; margin irregular, surface scaly, scales appressed, fibrillose, arranged in fascicles. Flesh turning yellow at exposure. Lamellae free, moderately broad, light pink, becoming greyish brown at maturity. Spore print greyish brown. Stipe upto 3.7 × 0.7 cm, bulbous, yellowish white, yellow where bruised. Annulus white, sheathed above, cortina present. Gill trama made up of irregular hyphae.

Distribution

Reported from Patiala, Punjab by Atri *et al.* (1992).

15. *Agaricus cupreo-brunneus* (Schäffer and Steer) Moller

Habitat

Grows Scattered to gregarious in disturbed ground; along paths, in sparse grass, vacant lots etc.

Habit

Cap 2.5-6 cm broad, convex, expanding to nearly plane; margin at first inrolled, then decurved, plane to upturned at maturity; surface dry, the disc sometimes tomentose, becoming appressed squamulose, brown to greyish-brown, occasionally tinged vinaceous, the squamules often raised near the margin giving a shaggy aspect, obscurely squamulose to merely appressed fibrillose in age or from weathering; flesh thin, soft, pallid, discoloring slowly to pale dull-brown when injured; not yellowing in KOH. Gills free, close, moderately broad, dingy-pink when young, blackish-brown at maturity. Stipe 1.5-3.5 cm long, 1-1.5 cm thick, straight, round, equal or narrowed at the base, stuffed to hollow at maturity; surface pallid, smooth with scattered flattened scales at the apex, minutely scaly below, often nearly glabrous in age, sometimes discoloring brownish where handled; veil cottony-membranous, thin, whitish, forming an inconspicuous, short, erect, medial to superior ring. Spore print blackish-brown.

Distribution

Reported from Poona (Patil *et al.*, 1995).

16. *Agaricus cylinderacea* (DCex Fr.) Maire

Synonym

Pholiota aegerita (Brig) Quel

Common Name

Poplar Fiddcap

Habitat

Grows on dead and drying old Poplars in large clump.

Habit

Cap 4–10 cm across, hemispherical, convex becoming flattened and sometimes cracked at centre and often wavy near the margin, pale buff to almost white with rust flush at centre when young, becoming darker brown with age. Stipe 50–100 × 10–15 mm, cream at first, darker brown with age, with persistent ring which soon becomes dusted brown by the spores. Flesh white in the cap and stipe, brown in the stipe base. smell of old wine casks. Gills adnate or slightly decurrent, cream at first then tobacco brown due to the spores. Spore print tobacco brown. Spores ovoid-ellipsoid, 8.5–10.5 × 5–6 µm with minute germpore. Cap cuticle cellular.

Distribution

Reported from North West Himalayas (Watling and Gregory, 1980).

17. *Agaricus diminutivus* Peck

Synonym

Agaricus amethystina; A. dulcidulus

Habitat

Grows scattered in open grasslands.

Habit

Carpophore upto 7 cm in height, placomycetoid. Pileus 2.5-5 cm broad, convex then expanded or centrally depressed, whitish or pinkish brown, thin, fragile, surface dry with small, silky and brownish scales. Stipe, tapering upwards, glabrous, whitish or pallid, 4-5 cm long, annulated, annulus thin, white and persistent. Spores (4-5 µm) ellipsoidal, brown.

Distribution

Reported from Kerala (Bhavani Devi, 1995) and from Punjab (Saini and Atri, 1989).

18. *Agaricus edulis* (Vitt) Moll.

Habitat

Grows solitary or gregarious on soil.

Habit

Pileus 8.5-15.5 cm broad, dull white, hemispherical with depressed centre, surface squamulose, margin irregular, inrolled. Gills free, greyish red to reddish brown to dark brown. Stipe upto 6 cm long, concolourous, bulbous, robust, solid, doubly annulate. Spores (5-8.5×3.7-6.5 µm) ellipsoidal, smooth, double walled, apiculate.

Distribution

Reported from Punjab (Saini *et al.*, 1992).

19. *Agaricus endoxanthus* Berk and Br.

Habitat

Grows solitary to scattered on soil.

Habit

Pileus 2.5-5 cm broad, convex, finally expanded, surface greyish brown, covered with appressed redially fibrillose squamules, sometimes splitting at places to reveal the white context below, margin incurved then straight. Lamellae free, pinkish white, darkening to chocolate brown, crowded with lamellulae of different lengths. Stipe 6-12 cm × 4-7 mm, central, cylindrical, equal with a marginate basal bulb, solid becoming hollow with age. Annulus superior, white, large, pendent. Context white upto 5 mm wide, composed of thin walled hyphae. Cheilocystidia in fascicles. All hyphae lacking clamp-connections.

Distribution

Reported from Kerala by Vrinda *et al*. (1995).

20. *Agaricus fusco-fibrillosus* (Moller) Pilat.

Habitat

Grows gregariously on lawn or below the hedge of *Clerodendron inerme* (Linn.)

Habit

Sporophore upto 4.5 cm in height, pileus upto 3 cm broad, convex to convex-flattened when young, infundibuliform at maturity. Pinkish white when young, umbo broad, margin irregular, striate, surface dry, scaly throughout, scale appeared fibrillose, cottony white to reddish white in young. Carpophores, greyish with maturity, flesh 3 mm thick in the centre, white, turns orange with reddish tinge in exposure. Gills free, pink in young. Stipe upto 4.2 × 0.3 cm (0.5 cm broad near the base).

Distribution

Reported from Punjab plains (Patiala) by Saini *et al*. (1991).

21. *Agaricus impudicus* (Rea) Pilat

Habitat

This uncommon mushrooms grows in lucidness or coniferous forest.

Habit

As with all *Agaricus* species, gills are free, colour progresses with age from pale-pink to a chocolate colour, and spores are dark brown. Cap 4–15 cm wide, and appears brownish due to numerous brownish scales on a white background. The stipe with clear annulus ring, white, 6–12 cm tall and 0.8–2 cm thick, cylindrical and wider towards the bottom, or ending in a bulb. It is distinguished from similar forest growing *Agaricus* mushrooms in that it does not bruise yellowish or reddish when cut and the widening stipe.

Distribution

Reported from Poona (Patil *et al.*, 1995).

22. *Agaricus langei* (F.H. Moller) Moller

Common Name

Named after Danish mycologist Lange.

Habitat

Sporophores gregarious or scattered, grows under the trees.

Habit

Cap 6-12 cm, semiglobose then expanded, covered with brownish, fibrillose scales, thinning out toward edge, cuticle joined and darker at the center, dry. Gills free, pinkish, finally blackish brown, crowded, short, margin sterile and whitish. Stipe 7-12×1.5-2.5 cm almost cylindrical or faintly enlarged at base, smooth above ring, white and floccose below, tinged with red and then darkening if touched, hollow. Ring descending with small brown scales below. Flesh at once crimson when cut.

Distribution

Reported from Trivendrum, Kerala (Bhavani Devi, 1995).

23. *Agaricus micromagathus* Peck. Bull.

Synonym

Agaricus pusillus Peck.

Common Name

Almond Mushroom.

Habitat

Gorws in variuos kinds of soil in shaded or exposed places, solitary or caespitose.

Habit

Pileus fleshy but thin, convex, becoming plane, sometimes slightly depressed at the center, 1-5 cm broad, surface dry, silky–fibrillose or fibrillose–squamulose, grayish brown, darker or brown on the disk, often with yellowish or ferruginous stain, context fragile, white or whitish, not changing colour when wounded, with taste and odour of almonnds, lamellae thin, crowded, free, grayish, soon pinkish, finally brown. Stipe equal or slightly tapering upward, sometimnes bulbous, stuffed or hollow, slightly fibrillose, white, 1-2.5 cm long, 2-6 mm thick, annulus slight, often evanescent.

Distribution

Reported from Poona (Maharashta) by Sathe and Deshpande (1979) and from south west regions of India (Sathe and Rahalkar, 1978).

24. *Agaricus nivescens* (Moller) Moller.

Habitat

Grows on pine needles in coniferous forest.

Habit

Pileus upto 9 cm in diameter, convexo-applanate at maturity; surface dry, non-hygrophanus, white to pale white, dark centre, glabrous, margin appendiculate, cuticle fully peeling, flesh upto 6 mm in thickness, unchanging. Lamellae free, densely crowded, unequal, 4-5 sized, dark purplish brown at maturity, 5-7 mm in breadth, edge smooth. Stipe upto 9 cm. in length, 7-13 mm in thickness, tapering upwards, fistulose, base sub bulbous with rhizoids, fleshy, surface white concolorous.

Distribution

Reported from Shimla, Himachal Pradesh (Upadhyay *et al.*, 2007)

25. *Agaricus pattersonae* Fr.

Habitat

Grows singly or in groups associated with *Cupressus macrocarps* or with mature tree of the age 40 years.

Habit

The cap, 5–19 cm in diameter, initially almost hemispherical in shape, transforming to broadly convex and finally to flattened or with edges upturned in age. The cap surface dry, with fibrils when young, but later the fibrils form large, dark brown appressed squamules (2–9 mm long x 2–5 mm broad). The cap colour may be various shades of brown depending on the maturity of the specimen. The cap flesh typically 1–3 cm thick, firm, white, and stains deep red 20–30 seconds after injury or bruising. The gills are free in attachment, close, 5–15 mm broad, and marginate. The light-cinnamon colour in young specimens turns to a dark blackish brown in age after the spores develop. Bruised gills stain a vinaceous (wine-colored) red. The stipe usually 8–18 cm long x 2.5–4 cm thick with a bulbous shape. The interior is hollow, with the internal cavity being between 5–9 mm thick. The stem context somewhat fibrous and white in colour, except for the basal section which is yellowish. Bruising or cutting results in a red stain after a minute. The surface of the stipe barely striate above the annulus, and smooth below except for fragments of the universal veil. During development the veils rupture and form an upper veil (partial veil), which initially hangs from the pileus edge and a lower veil. As the partial veil disintegrates, it often leaves fragments 2–3 mm in size attached to the margin.

Distribution

Reported from Poona (Maharashtra) by Sathe and Deshpande (1979).

26. *Agaricus placomyces* Peck.

Synonym

Psalliota meleagris Schaeft.

Common Name

Flat capped *Psalliota*, wood mushroom.

Habitat

It grows in scattered groups on the ground, soil, among grasses, below trees and between the heaps of rocks.

Habit

Pileus upto 10 cm in diameter, ovate then convex and finally flattened with age, white but brown at the centre due to concentration of brown scales towards the centre. Gills free, white to pink to blackish brown with age, stipe 5-15 cm long, cylindrical with slightly swollen base, pale brown and annulated. Spores (4.5-6.5 × 3-4 μm) purple-brown, ellipsoidal and without ornamentation.

Distribution

Reported from South-West regions of India (Sathe and Rahalkar, 1978), Himachal Pradesh (Sharma *et al.*, 1978), Patiala, Punjab (Saini *et al.*, 1991), Kerala (Bhavani Devi, 1995), from Lucknow (Ghosh *et al.*, 1974), from Solan, Himachal Pradesh (Sharma and Thakur, 1978) and from Maharashtra (Sathe and Deshpande, 1979).

27. *Agaricus rhoadsii* Murrill

Habitat

Grows solitary on humus soil below hedge of *Clerodendron inerme.*

Habit

Pileus upto 6.5 cm broad yellowish white with light brown disc, applanate, margin irregular, slightly raised, brown on drying, surface moist with fibrillose scales, scales more towards centre. Gills free, crowded, unequal, not in series. Stipe 7.7×0.5 cm, concolourous solid, bulbous, annulated, annulus white and movable. Spores (5-6.8×3.8-4.5 μm) ellipsoidal, double walled, pigmented, apiculate and guttulate.

Distribution

Reported from Punjab (Saini *et al.*, 1991)

28. *Agaricus rodmanii* Peck

Synonym

Agaricus bitorquis

Habitat

Grows solitay to scattered below angiospermic tree.

Habit

The semiunderground fruiting is the characteristic of this species. Cap upto 15 cm across, glabrous, margin inrolled when young often extending beyond the gills in maturity. Ring often pulling away from the stalk at its top edge and its bottom edge, often flaring, usually barely breaking the surface of the ground.

Distribution

Reported from Patiala (Punjab) by Saini *et al.* (1993) and from Poona by Sathe and Deshpande (1979).

29. *Agaricus silvaticus* Schaffer. ex Secc.

Habitat

Sporophores solitary or scattered on the ground under the *Casuarina* plant, in forests and in the field adjacent to woods.

Habit

Pileus 5.0-11.0 cm in diameter, convex or campanulate when young, later expanded, whitish-brown, brown or grey, soft, fleshy, fibrillose or with brownish or reddish brown appressed scales. Stipe central, uniform (1.0-2.5 cm thick). Flesh white, turns reddish when cut. Fresh gill reddish-brown. Volva absent.

Distribution

Reported from Trivandrum, Kerala (Bhavani Devi, 1995); Arunachal Pradesh (Mishra, 1999); Darjeeling, West Bengal (Berkeley, 1856); Darjeeling, West Bengal, 2500ft (Butler and Bisby, 1931; Smith, 1949) and from Maharashtra (Sathe and Deshpande, 1979).

30. *Agaricus silvaticus var. rubribrunescens* (Murr.) Heinem

Habitat

Sporophores solitary on grassy soil.

Habit

Carpophores smaller in size ranging from 4.5-7.8 cm in height, stature compestroid, universal veil, tissue hyphal. Pileus 2.9-6.5 cm broad, pinkish white surface with brownish centre, scales cottony white, flesh pinkish white, first becoming red finally deeper on exposure, caulocystidia absent.

Distribution

Reported from Punjab (Saini *et al.*, 1997).

31. *Agaricus silvaticus var. silivaticus* Schaeff.

Habitat

Grows scattered on humicolous soil in *Abies* among *Onichium*, *Fragaria* etc.

Habit

Sporophores upto 12.5 cm high, placomycetoid. Pileus upto 9 cm broad, conical when young and umbrella like at maturity with broad umbo, surface dry, orange yellow, yellowish brown towards the margin with grayish white back ground, scales appressed, fibriller with few separating from the surface, margin irregular, splitting at maturity, culticle fully pealing except at the umbo. Flesh upto 7 mm thick, white underneath, turns pinkish on exposure. Gills free, subdistant to close, unequal, not in series. Stipe upto 10.9×2 cm, distinctly bulbous, white turning to grayish-yellowish brown above annulus but light coloured below at maturity, surface turns brownish on bruising, smooth, hollow.

Distribution

Reported from Narkanda, Himachal Pradesh (Saini *et al.*, 1997); Darjeeling, West Bengal (Berkley, 1856); Pune, Maharashtra (Sathe and Rahalkar, 1978).

32. *Agaricus silvicola* Vittr. Fr.

Synonym

Psalliota silvicola

Common Name

Wood mushroom.

Habitat

Grows in group or scattered on humicolous soil and among dead leaves near hedge of *Clerodendron* or *Morus alba* or under *Diospyros*.

Habit

Cap 5-12 cm, first globose then campanulate becoming expanded, cuticle dry, shiny, whitish yellowing with age or when touched. Gill crowded, free, initially, dirty white then pinkish, sepia brown when mature. Stipe (6-11×1.5-2.5 cm) slender, hollow, fragile, basal bulb white turning to yellow, pinkish above ring, ring double with darkening fringe.

Distribution

Reported from South west India (Sathe and Rahalkar, 1978); Punjab, Patiala (Atri *et al.*, 2001); by Atri from H.P. and Gupta from Dehradun (in Atri *et al.*, 2001). And from Maharashtra (Sathe and Deshpande, 1979).

33. *Agaricus subrufescens* Peck

Synonym

Agaricus blazei Murrill; *Psalliota subrufescens* (Peck) Kauffman.

Common Name

Reddish *Psalliota*.

Habitat

Grows in forests or woodlands on soil, generally in group but solitary also.

Habit

Pileus 9.5-11.5 cm in diameter, plano-convex, convex, finally broadly expanded, white to reddish brown, fleshy, soft, surface smooth covered with reddish brown scales. Stipe stout and whitish, slightly swollen at the base. Gills crowded and free, dark brown.

Distribution

Reported from Arunchal Pradesh (Mishra, 1999) from Chambaghat, Solan, Himachal Pradesh (Sharma and Thakur, 1978).

34. *Agrocybe broadwayi* (Murr.) Dennis

Synonym

Hebeloma broadwayi Murill.

Habitat

Grows solitary on soil

Habit

Fruit body with narrow stipe and dull colour. Pileus fleshy, convex to expanded, margin entire, 2- 4 cm. broad, surface smooth and white, glabrous, subviscid, not striate, lamellae adnexed crowded, not radially separable from the context. Volva and annulus absent. Veil present at time in young stage but evanescent. Pileus centrally stipitate. Stipe fleshy. Universal veil not arachnoid. Spore print-dark brown. Cheilocystidia clavate to lageniform, thin walled hyalin. Pleurocystidia absent.

Distribution

Reported from Kolhapur (Maharashtra) by Patil *et al*. (1995).

35. *Agrocybe cylindracea* (De. ex Fries) Maire

Synonym

Agrocybe aegerita (Brig.) Singer.; *Pholiota aegerita* (Brig.) Quel.

Common Name

Southern poplar mushroom.

Habitat

Grows in tufts on stumps and old trunks of poplar or elm.

Habit

Pileus 3-14 cm in diameter, convex to expanded or plane with age, central region depressed, light brown but fading to white towards margin, surface viscid, margin curved. Stipe slender, white, brownish at base, surface fibrillose and striate, annulated. Spores elliptical, smooth and brown.

Distribution

Reported from Jammu and Kashmir (Watling and Gregory, 1980).

36. *Agrocybe molesta* (Larch) Singer

Synonym

Agrocybe dura (Bolt. ex Fr.) Singer.; *Pholiota dura* (Bolt. ex. Fr.) Kummer.

Common Name

Roger's Mushroom

Habitat

Grows solitary in grassy land sometimes in hop yard.

Habit

Pileus upto 7 cm in diameter, first convex then expanded, white to cream. Stipe upto 8 cm long, whitish, annulate, annulus cottony. Spores ovoid to ellipsoidal, thick and coloured wall.

Distribution

Reported from Srinagar, Jammu and Kashmir (Watling and Gregory, 1980).

37. *Agrocybe praecox* (Pers. ex Fr.) Fayod.

Synonym

Pholiota praecox (Pers.ex Fr) Kummer; *Agaricus praecox.*

Habitat

Sporophores solitary or in group or scattared on the ground or in lawns, pastures or in open woods.

Habit

Sporophores usually centrally stipitate. Pileus 2.0-6.0 cm in diameter, convex young and expanded when old, sometimes with an umbo, whitish or yellowish at first, later yellowish brown, margin slightly incurved when young but later upturned, sometimes, surface uneven with numerous shallow pits. Gills crowded, adnate or slightly decurrent, sinuate, at first whitish, then greyish and finally turns rusty brown, broader in the middle, edge white and finely wavy. Stipe more or less slender, cylindrical, 4.0-8.0 cm long, uniformly thick, slightly narrowed at the base, whitish, greyish or flesh coloured, solid or hollow. Veil forming ring at the top of the stipe or sometimes remaining as fragments at the margin of the pileus, disappearing readily with age.

Distribution

Reported from Saharanpur, U.P. (Butler and Bisby, 1960; Nita Bahl, 1988) and from Lucknow, U.P. (Pathak and Gupta, 1982).

38. *Agrocybe semiorbicularis* (Fr. ex St. Amous) Fayod.

Synonym

Naucoria semiorbicularis Bull., *Agrocyby pediades*

Habitat

Sporophores scattered or in groups on the ground, in pastures, lawn, road sides, grassy lands, in waste places and on dung.

Habit

Pileus 1.0-4.0 cm in diameter, convex but expanded when old, tan to yellowish, smooth, occasionally cracks on the surface, wrinkled when old, surface sticky when moist, Gills crowded, adnate to sinuate, pale tan when young, turning to rusty brown with age, broader than the thickness of the pileus, edge often delicately fringed. Stipe slender, uniform 4.0-6.0×0.2-0.6 cm, light reddish brown, surface smooth, and shining, tough but pliable, sometimes with striations, stuffed with white pith, without ring and volva. Flesh light tan.

Distribution

Reported from India (Butter and Bisby, 1960), Chandigarh, Punjab (Rawla *et al.*, 1982; Saini and Atri, 1995).

39. *Albatrellus confluens* (A. and S. Per Fr.) Kotl. and Pouzar

Synonym

Polyporus confluens Alb and Schw. ex Fr.

Habitat

Grows solitary on soil in oak forest.

Habit

Sporophores laterally stipitate, overlapping and confluent, fan-shaped. Stipe solid, unbranched, cylindrical and glabrous. Upper surface creamish yellow, scaly, margin acute, uncurved. Hymenial surface white when fresh, dark brown when dry, pores round and angular. Spores (2.8-4×2.4-3.2 µm) globose or ovoid, hyaline, smooth, apiculate, uniguttulate.

Distribution

Reported from Uttarakhand (Thind and Chatrath, 1960; Thind and Dhanda, 1979).

40. *Aleuria aurantia* (Pers ex Fr.) Fuckel

Synonym

Peziza aurantia Fries

Common Name

Orange Peel fungus or Scarlet fungus or Golden cup fungus or Orange cup fungus.

Habitat

Grows on ground or bare sandy soil.

Habit

Sporophores in groups, gregarious, short, 1.5-4.0 cm in diameter, cup shaped but flattened with age, sessile or laterally stipitate, sometimes split along the margin. Cup upto 10 cm in diameter irregularly saucer shaped but when caespitose they may be compressed from pressure. Flesh thin and brittle turning green by iodine. Hymenium bright orange, asci 8 spored not turning green by iodine 132.0-171.0×14.0 µm. Ascospores elliptical, surface reticulate, oil drops present (13.2×10.5 µm), paraphyses clavate projecting beyond asci.

Distribution

Reported from Sikkim, Darjeeling, West Bengal (Barkeley, 1856; Currey, 1876), Mussorie, U.P. (Thind *et al.*, 1957; Thind and Sethi, 1957), Pangim Tuting in Arunachal Pradesh (Mishra, 1999), Mongpoo, W.B. (Kar and Dewan, 1975).

41. *Amanita caesarea* (Scop. ex. Fr.) Pers. ex Schw.

Common Name

Caesar's mushroom, Royal agaric, Orange amanita.

Habitat

Sporophores solitary or in small groups. Grows under tree, in forests and on the ground.

Habit

Sporophores usually larger in size, yellow or orange with distinct veilar remnants. Pileus 5.2-10.2 cm or more in diameter, ovate when young, then convex and finally flattened, yellow or orange, deeper at the center. Young fruit-bodies pale, surface smooth, margin with prominent striations, sometimes curved downward at maturity. Gills free, yellow or orange. Stipe centrally placed, cylidrical, 12.7-20.3 cm. long, yellow or orange, stuffed or hollow, annulate, volvate, scaly upto annulus, volva white, large, loose, sac like, fleshy, sometimes very reduced, volva covering entire fruit body at button stage. Annulus large, yellow or orange, membranous, having from the top of the stipe.

Distribution

Reported from Arunchal Pradesh (Mishra, 1999), Baroda (Moses, 1948), Assam (Butter and Bisby, 1960; Nita Bahl, 1988); Meghalaya (Berkeley, 1856); Boroda, (Nita Bahl, 1988) and from Kerala (Bhavani Devi, 1995).

42. *Amanita cociliae* (Berk and Br) Bas.

Synonym

Amanita inaurata (Gillet) Fayod.

Habitat

Grows solitary, scattered or gregarious, mixed under conifers mycorhizal with *Cedrus deodara*.

Habit

Cap 5–12 cm across, convex to flat with an upturned deeply lined margin and a low umbo, brownish black to brownish grey, darker at the disc, paler towards the margin, smooth, slightly sticky when moist, with loose, charcoal grey patches of scales cuplike volvar remnants, dotted around the cap. Gills free, close, white. Stipe 50–160 × 7–15 mm, hollow to lightly stuffed, tapering slightly towards the top. Spore deposit white.

Distribution

Reported from North West Himalayas (Kumar *et al.*, 1990).

43. *Amanita flavoconia* Atte.

Common Name

Yellow patches, Yellow wart, Orange amanita or Yellow dus *Amanita*.

Habitat

Grows solitary, scattered gregarious in coniferous forests, mycorhizal with *Cedrus deodara*.

Habit

Deep Yellow to deep Orange pileus, Yellow floccose patches. The cap initially ovoid, beome convex and eventually flattenad with age, orange to bright yellow–orange reaching a diameter of 3–9 cm, covered with chrome yellow wart when young which may be rubbed off or washed away with rain. The cap surface smooth and sticky beneath the warts. The flesh white or tinged yellow on the edges, initially covered with yellowish partial veil. The stipe typically 5.5–11.5 cm long by 0.7–1.4 cms thick, equal or slightly tapered upward from a small rounded bulb at the base, white to yellowish orange in colour, surface smooth or covered with flake. The partial veil leave a skirt like ring (annulus) on the upper stipe. The spore print white.

Distribution

Reported from North West Himalayas (Kumar *et al.*, 1990).

44. *Amanita fulva* (Schaeff. ex) Pers.

Common Name

The Tawny Grisette or Orange Brown ringless *Amanita*.

Habitat

Grows solitary or scattered, on the ground in coniferous and mixed woodlands associated with *Cedrus deodara*, *Pinus roxburghii*, *Quercus incana* and *Rhododendron arboreum*.

Habit

Pileus 4-10 cm in diameter, convex to plane with age, umbonate, reddish orange to reddish brown, slightly viscid, smooth with striated margin. Stipe 8-13 cm long, central, cylindrical, white or pale yellow, finely squamulose, broad at the base and tapered above, without annulus, with a white, free, saccate volva present at the base. Spores (9-12 µm) globose, smooth, hyaline, apiculate, non-amyloid and guttulate.

Distribution

Reported from North-Western Himalayas (Bhatt and Lakhanpal, 1988a) and from Orissa (Das and Sinha, 1990).

45. *Amanita gemmata* (Fr Bertillum)

Common Name

Gemmed Mushroom

Habitat

Grows solitary, scattered on humicolous soil, mycorhizal with *Pinus roxburghii* and *Cedrus deodara*.

Habit

Sporophoe creamy to pale yellow, golden yellow often slightly darks at the center. Cap 4–10 cm, in diameter, pale yellowish white, striated, thin, yellowish white annulus, rimmed volva broad becoming convex to plane, surface slightly viscid when moist. The whitish universal veil covering entirely the younger mushroom, form whitish spots or warts on mature mushroom, and may eventually wash off with age. Stipe 5–13 cms. usually smooth above the ring and sometimes scaly below, narrowing slightly towards the cap, dry white or tinged yellowish in colour. Gills free or slightly attached, crowded, broad to narrow and white. Spore print white.

Distribution

Reported from North West Himalayas (Kumar *et al.*, 1990).

46. *Amanita hemibapha* (Berk and Br.) Sacc.

Synonym

Agaricus hemibapha Berk.

Habitat

Grows solitary or gregarious on humus, always associated with forest trees.

Habit

Pileus upto 11 cm in diameter, hemispheric then expanding to broadly convex, often becoming depressed with the margin upturned with age, without umbo, red, smooth, glabrous, margin striate. Stipe upto 15 cm long, central, yellow, slightly narrowed upward, veil present. Spores (8.3-10.5×5.3-6 µm) ellipsoid to elongate, hyaline, thin walled, smooth inamyloid, apiculate, guttulate.

Distribution

Reported from Assam (Berkley, 1952), Gujarat (Moses,1948), Uttar Pradesh (Bakshi, 1974), Himachal Pradesh (Kumar *et al.*, 1967), Kerala (Vrinda *et al.*, 2005) and from Orissa (Das and Sinha, 1990).

47. *Amanita muscaria* (L. ex Fr.) W.J. Hooker

Common Name

Fly Agaric Mushrooms.

Habitat

Grows associated with *Betula utilis*

Habit

Fruiting body grows 25 cm and forms a cap as large as 20 cm in diameter. The remmants of veil-like membrane of mature mushrooms extending from the margin of cap to the stipe and tom by growth revealing the gills of mature sporophores. Pileus orange to orange-red with white floccose patches.

Distribution

Reported from North West Himalayas (Watling and Gregory, 1980; Manjula, 1983 and Kumar *et al.*, 1990) and from South India (Natarajan, 1977b).

48. *Amanita pachycolea* Stuntz

Habitat

Grows solitary-scattered, on soils, rich in humus under the trees of *Lyonia ovalifolia, Myrica esculenta, Quercus leucotrichophora* and *Rhododendron arboreum*.

Habit

Pilues 7-14 cm in diameter, convex becoming plane at matuarity, unbonate, brown on the disc, fading towards margin, surface viscid, remnants of veil when present solitary, white to greyish white patch on the centre of the pileus. Margin decurved, becoming plane to uplifted with age, conspicuously striate or tuberculate striate. Pileus context white, thick near the disc, unchanging on cutting and bruising. Lamellae upto 0.5 cm broad, fleshy, thin, free or adnate, sometimes decurrent, subdistant to close, white becoming irregularly orangish with age. Stipe 8-19 cm long and 1-2.5 cm wide, fleshy to subfleshy, central, cylindrical, tapering above, fibrillose, fibrils appressed to scaly surface, stuffed, hollowing with age. Annulus absent. Volva thick, membraneous, cuplike, saccate, persistent, extending upto 1.5 cm up stem.

Distribution

Reported from Uttar Pradesh, Garhwal by Bhatt *et al.* (1999).

49. *Amanita rubescens* (Fr.) S.F. Gray

Habitat

Grows solitary, scattered, occasionally gregarious, mycorhizal with *Cedrus deodara* and *Quercus incana*.

Habit

Pileus greyish red to reddish brown, beset with red warts, convex, upto 15 cm across, sometimes covered with an ochre–yellow flush which can be washed by the rain. The flesh of the mushroom white, becoming pink when bruised or exposed to air. Stipe ridged, white with flushes of cap colour, measuring upto 15 cm. long. The gills white and free of the stipe and display red spots when damaged. The ring striate on the upper side. Spore print white.

Distribution

Reported from North West Himalayas (Kumar, 1987).

50. *Amanita vaginata* (Bull, ex Fr.) Vitt.

Synonym

Amanitopsis vaginata (Bull. ex Fr.) Roze.

Common Name

Grisette.

Habitat

Sporophores grow solitary or scattered to gregarious on soil rich in humus in mixed forest and in open area and wayside, mycorhizal with *Pinus roxburghii*. Also associated with. *P. wallichiana, Quercus incana, Cedrus deodara, Picea smithiana* and *Abies pindrow.*

Habit

Sporophores centrally stipitate. Pileus 5.0-10.0 cm in diameter, ovate when young turning convex to expanded when old, white, grey or light orange yellow, usually smooth, rarely covered with the fragments of volva, margin thin and with prominent striations, slightly sticky when young. Gills distinctly formed, free whitish. Stipe central, cylindrical or slightly narrow at the top, slender, 9.0-15.0×0.6-1.0 cm, thick, fragile, glabrous or scaly, stuffed when young or hollow, base not bulbous. Volva white, cup like adhering to stipe. Flesh white, thin and soft.

Distribution

Reported from Arunachal Pradesh (Mishra, 1999), Meghalaya and Assam (Berkeley, 1954), Solan, Himachal Pradesh and Kashi Hills Assam (Sohi *et al.*, 1964), Nagpur (Trivedi, 1972), Lucknow (Ghose *et al.*, 1974), Assam (Butler and Bisby, 1960), Below Nurklow, Khasi Hills, Assam, Solan, Nagpur (Nita Bahl, 1988); Punjab Pathankot, Katori Bangalow (700 m) by Atri and Kour (2005) from Birbhum (W.B) by Gupta (1984), and from Himachal Pradesh, (Kumar *et al.*, 1990). Maharastra (Patil *et al.*, 1995).

51. *Amyloporia xantha* (Fr.) Bond and Sing.

Synonym

Poria xantha (Fr.) Cooke.

Habitat

Grows on logs of *Pinus excelsa*, *Quercus semicarpifolia* or sometimes on charred logs.

Habit

Sporophores broadly effused or resupinate, brittle, freely splitting when dry. Hymenial surface creamish, chalky, poroid, pores circular. Spores (3.6-4.3×1.1-1.4 µm) cylindrical, sometimes allantoid, small, hyaline, smooth.

Distribution

Reported from Himachal Pradesh (Puri, 1956).

52. *Armillaria mellea* (Vahl ex Fr.) Kummer

Synonym

Armillaria mellea (Fries) Karsten.

Common Name

Honey Fungus, Boot Lace Fungus, Honey Tuft.

Habitat

Sporophores solitary, gregarious or in tufts at the base of the old trees and stumps of oaks or on the roots of living tree or on the dead trees of other hard woods and conifers on the ground, in forests or in orchards. It is a distinctive parasite of woody plant.

Habit

Sporophores more or less larger in size, generally honey colured, centrally stipitate, ring or obvious veil present; attached to black rhizomorph. Pileus 2.5-15.2 cm in diameter at first spherical, then convex, later expanded and subumbonate, honey coloured to dull reddish brown, dark colour at the center, surface covered with blackish to brownish evanescent scales especially at the center, margin glabrous, occasionally adorned with loose floccose scales, sometimes scales entirely absent, usually dry but sticky when young, margin often striate when old. Gills subdistant, fully formed, adnate or decurrent, whitish when young, coloured with age, 0.6-1.1 cm broad. stipe central uniform in thickness, 2.5-15.2×1.0-2.5 cm thick, yellowish or brownish, whitish at the top, smooth or floccose scaly, stuffed or hollow, ring prominent, white, not movable, usually it disppears. Flesh white.

Distribution

Reported from Kerala (Bhavani Devi, 1995), Uttar Pradesh (Hole, 1927), Baroda (Moses, 1948), North Eastern and Western Himalayas (Butler and Bisby, 1931, Chandra and Watling, 1982), Baroda, Deoband, Jaunpur, U.P. (Nita Bahl, 1988).

53. *Armillaria tabescens* (Scop.:Fr) Emel

Synonym

Clitocybe tabescens Scop.

Habitat

Sporophores in dense clusters at the base of the tree stumps, deciduous timber especially oaks.

Habit

Remarkably similar to *Armillaria mellea* but more densely tufted, stems fused, not bulbous, entirely without ring. Spores white, ovate 8-10×5-7 µm.

Distribution

Reported from Trivandrum, Kerala (Bhavani Devi, 1995).

54. *Astraeus hygrometricus* (Pers.) Morg.

Synonym

Geastrum hygrometricum Pers.

Common Name

Hygroscopic Earth Star.

Habitat

Grows on the ground subterranean but comes out of soil at maturity.

Habit

Sporophores upto 5 cm in diameter, astipitate and globose. Peridium two layered exoperidium, ochraceous dark brown, hygroscopic, splitting stellately at maturity. Endoperidium, light grey and opening by ostiole. Gleba brown with age. Capillitium threads attached to the side of the endoperidium. Spores spherical or globose, cinnamon brown and finely warted.

Distribution

Reported from West Bengal (Thakur, 1980) and Das *et al*. (2002).

55. *Astroboletus gracilis* (Pk.) Wolfe.

Synonym

Boletus gracilis Peck.

Habitat

Grows solitary to scattered on soil humus and leaf litter in *Cedrus deodara* and *Quercus incana*.

Habit

Pileus upto 9 cm broad, convex to plano-convex with age, reddish brown to deep reddish brown, surface dry, finely granular, margin fertile, recurved with age. Stipe upto 7 cm long, unequal, base smooth, deep reddish brown fading towards the base, whitish at the base. Spores (10-16×5-7 µm) ovoid to elliptical, brownish, thick walled, apiculate.

Distribution

Reported from Himachal Pradesh (Sharma *et al*., 1978).

56. *Auricularia auricula* Judae

Synonym

Hirneola auricula Judae.

Common Name

Jew's ear, Tree ear, Wood ear.

Habitat

Grows on wood of broadleaf, dead wood or logs of *Shorea robusta, Bamboosa* sp, *Ficus religiosa* or on tree trunk and on injured Tea stem.

Habit

Carpophore 6-10 cm, ear shaped, sessile or with a short attaching peduncle bony hard when dry, outer surface velvety, sterile, pubescent, with slight venations, inner surface fertile, yellowish brown to reddish brown at first at first almost smooth then venose, pruinose because of the spores. When drying out it tends to turn violet and become increasingly circumvolute. Flesh soft, gelatinous, translucent, fragile when dry. Spores white, cylindrical, smooth, 12-17×4-7 µm.

Distribution

Sohi and Upadhyay (1990) reported eight species of *Auricularia* including *A. auricula* from North Eastern states of Meghalaya, Assam and Manipur (in Verma and Upadhyay, 2000); Arunachal Pradesh (Bisht and Harsh, 2001); Himachal Pradesh (Sohi and Upadhyay, 1990) and from West Bengal, Musoorie, Khandala (Maharashtra), Sonemarg, Kashmir (Banerjee, 1946, Banerjee and Ghosh, 1942; Butler and Biisby, 1931).

57. *Auricularia delicata* (Fr.) P. Henn

Synonym

Hirneola deliacta (Fr.) Bres.

Habitat

Grows on wood.

Habit

Fruit bodies 5.0-8.9 cm wide, ear shaped, mostly sessile, vinaseous grey in colour, margin slightly curved, rubbery texture, gelatinous, tough, surface with hyaline hairs, six distinct hyphalzones in transverse section, hymenium merculoid to strongly poroid, reticulate, 80.0-95.0 µm wide, basidium cylindrical, transversely septate, basidiospores allantoid, 12-14×5-7 µm.

Distribution

Reported from Arunachal Pradesh (Mishra, 1999); Lebong, Darjeeling, West Bengal (Berkeley, 1856), North Eastern Hills of Meghalaya (Verma and Singh, 1981).

58. *Auricularia fuscosuccinia* (Mont.) Farl.

Habitat

Tenaceously attached to dead branches of the trees, solitary or gregarious.

Habit

Fruit bodies 8.0-12.0 cm wide, cup shaped, rosy to vinaceous when fresh, translucent when dry, sessile, sometimes substipitate, cartilaginous, surface pileate, upper surface grey and hairy, hyaline hairs with round tips, hymenium smooth, 70-80 µm wide, basidia cylindrical, 45-60×4-6 µm, basidiospores allantoids 14-16×6-7 µm in size.

Distribution

Reported from West Bengal (Banerjee, 1947 and Low, 1951), Arunachal Pradesh (Mishra, 1999).

59. *Auricularia mesentricha*

Common Name

Tripe Fungus

Habitat

In groups on broadleaf of wood.

Habit

Carpophores 5-12 cm, bracket shaped, sessile edge lobate, grey-brown, outer surface sterile with whitish zonations, quite villose, pale blue-grey, slightly velvety, lower surface fertile, very pleated, purple-brown, pruinous because of the spores. Flash gelatinous, quite thick and elastic, leathery when dry. Spores white, slightly incurved, smooth, 15-18×7-8 µm.

Distribution

Reported from Jammu and Kashmir (Pandotra, 1966) and from Maharashtra (Lloyd, 1904–1919; Theissen, 1911).

60. *Auricularia peltata* Lioyd.

Habitat

Grows on wood of *Jacaranda mincosaefolia*.

Habit

Sporophores upto 5 cm wide, disc like, gelatinous, recupinate with hairy margin. Sporophores exhibit six distinct hyphal zones in cross section. Zona pilosa, the uppermost, consist of hyaline to light brown hairs with rounded apices. Hymenium, the lower most, is characterised by the presence of scattered amorphous crystals. Spores allantoid.

Distribution

Reported from Rohtak, Punjab (Ahamad, 1945).

61. *Auricularia polytricha* (Mont.) Sace.

Synonym

Hirneola polytricha Mont.

Common Name

Ear Fungus or "Mou Leh"

Habitat

Grows solitary or gregarious on dead angiospermic woods or logs or on decaying wood.

Habit

Sporophores upto 10 cm in diameter, leathery, mostly sessile, ear-shaped, red brown when fresh, grey or tan on drying, eight distinct hyphal zones in transverse section. Hymenium 80-90 µm wide, dark, smooth and sometimes papillate.

Distribution

Reported from various states of India including Punjab (Ahmad, 1945), West Bengal (Banerjee, 1947), Maharashtra, Himanchal Pradesh, Timil Nadu (Butler and Bisby, 1931), Sikkim (Verma *et al.*, 1995) and from Kerala (Butler and Bisby, 1931)

62. *Auricularia tenuis* (Lev.) Farlow.

Habitat

Grow solitary or gregarious on dead branches of trees.

Habit

Sporophores 1-2 mm thick, sessile, papery, orbicular, transluscent to opaque, horny, brittle on drying, upper surface smoky, velvet to smooth. Hymenium gelatinous with reddish ting. Spores cylindrical to allantoid.

Distribution

Reported from Kolkata, West Bengal (Banerjee, 1947).

63. *Battarraea albicans* (Libosch.) Fr.

Habitat

Grows solitary on the ground with volva hidden under the soil.

Habit

Sporophores 30-35 cm high, peridium depressed, globose, seated at the apex of the stipe. Stipe woody, 2.5-3 cm thick, covered with scale. Volva persists as a cup at the base of the stipe. Gleba rust brown, coloured and consists of hyaline thread of capillitium and annulated cells 60-80 µm long and 6-8 µm thick, epispore 6-7.2 µm in diameter.

Distribution

Reported from Ganeshkhind, Poona, Maharashtra by Nair and Patil (1978).

64. *Battarraea stevenii* (Libos.) Fr.

Synonym

Lycoperdon phalloides/Batterea phalloides

Common Name

Scaly stalked puffball or sandy still ball.

Habitat

Grows solitary to scattered, in decideous or coniferous forest.

Habit

Cap with gills. This species has a spore sac atop the stipe. When young, the fruit body is roughly spherical and completely encased in an outer wall which later splits in a circumscissile fashion (along a circular or equatorial line), the lower part forming a volva and the upper part forming scales that cover the inner wall. The upper part rolls upward and backward and eventually falls away in one piece, exposing a spore sac lined with a narrow ring of capillitium and spores. The spores are sticky. As these are carried away by the wind, the drying action of the latter cause the edges of the peridium to shrivel and roll up more, exposing more spores. This is continued until the upper half of the peridium has shriveled and blown away and there remains only a few spores, which may be washed away by rain.The fruit body develops rapidly; when mature, it is rust-colored, with a hemispherical to somewhat conical "head" 1 to 3 cm in diameter, and with a stalk up to 40 cm long by 0.4 to 1.5 cm thick.

Distribution

Reported from Poona (Maharashtra) by Patil *et al.* (1995).

65. *Bolbitius vitellinus* (Pers. ex Fr.) Fr.

Common Name

Yellow cow-pat load stoop.

Habitat

Grows usually on horse-dung, characterized by the fragile and suddenly collapsing nature of the fruit-body.

Habit

Pileus 1.5-3 cm in diameter, yellowish brown, campanulate then convex, surface viscid, smooth, margin striate, sometimes split. Gills crowded, adnexed or free, thin, ochraceous then brown. Stipe upto 9.5 cm long, whitish, covered with floccose scales. Spores smooth, yellow to reddish brown, broadly elliptical, with germ pores.

Distribution

Reported from Punjab (Manju, 1933).

66. *Bolbitius flavellus* (Murr.) Singh and Tewari

Synonym

Pluteolus flabellus Murr.

Habitat

Grows scattered to aggregated on decaying organic matter and dung.

Habit

Carpophore large upto 9 cm. Pileus 6-9 cm broad, convex, subumbonate to expanded, viscid when young and moist, fragile, pale ochraceous-tan, glabrous, margin striate, splitted and pallid, flesh soft, moderately thick, watery, white, light yellow with age. Lamellae adnexed, crowded, moderately broad, wedge shaped, changing to pale ochre colour, edge minutely fimbriate. Stipe 7-8 × 0.5-1.5 cm, central, fragile, hollow, thicker towards the base, white becoming yellow on drying. Spore print yellow.

Distribution

Reported from Varanasi by Singh and Tewari (1976).

67. *Boletus bicolor var. bicolor* Peck.

Habitat

Gregarious in mixed forests.

Habit

Pileus 6-12 cm in diameter, broadly convex to parabolic becoming plano-convex at maturity, reddish brown to pinkish with age, surface dry, areolately cracked and pruinose. Pores round. Tubes depressed to subdecurrent, bright yellow. Stipe 6-12 cm long, cylindrical, glabrous, reddish at base and yellow at apex. Spores (9.5-14.5×3.5-5.5 µm) olive brown, oblong, smooth and thin walled.

Distribution

Reported from Himachal Pradesh (Lakhanpal, 1996; Lakhanpal and Sharma, 1988) and from North West Himalaya (Sagar and Lakhanpal, 1989).

68. *Boletus bicolor var. subreticulatus* Smith and Thiers.

Habitat

Solitary or scattered, sometimes in tufts on soil in mixed canopies of Oak and Deodar.

Habit

Pileus 5-14 cm in diameter, convex to plano-convex to flattered with age, deep reddish to rusty red, surface dry, velvety, finely areolate with age, margin regular, smooth and incurved when young. Pores small, yellow orange to olive brown, blue on bruising. Tubes adnate to subdecurrent, pale yellow to greenish yellow, blue on bruising. Stipe 6-9 cm long, cylindrical, solid, whitish pale at base and yellow to yellowish orange at apex, reticulate. Spores (9.5-13×3.5-4.5 µm) yellowish, elliptical to oblong, smooth, obscurely unequilateral in profile.

Distribution

Reported from Himachal Pradesh (Lakhanpal, 1996).

69. *Boletus caespitosus* Peck.

Synonym

Pulveroboletus caespitosus.

Habitat

It grows in clusters on ground in the vicinity of Oaks and *Rhododendron*.

Habit

Pileus 3-7 cm broad, plane or plano-convex, deep reddish brown to rusty brown, surface viscid, areolate and tomentose, margin turning upwards and wavy at maturity. Pores angular, boleti like. Tubes depressed to subdecurrent, light yellow to dark yellow. Stipe 2-5×1-1.5 cm at base, smooth, unequal, narrowing upwards, yellowish brown at apex. Spores (8-12×3-5 µm) yellowish brown, smooth long, oval to oblong.

Distribution

Reported from Himachal Pradesh (Lakhanpal, 1996; Lakhanpal and Sharma, 1988; Sagar and Lakhanpal, 1993).

70. *Boletus edulis* Bull ex Fr.

Common Name

Cap, penny bun bolete.

Habitat

Sporophores scattered to gregarious on the ground in open forest or in gardens under the perennials like Jack Fruit tree (*Atrocarpus heterophyllus*) and mango (*Mangifera indica*), conifers; mycorhizal with *Cedrus deodara* and *Pinus wallichiana.*

Habit

Pileus 9.0-18.0 cm, broad, hemispherical then convex when young, broadly convex with age, light yellowish brown to darker, dusted with whitest bloom, context firm, pure white and all parts unchanging on exposure, surface dry, viscid when wet, glabrous, smooth, uneven, wrinkled to somewhat shallowly pitted. Stipe central 40-90 mm long, 10-20 mm across, often swollen in the middle or below, irregularly shaped, reticulate in the upper half. Flesh firm, white, tubes adnexed, long and thin, easily detached from cap, pores minute and roundish.

Distribution

Collected from Arunchal Pradesh (Mishra, 1999), North West Himalaya (Lakhanpal *et al.*, 1988), Himachal Pradesh (Sharma *et al.*, 1978; Lakhanpal, 1996; Lakhanpal and Sharma, 1981; Lakhanpal *et al.*, 1988), Mahabaleshwar (Patil *et al.*,

1995); Trivandrum and Wynad districts of Kerala (Bhavani Devi, 1995) and from South India (Verma and Upadhyay, 2000).

71. *Boletus edulis var. clavipes* Peck.

Synonym

Boletus clavipes (Peck) Pilat and Dermek.

Habitat

Gregarious and mycorrhizal with oaks.

Habit

Pileus 6-12 cm broad, convex to plano–convex to plane with age, light brown when young and dark brown at maturity, surface dry, margin with fragments of veil when young which disappears at maturity leaving a prominent band of sterile tissue at the margin. Pores roundish. Tubes adnate to slightly depressed, pinkish white or light orange. Stipe 5-7 cm long, unequal, swollen and light brown at base, fibrillose, medium brown and reticulate at the apex. Spores (11-15×4-5 µm) yellowish brown, subfusiform.

Distribution

Reported from Himachal Pradesh (Lakhanpal, 1996; Lakhanpal and Sharma, 1988) and North West Himalaya (Sagar and Lakhanpal, 1993).

72. *Boletus edulis var. ochraceus* Smith and Theirs.

Habitat

Grows solitary or gregarious on ground in angiosperimic forest.

Habit

Pileus 6-8 cm in diameter, convex and medium yellowish brown when young to broadly convex and light yellowish brown at maturity, surface dry, even and smooth. Pores roundish. Tubes deeply depressed at maturity, medium to light greenish yellow. Stipe 10-15×1-2 cm at base which is swollen, unequal, narrowed at the apex, solid, surface dry, reticulate all over but less distinct at the apex. Spores (14-16×3.5-5 µm) yellowish brown and subfusiform to oblong.

Distribution

Reported from Himanchal Pradesh (Lakhanpal, 1996; Lakhanpal and Sharma, 1981) and North West Himalaya (Sagar and Lakhanpal, 1993).

73. *Boletus erythropus* (Fries) Kromb.

Common Name

Dotted stem bolete.

Habitat

Grows gregarious on soil, humus and leaf litter in mixed forests of *Cedrus deodara*, *Pinus wallichiana* and *Quercus incana*.

Habit

Pileus 1.5-15 cm broad, convex to planoconvex with age, greyesh brown to reddish brown with maturity, surface dry, finely tomentose, margin yellowish, regular, smooth and incurved. Pores minute, roundish, orange red near the stipe and yellowish to orange near the margin, instantly bluish on bruising. Tubes adnate to subdecurrent, depressed around the stipe, yellowish. Stipe 3-12 cm, massive, bulbous when mature, solid, orange yellow at the top and orange brown near the base, surface finely pruinose. Spores (12-17×5-6 µm) olive brown, elliptical to subfusiform, smooth.

Distribution

Reported from Himachal Pradesh (Lakhanpal, 1996).

74. *Boletus aestivalis* Fries

Habitat

Solitary or gregarious on the ground of mixed forests, in the vicinity of Oaks and *Rhododendron*.

Habit

Pileus 8-24 cm in diameters, parabolic to broadly convex to almost flattened with age, yellowish white to yellowish brown, surface viscid when fresh and moist, pores roundish and stuffed when young. Tubes adnate, pale yellow to light grey. Stipe 6-15 cm long, unequal, swollen, light yellow at apex and yellowish brown at base. Spores (12-14.5×5 µm), deep yellowish brown, subelliptical to subfusiform.

Distribution

Reported from Himachal Pradesh (Lakhanpal, 1996; Lakhanpal and Sagar, 1989).

75. *Boletus flammans* Dick and Snell

Habitat

Grows on soil, solitary, scattered to gregarious, under Conifers, *Cedrus deodara* and *Picia smithiana*.

Habit

Pileus 4-10 cm broad, convex, deep reddish brown, surface dry but subviscid when wet, velvetty when young and glabrous on maturity, margin entire and incurved when young. Pores small, roundish, deep reddish to bright red, blue on bruising. Tubes deep, depressed around the stipe, pale yellow to greenish yellow. Stipe 5-8 cm long, equal, reticulate in the apical part, pruinose below, reddish orange above and reddish brown below but base yellow orange. Spores (9.5-14×3.5-4.5 µm) smooth, thin walled, ellipsoid to subfusoid.

Distribution

Reported from Himachal Pradesh (Lakhanpal, 1996).

76. *Boletus gertrudiae* Peck

Habitat

Grows solitary or scattered associated with *Quercus incana* in a mixed canopy with *Quercus, Rhododendron* and *Pinus wallichiana*.

Habit

Pileus 5-15 cm broad, convex to flattened, yellowish to brownish yellow, surface dry, soft, glabrous, bright and shining, margin entire and smooth. Pores minute. Tubes adnate but depressed around stipe, pale white or yellow when young and olive yellow or yellowish brown with maturity. Stipe 6-10 cm long, almost equal, slightly swollen at base, solid, white to pale orange when young and yellowish with age, finely reticulate, reticulum orange brown. Spores (11-17×4-5.5 µm) smooth, subfusiform to ellipsoid, light brownish.

Distribution

Reported from Himachal Pradesh (Lakhanpal, 1996; Lakhanpal and Sharma, 1988) and North West Himalaya (Lakhanpal, 1994).

77. *Boletus protentosus* Berk and Br.

Synonym

Phlebonus protentosus (Berk and Br.) Boedjin; *Phaeogyroporus protentosus* Berk and Br. Mc. Nabb.; *Polyporus olivaceofuscus* Berk and Br. Mc. Nabb.

Habitat

Sporophores solitary to scattered on ground.

Habit

Pileus 8-24 cm in diameter, first convex becoming plano-convex, often with shallow depression at the center, surface olive brown to sepia-brown at the center, paler elsewhere, slimy when wet, otherwise dry with a nonseparable cuticle, smooth, glabrous, margin first involute projecting beyond hymenophore, undulate. Hymenophore adnaxed to adnate, tubes lemon yellow. Upto 14 mm long, darkening on cutting. Stipe 6-174-10 cm, central, robust with swollen base, solid. Context upto 5 cm wide at center, spongy butter yellow, bluing on bruising when fresh and young, odour pleasant, pileal surface turns violet in NH_4OH.

Distribution

Reported from different localities in Kerala (Vrinda *et al.*, 2000).

78. *Boletus scaber* Fr.

Synonym

Leccinum scabrum L.

Habitat

Sporophores solitary to scattered.

Habit

Pileus 60-100 mm, convex to parabolic, surface dry, viscid when wet, glabrous, often with depressions in age, light greyish brown at center, fading to yellowish brown towards the periphery, margin even, not projecting beyond the tubes, context 10-12 mm, white, tubes 6-15 mm deep, deeply depressed around the stipe, pores roundish, minute (2-3 mm), whitish first later brownish. Stipe 60-120x15-25 mm, gradually tapering upward, yellowish white with black scabers. Flesh whitish yellow to pale yellow.

Distribution

Collected from Mariyang and Tuting, Arunachal Pradesh (Mishra 1999); Himachal Pradesh (Sharma *et al.*, 1978 and Lakhanpal, 1996).

79. *Bovista gigantea* (Batsch.) Gray

Synonym

Lycoperdon gigantea, Langarmannia gigantea, Clavaria gigantea.

Habitat

Grows on ground.

Habit

Carpophore very large ranging from 12 cm to 60 cm in diameter, globose, outer peridium at first white but turn yellowish with age, glabrous, at length peeling off in large irregular flakes. Inner peridium thin olivaceous, bursting irregularly. The interior composed of homogenous white substance when young, afterwards bacoming spongy, greenish and at last minute globose sporidia are developed with pedicels by which they are attached to numerous intermingled filaments.

Distribution

Reported from West Bangal (Bose and Bose, 1940).

80. *Bovista plumbea* Pers.

Synonym

Lycoperdon plumbeum Vittad

Habitat

Grows on humicolous soil or in old pastures without anchoring rhizoids and at maturity freeing from substrate.

Habit

Sporophores 2-4 cm in diameter, globose or depressed globose, whitish, glabrous when young, splitting irregularly into white granules and finally falling off. Peridium red coloured, papery. Gleba at first white, then red then dark brown to purple. Capillitium threads free. Spores (5-5.5 µm) subglobose to ovate, smooth or warted, pedicellate, yellowish to purple brown.

Distribution

Reported from Uttaranchal (Hennings, 1901), Jammu and Kashmir (Ahmad, 1941; Thind and Thind, 1982) and Himachal Pradesh (Gupta *et al.*, 1974; Thind and Thind, 1982), Arnigadh, Mussoorie; Sonmarg, Kashmir, Khadreala, Himachal Pradesh (Nita Bahl, 1988).

81. *Bovista pusilla* (Batsch. ex Pers.) Pers.

Synonym

Lycoperdon pusillum (Batsch ex Pers) Schum.

Habitat

Grows on the ground.

Habit

Sporophores small, 2 cm in diameter, with a cord like rooting base. Exoperidium fugacious, often squamulose or warted. Endoperidium buff or light coloured, brown at maturity, membranous, smooth with a small and irregular pore. Gleba brown at maturity. Capillitium branched, brown, usually pitted. Spores globose, thick walled, verrucose, sometimes smooth, minutely pedicellate.

Distribution

Reported from Dehradun, Uttaranchal (Ahmad, 1942) and Pune, Maharashtra (Nair and Patil, 1978).

82. *Calocera cornea* (Batsch. Fr.) Fr.

Habitat

Grows isolated or in groups on fallen broadleaf and coniferous tree stump or on rotting woods.

Habit

Sporophores long, erect, unbranched approximately 1 mm thick and 1–1.5 cm tall, jelly like. Light–yellow when young and creamish white when old, tapered to a narrow, obtuse tip, single or several fused to a common base, surface viscid when moist, yellow to yellowish–orange, context stiff–gelatinous.

Distribution

Reported from Jaintia Hills district of Meghalaya (Verma *et al.*, 1995) and from West Bengal (Banerjee, 1947).

83. *Calocera viscose* (Pers.) Fr.

Common Name

Yellow Staghorn Fungus, Yellow tuning Fork, Yellow False Coral.

Habitat

Grows on wood, saprobic, several to many in groups on dead coniferous, logs and stumps.

Habit

Sporophore 3-10 cm tall and branched. The branches yellow to yellow orange. The fruitbody arising from a common base. The upper branches usually forked. Its tough gelatinous texture distinguishes it from coral fungus which are brittle and easily broken. Sporophores greasy and viscid with anther like branches, can be confused with some of the *Ramaria* species of coral fungi, but be greasy, viscid surface is an immediately obvious distinguishing feature. Spores white.

Distribution

Reported from Mysore (Cooke, 1880) and from West Bengal (Currey, 1874).

84. *Callocybe gambosa* (Fr.) Singer.

Synonym

Tricholoma gambosum.

Common Name

St. Geogre's mushroom, Spring mushroom.

Habitat

Grows in grassland (hilly and mountaineous areas), prefer calcareous soil, in broadleaf woods.

Habit

Cap 6-10 cm or more, hemispherical then convex, very fleshy and thick, then distended, normally humped, dry, finely velvety, colour from white to buff to yellowish, margin smooth, pruinous, curled, later expanded and often sinuate. Gills crowded, thin whitish then cream coloured, adnate, stipe 3-6 x 1-2 cm or more, cylindrical but often enlarged at base, ochreous marking toward bottom, fibrillose, pruinous at the top, solid. Flesh compact, thick, white, strong, pleasant smell of fresh meal.

Distribution

Collected from different localities in Trivandrum district of Kerala (Bhavani Devi, 1995), Nagpur, Maharashtra (Patil *et al.,* 1995) and from Western Ghats of India (Anand and Prakasham, 2000).

85. *Calocybe indica* (P and C)

Habitat

Grows solitary in soil.

Habit

Sporophores robust in size, and fleshy, white or pale colourad. Pileus 10.0-14.0 cm in diameter, convex, later expanded and flattened, white, non-hygrophorus, cuticle easily peeled, mat polished, sometimes appressed scale present, at or around the center, margin regular, incurved, smooth, non-striate. Gills crowded, distantly formed, emarginate, separable, white, unequal, pliable, attenuated toward the margin of the pileus, entire. Stipe central, sometimes eccentric, cylindrical with sub bulbous base

10.0 cm long, white cartilaginous surface dry and fibrillose, base not hollow, without annulus and volva. Flesh white.

Distribution

Reported from Trivandrum, Kerala (Bhavani Devi, 1995), West Bengal (Purkayastha and Chandra, 1974, Das *et al.*, 2002) and from Western Ghats of India (Anandh and Prakasham, 2000).

86. *Calvatia caelata* (Massee) Morgan

Synonym

Calvatia saccata var. *elata* (Massee) Hollos, *Clavatia excipuliformis* (Schaeff. ex Pers.) Perdeck.

Habitat

Grows solitary, sometimes scattered in coniferous woods on ground.

Habit

Sporophores subcylindrical, stalked, 7.5 cm long, at first white later becoming brown. Peridium consisting of an outer thick and inner thin layer, cracks when old and falls off. Gleba brown. Capillitium dichotomously branched, fragmented, moderately pitted. Basidiospores (4.2-50.0 µm) brown, globose, minutely verrucose short, pedicellate.

Distribution

Reported from Himanchal Pradesh (Gupta *et al.*, 1974) and from Western Himalayas (Ahmad,1941).

87. *Calvatia cyathiformis* (Bosc) Marg.

Synonym

Bovista lilacina Berk and Mont., *Lycoperdon lilacinum* (Berk.) Massee.

Habitat

Sporophores grows in groups on soil, grassyland and sometimes on the cultivated fields; sometimes solitary on sandy soil.

Habit

Sporophores 7.5-15.0 cm in diameter, round to pear shaped with stalk, white when young but turning into brown with pinkish tinge at maturity. Peridium with thick tapering base composed of spongy mycelium, cracking into irregular and brittle fragments, subsequently falling off exposing the spore mass within the cap shaped base. Flesh white when young and purplish at maturity. Capillitium purple brown. Spore mass dark purple with age.

Distribution

Reported from Kerala (Bhavani Devi, 1995); Punjab (Ahmad, 1941; Nita Bahl, 1988), Himachal Pradesh (Gupta *et al.*, 1974; Nair and Patil, 1978; Sharma and Thakur,

1978), Maharashtra (Nair and Patil, 1978; Sharma and Thakur, 1978; Anand, 1964), Gujarat (Nita Bahl, 1988) and from U.P (Khare, 1976).

88. *Calvatia fragilis* (Vitt.) Morgan

Habitat

Grows scattered on humicolous soil in the coniferous woods.

Habit

Sporophores usually 4.0-5.0 cm high, but the base region abruptly ending into a blunt tip, colour varies from brownish to straw. Peridium two layered, inner layer thick when young and fresh and thin with age, usually deep purple brown, brittle and sometimes withered irregularly from upper region. Gleba soft, purple brown with age. Capillitium branched, tapering at both ends, short, segmented, brown. Basidiospores (3.1- 4.2 µm) usually spherical, dark purplish brown, smooth, sometimes minutely verrucose, short pedicellate.

Distribution

Reported from Himachal Pradesh (Gupta *et al.*, 1974).

89. *Calvatia gigantea* (Pers.) Lloyd.

Synonym

Lycoperdon maxima, Lycoperdon gigantea Batsch ex Pers., *Lagermania gigantea.*

Common Name

Giant Puffball, Local name in Kerala is "Muttakoonu"

Habitat

Grows solitary or in arcs or gregarious on ground, woods, meadows, under bush and in fields and gardens along small drains.

Habit

Fruit body giant in size (15-30 cm in diameter). It has been reported to attain more than 2 feet in height weighing 45 pounds in Minnesota. Carpophores globose to subglobose, sessile, leathery, smooth, whitish but brownish or greenish yellow with age. Fruit body is somewhat greater in height than in diameter.

Distribution

Reported from Kerala (Bhavani Devi, 1995). Manipur and Meghalaya (Verma *et al.*, 1995); Jammu and Kashmir (Butler and Bisby, 1931).

90. *Calvatia lilacina* (Berk.) P. Henn

Synonym

Bovista lilacina Mont. and Berk. *Globaria lilacina* (Mont. and Berk.) Speg. (*Lycoperdon lilacinum* (Mont. and Berk.) Speg.

Habitat

Grows solitary on the ground, usually in sandy areas.

Habit

Peridium up to 15 cm diameter, subglobose or pyriform, tapering abruptly into a large, well-developed, strongly crenulate rooting base; exoperidium smooth or more frequently floccose, often areolate, cream to bay brown, thin, fragile, fugacious; endoperidium brown, thin, fragile, breaking away irregularly from the apical portion; sterile base well developed, persistent, cellular at the periphery, hemicompact within, separated from the gleba by a prominent diaphram. Gleba some shade of purple, sometimes with a greyish tinge, at first compact, soon pulverulent; capillitium threads long, branched, septate, equal, pallid olivaceous. *Spores:* globose, 5.5–7.5 mm. diam., occasionally apiculate; epispore strongly verrucose, violaceous.

Distribution

Reported from Punjab (Ahmad,1941) and from Poona, Maharashtra (Cunningham,1943; Butler and Bisby, 1931).

91. *Calvatia pachyderma* (Peck.) Morgan

Habitat

Sporophores grow scattered on humicolous soil.

Habit

Sporophores 3.6-6.0 cm high and 4.0-7.5 cm in diameter, subglobose to subcylindrical, stalk tapering towards the base, pale to brown. Peridium single layered, thick at first and thin with age, glabrous or grannulated, crack starts at the top and later extends downwards. Gleba bright, olivaceous when old. Subgleba present as a sterile basal part. Capillitium often branched, irregularly pitted, olivaceous yellow in colour. Basidiospores (4.2-6.3 μm) olivaceous yellow, globose to oval, smooth, pedicellate.

Distribution

Reported from Himachal Pradesh (Gupta *et al.*, 1974).

92. *Calvatia pistillaris* Fr.

Synonym

Clavariadelphus pistillaris; Clavaria pistillaris

Habitat

Grows singly or scattered on the ground, in beech woods.

Habit

This is the largest of the simple club fungi, the club and pestle-shaped fruit bodies reaching heights of 10-25 cm, the wrinkled surface yellow ochre to orange, then reddish brown. Flesh whitish. Spores pale yellow, ovate, 11-16×6-9 μm.

Distribution

Reported from Kerala (Bhavani Devi, 1995).

93. *Calvatia utriformis* (Bull ex Pers.) Toap.

Synonym

Calvatia caelata (Bullo) Morg., *Clavatia bovista* (Pars.) Kamblee and Lee.; *Lycoperdon caelatum.*

Common Name

Mosaic Puffball

Habitat

Sporophores grows solitary or scattered on the ground, in meadows, pastures or grassy lands.

Habit

Sporophores with distinct stem like base, 6.0-14.0 cm high, surface rough peridium 5.0-12.0 cm wide, spherical but becoming flattened or depressed with age, narrowed below into a short, cylindrical, stout cup shaped base, white at first, yellow or brown when old, surface with warty patches, spines or cracks. Subgleba present as a sterile basal part. Basidiospores greenish yellow to umber coloured, 4.0 µm in diameter.

Distribution

Reported from Himachal Pradesh (Ahmad, 1941; Nita Behl, 1988), Kerala (Bhavani Devi, 1995).

94. *Camerophyllus niveus* (Scoop. ex Fr.) Karst

Synonym

Cuphophyllus niveus (Fr.) Bon.

Habitat

Grows on dead and decaying twigs of *Shorea robusta* in woods or amongst short grass in pasture and open woodland.

Habit

Cap 1-3 cm across, convex becoming flattened and depressed, white then ivory or tinged ochraceous with age especially at centre, striate when moist. Stipe 25-50 × 2-4 mm, tapering towards the base, white. Flesh thick at the centre of cap, whitish. Gills decurrent, widely spaced, whitish. Spore print white. Spores ellipsoid or pip-shaped, 7-9 × 4-5.5 µm in four spored forms but 10-12 × 5-6 µm in two spored forms.

Distribution

Reported from Kalinga, Orissa (Dhancholia and Sinha, 1988).

95. *Camerophyllus pratensis* (Berk and Br.) Pegler

Synonym

Hygrophorus pratenis (Pers, Fr) Murril

Common Name

Bull Cap.

Habitat

Grows gregarious in pastures and meadows, solitary to scattered in mixed hardwood, occasionally in grassy or open areas.

Habit

Sporophores 3-7 cm broad, convex, expanding to nearly plane, slightly umbonate, margin entire or wavy, surface moist, smooth, pale orange, to apricot orange, sometimes spotted, darker orange, fading when dry to pale peach and becoming slightly fibrillose, flesh thick at the disc, otherwise thin, white unchanging. Gills subdecurrent at maturity, broad subdistant, thick waxy. Stipe 2.5 to 7 cm tall, 1-2 cm thick, stuffed, equal to narrowed at the base, surface dry, smooth to fibrillose striate. Flesh white unchanging. Veil absent.

Distribution

Reported from Kerala (Bhavani Devi, 1995) and from Khasi Hills, Meghalaya (Shajahahn *et al.*, 1988).

96. *Cantharellus cibarius* Fries

Common Name

Golden cantherella

Habitat

Grows scattered or in semicircular rings on forest floor of frondose and coniferous woods and on the ground under *chir* and *sal* forests.

Habit

Sporophores top-shaped usually solid or stuffed, chrome yellow coloured, asymmetrical occasionally symmetrical in form. Pileus 3-12 cm in diameter, convex when young, then flattened and finally shallow funnel shaped, yellow in colour, firm, fleshy, surface smooth, margin involute to spreading sometimes wavy. Gills not crowded, docurrent, yellow, forked (often dichotomously) thick, narrow, edge blunt. Stipe central, cylindrical to compressed usually short, equal or shortly tapering at the base, 3-6 cm or more long, rather thick, yellow, firm, not hollow, glabrous. Flesh white, 1-2 cm thick, fibrous, consisting of thin walled hyaline 3-10 μm wide. Hyphae lacking clamp connection. Annulus none.

Distribution

Reported from Meghalaya and Arunachal Pradesh (Verma *et al.*, 1995), Kerala (Abraham *et al.*, 1995), West Bengal (Bose and Bose 1940), Jammu and Kashmir and Uttaranchal (Chopra and Chopra, 1955), Uttaranchal (Butter and Bisby, 1960), Himanchal Pradesh (Sohi *et al.*, 1964), Uttaranchal and Meghalaya (Butter and Bisby, 1931; Ghosh *et al.*, 1974), West Bengal, Jammu and Kashmir, Uttaranchal, Himachal Pradesh (Nita Bahl, 1988), Jammu and Kashmir (Abraham *et al.*, 1980) and U.P. (Hennings, 1901).

97. *Cantharellus cinnbarinus* Schweinitz.

Common Name

Red chantherelle.

Habitat

Solitary or scattered in soil in open places.

Habit

Pileus 4.5-9.0 cm diameter, convex soon depressed; surface reddish-orange, fading with age and exposure, nonhygrophanus, margin distinctly incurved, undulate or lobed, not striate, sometimes crisped and irregular or incised. Hymenophore deeply decurrent with false lamellae, greyish orange, discolouring faun brown on bruising, well separated, intervein and forked. Stipe 4.0-5.0×7.0-10.0 mm, central, cylindrical to slightly compressed, equal or tapering below, often curved at the base, solid, surface smooth. Annulus none. Odour fruity. Context thin, white with pinkish tinge, composed of thin walled, 5-12 µm, wide hyphae with clamp connections.

Distribution

Reported from Kerala (Abraham *et al.*, 1995).

98. *Cantharellus infundibuliformis* Scop ex Fr.

Synonym

Chatharellus tubaeformis Fr.

Common Name

Trumpet shaped chantherelle

Habitat

Sporophores grows scattered or in tufts or sometimes gregarious on ground under damp forests or in swamps.

Habit

Sporophores usually smaller in size. Pileus 1.5-5.0 cm, in diameter, convex when young, umbilicate or funnel shaped when old, often depressed in the centre, depression extending into the stipe, hygrophanous, yellow to brown when moist, greyish to greyish brown when dry, thin, surface smooth, floccose to fibrillose when dry, margin wavy or irregular. Gills not crowded, decurrent, yellowish to lilac, narrow, dichotomously or irregularly forked. Stipe 3.0-8.0 cm long, yellowish, smooth, hollow.

Distribution

Reported from Assam (Berkeley, 1856; Butler and Bisby, 1960).

99. *Cantharellus luteocomus* Bigelow.

Habitat

Grows gregarious or subcaespitose on forest floor associated with *Abies pindrow, Cedrus deodara, Picia smithiana, Pinus wallichiana* and *Quercus incana*.

Habit

Pileus 3-15 mm diameter, convex then plane, margin uplifted when old, wavy, often crenate, sometimes infundibuliform, surface melon yellow, glabrous and smooth. Lamellae decurrent, smooth or very slightly wrinkled, orange white. Stipe 1-3.2 cm × 1-2 mm, central, cylindric, equal, stuffed, surface concolourous with the pileus, glabrous. Context thin, soft, concolorous with the pileus, composed of cylindric to slightly inflated hyphae with clamp connections.

Distribution

Reported from Kerala by Vijay Joseph *et al.* (1995) and from Himachal Pradesh, Simla by Bhatt and Lakhanpal (1988 b).

100. *Cantharellus minor* Peck

Habitat

Grow scattered or in clusters on ground covered with decaying organic matter, in woods or in open places.

Habit

Sporophores usually smaller, stipitate, fleshy. Pileus 1.5-2.5 cm in diameter, convex when young, later expanded, centrally depressed, yellow, thin, fleshy, surface glabrous, margin wavy or irregular. Gills distant, decurrent, dichotomously branched, yellowish, narrow. Stipe slender, smooth, reddish to yellowish, solid at first, hollow with age. Flesh pale yellow.

Distribution

Reported from Kerala (Abraham *et al.*, 1995), India (Bose and Bose, 1940), Uttaranchal (Thind and Anand, 1956) and from Himachal Pradesh (Sohi *et al.*, 1964).

101. *Catharellula umbonata* (Fr.) Singer.

Synonym

Catharellus umbonata (Fr.), *Clitocybe umbonata* (Fr.) Konrad

Habitat

Grows scattered or in groups in forest or in open places, rarely on bare soil and not on woods.

Habit

Cap 2-5 cm across, convex becoming funnel-shaped with a small central umbo and sometimes a wavy margin; grayish brown to smoky or violaceous gray; dry and minutely hairy. Gills decurrent, crowded, narrow, thickish, regularly forked; whitish on bruising red or yellow. Stipe 25-80 × 3-7 mm, central, equal,slightly tough, stuffed at the top; whitish to grayish; silky. Flesh white, on bruising red where cut or handled. Odor scented. Spores subfusoid, smooth, amyloid, 8-11 × 3-4.5 µm. Spore deposit white.

Distribution

Reported from Poona (Maharashtra) by Sathe and Deshpande (1979).

102. *Chlorolepiota mahabaleshwarensis* Sathe and Deshpande

Habitat

Grows on the ground.

Habit

Pileus 5-8 cm broad, conical to convex, umbonate, buff or straw coloured, covered with umber coloured scales, crowded at the centre, smooth with mostly entire margin. Stipe central, concolorous, cylindrical, hollow, base bulbous, annulated, annulus fixed then mobile and complex. Spores (14.30×8.5 µm) ellipsoidal, greyish–yellow–green, laterally apiculate.

Distribution

Reported from Maharashtra (Sathe and Deshpande, 1979).

103. *Chlorophyllum molybdites* (Meyer ex. Fr.) Mass.

Synonym

Agaricus morganii Peck., *Lepiota morganii* (Peck) Sacc., *Agaricus guadelupensis* Pat., *Chlorophyllum morganii* Graff.

Common Name

False parasol, Green spored Parasol.

Habitat

Grows on soil.

Habit

Pileus upto 20 cm in diameter, convex then expanded, sometimes campanulate, buff to cinnamon coloured, covered with fuscous to crustose scales. Gills distant, free, white or cream then green, broad. Stipe upto 20 cm long, narrowed towards the apex, base bulbous in appearance, annulated, annulus white and double, volva absent. Spores (30-46×8-14.5 µm), smooth, thick walled, yellowish green, ovoid to subovoid, with broad germ-pore.

Distribution

Reported from Uttar Pradesh (Ghosh and Pathak, 1965; Ghosh *et al.*, 1976), Maharashtra (Sathe and Rahalkar, 1976; Sathe and Deshpande, 1979 and 1980) and from India (Verma and Upadhyay, 2000).

104. *Clavaria compressa* Schw.

Synonyms

Clavaria ceranoides Pers., *Clavaria compressa* Schwein., *Clavaria fusiformis* Sowerby, *Clavaria fusiformis* var. *ceranoides* W.G. Sm., *Clavaria inaequalis* var. *fusiformis* (Sowerby) Fr., *Clavaria platyclada* Peck, *Ramaria ceranoides* (Pers.) Gray, *Ramariopsis fusiformis* (Sowerby) R.H. Petersen.

Common Names

Yellow spindle coral, Golden fairy spindle, Tongues of flame, Slender golden fingers.

Habitat

Typically in clusters on soil in grasslands, humus, in light forests under conifers, or in fields.

Habit

Fruiting body 2-8 cm high, 0.2-0.6 cm diameter, bright yellow, cylindrical, unbranched, often laterally compressed and grooved, tips blunt or pointed, becoming reddish to brown in age; no obvious distinction between fruiting body and stem. Flesh solid in younger specimens, hollow in age. Spores smooth, thin-walled, hyaline, globose to sub-globose with distinct apiculus, entire, non-amyloid, with droplets, 5-6.5 x 4.5-6 µm. Spore print buff. Chemical tests flesh turns green in ferric sulphate.

Distribution

Reported from West Bengal (Butler and Bisby, 1931).

105. *Clavaria fistulosa* (Helms) Fr.

Synonym

Clavariadelphus fistulosus (Fries) Corner.

Common Name

Early Morel.

Habitat

Grows singly or in small groups on the ground or on the fallen branches.

Habit

Fruit body 10-20 cm high 2 mm thick, yellow then rufescent cylindrical, subobtuse, straight or curved, often contorted at the base, hollow, root short, villose. Flesh yellowish, firm. Spores white verrucose, pip shaped or subfusiform, 12x7 μm with a large central gutta, basidia conspicuous about 40μm long with 4 erect sterigmata, contents fleshy grannular.

Distribution

Reported from various localities in several districts of Kerala (Bhavani Devi, 1995).

106. *Clavaria fumosa* Fr.

Synonym

Clavaria rubicundula.

Common Name

Smoky Clavaria

Habitat

Grows in close tufts on soil under Oak forest.

Habit

Sporophores upto 12 cm high, astipitate, greyish yellow at the base and yellowish at the top, cylindrical, attenuated at the top, usually smooth but sometimes brittle, tips blunt, longitudinal grooves present, fleshy. Clubs grows from a common base. Spores (7-8×2.5-6.5 µm) smooth, hyaline, broadly ellipsoidal, papillate, with granular contents.

Distribution

Reported from Uttaranchal (Thind and Anand, 1956; Thind, 1961).

107. *Clavaria mirus* (Pat.) Corner.

Synonym

Clavariadelphus mirus (Pat.) Corner (1950) accepted name.

Habitat

Grows on soil under oak forest.

Habit

Fruit bodies with fertile, rounded or pointed apex, very large flesh have brunnescent to vinescent stain. The basidiocarps light brown to brown and produce larger basidiosopores (10–13×6–8µm).

Distribution

Reported from N. W. Himalayas (Thind, 1961).

108. *Clavaria vermicularis* Fr.

Common Name

White spindles or worm like Clâvaria and Fairy fingers.

Habitat

Grows in groups or in tufts (sometimes solitary) on humicolous soil or on bare soil under Oak forest. Also common in pastures along the edges.

Habit

Sporophores upto 13 cm high, simple, consisting of 2-7 clubs, brittle, white then pale yellow and finally ochraceous when dried, hollow, tips pointed in young clubs and obtuse when old. Stipe simple (sometime forked), cylindrical, white and small. Spores (4-5.6×3-4 µm) hyaline, broadly ellipsoidal to subglobose, smooth, papillate and aguttulate.

Distribution

Reported from Uttaranchal (Thind, 1961 and 1973) and from Mussoorie (Thind and Raswan, 1958).

109. *Clavariadelphus mirus* (Pat.) Corner

Synonym

Clavaria mirus Pat.

Habitat

Grows solitary or in small clusters on forest soil or in leaf litters, twigs etc. in coniferous, Oak or mixed forest.

Habit

Sporophores cylindrical with obtuse, truncate or mamillate apex, simple, light brown but faded towards the base, smooth and longitudinally wrinkled towards the apex, solid. Stipe inconspicuous. Hymenium upto 87 µm wide. Spores broadly ellipsoidal, obovate or pyriform, smooth and aguttulate.

Distribution

Reported from Himalayas (Thind, 1973) and Upper Shillong in Meghalaya (Verma *et al.*, 1995)

110. *Clavariadelphus pistillaris* (Fr.) Donk

Synonym

Clavaria pistillaris Fr.

Common Name

Large club Clavaria or Giant Club.

Habitat

Grows scattered or gregarious on the ground in forests.

Habit

Sporophores upto 30 cm high, yellow at first and brownish with age, subcylindrical and acute, club shaped at maturity. Stipe white, hairy downwards. Hymenium compact. Spores white to yellowish, oblong or ellipsoid. Hyphae with clamp connections.

Distribution

Reported from Uttaranchal (Ahamad, 1949).

111. *Clavariadelphus truncatus* (Quel.) Donk

Synonym

Clavaria truncata Quel.

Common Name

Club coral.

Habitat

Grows scattered on ground amongst conifer needles, sometimes under *Picea smithiana* or *Abies pindrow*.

Habit

Sporophores 5-16 cm high, turbinate or subpileate with sterile truncate tips, orange red at the top and dark red at the bottom, smooth, club shaped, clubs simple, longitudinally rugose, hollow but solid downwards. Stipe inconspicuous, solid, sterile, pale yellow. Hymenium usually straight or incurved. Spores (9.5-12×5.5-7 µm) subhyaline, broadly ellipsoid or oval, papillate.

Distribution

Reported from Uttarakhand (Thind and Sukhdev, 1957; Thind, 1961) and Jammu and Kashmir (Watling and Gregory, 1980).

112. *Clavicorona pyxidata* (Fr.) Doty.

Synonym

Clavaria pyxidata Fr.

Common Name

Cup bearing Clavaria.

Habitat

Grows gregarious or in tufts on dead wood.

Habit

Sporophores upto 13 cm high, yellow to dark brown with age, stipitate, tough, gelatinous, horny on drying. Stipe smooth, whitish or brownish pink. Grows from an amorphous base or a resupinate, fine hairy, white or coloured mass. Hymenium upto 26 µm wide. Spores small, white, smooth, ellipsoidal and without guttules.

Distribution

Reported from Uttaranchal (Ramakrishnan *et al.*, 1952) and from Himachal Pradesh (Sharma *et al.*, 1977).

113. *Clavulina cinerea* (Fr.) Schoret.

Synonym

Clavaria cinera Fr.

Habitat

Grows in groups on floor of coniferous and broad leaf forests.

Habit

Fruiting bodies club like, erect, fleshy, branched, crested, dull white to grey with purple tinge, measuring upto 10 cm high; repeatedly branched; branches longitudinally ridged and usually blunt at tips; dark purplish-grey, whitish at base. Spores white, nearly round; smooth; 6.5×11 µm.This species might be confused with *Ramaria fennica* var. *violaceibrunnea*, which is also greyish-purple or violet with white base, however the latter has yellowish, warted, ellipsoid spores and unlike

Clavurina cinerea, pigmented surface of its flesh quickly turns orange-red upon treatment with 20 per cent KOH.

Distribution

Reported from Meghalaya (Verma *et al.*, 1995) and from Mussoorie (Thind and Sukhadev, 1956).

114. *Clavulina cristata* (Fr.) Schroet

Synonym

Clavaria cristata Fr.

Common Name

Crested Clavaria

Habitat

Grows on the forest floor along path in both coniferous as well as broadleaf forests. Sometimes the ground seems literally covered with this small white coral fungus.

Habit

Sporophores upto 7 cm high and 1.5 cm in diameter, white or cream, stipe very short or absent, smooth, simple or sparsely branched, cristate. Branches arising from a central thickened stem but ending in flattened greyish crested tips. Hymenium upto 10 µm wide. Spores (7-11×6-10 µm) globose or subglobose, papillate, uniguttulate, smooth and yellowish brown.

Distribution

Reported from Uttaranchal (Thind and Anand, 1956 and Thind, 1961) and Upper Shillong in East Khasi Hills and Nonstoin in West Khasi Hills districts of Meghalaya (Verma *et al.*, 1995).

115. *Clavulina rugosa* (Fr.) Schrot.

Synonym

Clavaria rugosa Fr.

Common Name

Wrinkled clavaria or Wrinkled club.

Habitat

Grows solitary, gregarious or in tufts on the ground in forests or in grassy fields.

Habit

Sporophores upto 12 cm high, white or cream, large, yellow on drying, simple but with a few anther like branches, often longitudinally wrinkled, hollow. Stipe not prominent. Hymenium upto 150 µm wide. Spores (9-14×8-12 µm) white, smooth, uniguttulate, ovoid-ellipsoidal.

Distribution

Reported from Murree Hills (Thind, 1961) and from Himachal Pradesh (Sharma and Munjal, 1977).

116. *Clavulinopsis fusiformis* (Fr.) Corner.

Synonym

Clavaria fusiformis Fr.

Common Name

Golden Spindles

Habitat

Grows solitary or in groups or in cluster on soil in forests. Sometimes on humus, in open grass, heathlands etc.

Habit

Sporophores usually erect, 6.5 cm in diameter, forming tufts of simple slender, unbranched clubs, spindle shaped and sharp pointed when in cluster, stipitale, orange at the top, gradually the colour fading to yellow, whitish at the base, fleshy, smooth, clubs cylindrical at first, flattened or grooved when old, tips acute. Stipe cylindrical in shape, not grooved.

Distribution

Reported from Uttaranchal (Thind and Anand, 1956), India (Bose and Bose, 1940 and Corner,1950).

117. *Clavulinopsis helvola* (Fr.) Corner

Synonym

Clavaria helvola Fr., *Clavaire janatre, Orngegelba keule.*

Common Name

Yellow club.

Habitat

Grows solitary or in small groups in wood on soil under Oak tree.

Habit

Sporophore 3-7 cm high 1.5-4 mm wide, yellow to orange yellow, simple. Spores white or faint yellow, subglobose to somewhat angular and bluntly echinulate, 4-7 X 3.5-6 μm.

Distribution

Reported from Mussoorie, U.P. (Thind and Raswan, 1958)

118. *Clitocybe clavipes* (Pers. ex Fr.) Kummer

Common Name

Clubfast

Habitat

Grows singly or in group on the ground under coniferous or sometimes deciduous trees.

Habit

Fruiting body white to cream, 5.15 cm and funnel shaped. Cap 2-8 cm across, flattened, convex with a slight umbo at first, later depressed; buff to gray-brown with an olivaceous tint, paling toward the margin; moist, smooth, with matted hairs and scurfy on the disc. Gills deeply decurrent, nearly distant, narrow to broad; pale creamy yellow. Stipe 30-70 × 5–15 mm, stuffed, greatly swollen at the base and tapering upward; whitish to ash; spongy, covered in silky fibers. Flesh thick, spongy then rather brittle when dry; white, but yellowish toward the base. Odor strong, sweet. Spores subglobose to ellipsoid, smooth, nonamyloid, 5-8 × 3-5 μm. Spore deposit white.

Distribution

Reported from North West Himalayas (Kumar, 1987).

119. *Clitocybe gibba* (Pers.) Kummer

Synonym

Clitocybe infundibuliformis (Schaeff ex Weinm) Qu

Common Name

Common Funnel Cap

Habitat

Grows in mycorrhizic association with *Quercus incana* in deciduous forest or on heaths.

Habit

Fruiting body funnel shaped, 5–15 cm in length. Pileus orange white to light orange having a shallow depression in the centre, context orange white, unchanging on bruising cap 3–8 cm across often with a wavy margin, pale pinkish buff to ochre, silky smooth. Stipe 30–80 × 5–10 mm, pale buff, tough and slightly swollen towards the base. Flesh whitish to buff. Smell faint and sweet. Gills decurrent, crowded, whitish. Spore print white.

Distribution

Reported from North West Himalayas (Kumar, 1987).

120. *Clitocybe nebularis* (Batsch ex Fr.) Quel.

Synonym

Lepista nebularis

Common Name

Cloudy clitocybe; Clouded funnel cap; Cloud funnel.

Habitat

Grows on the ground.

Habit

Pileus upto 6 cm in diameter, plano-convex to depressed at the centre, pale greyish, viscid, smooth. Stipe upto 7.5 cm long, cylindrical, solid, pale grayish in colour. Spores (5.5-8.5×2.8-4.2 µm) elliptic, smooth, hyaline, inamyloid.

Distribution

Reported from Chandigarh (Rawla and Arya, 1983) and from Mahabaleshwar, Maharashtra (Sathe and Deshpande, 1979).

121. *Clitocybe squamulosa* (Pers.) Fr.

Synonym

Infundibulicybe squamosa.

Common Name

Funnel Cap.

Habitat

Grows solitary, scattered or gregarioulsy on decaying needles of *Cedrus deodara* or on mosses.

Habit

Cap 2-11 cm; at first flat or with a central depression, becoming deeply vase-shaped; smooth or with fine fibers or scales; dry or slightly tacky; brown, cinnamon brown, or dark tan; fading with age; sometimes with a wavy margin in maturity. Gills running down the stem; close or nearly distant; white or pale cream. Stipe 3-7 cm long; up to 1 cm thick; equal; dry; fairly smooth; coloured like the cap; base often covered with white mycelium. Flesh thin, whitish or watery. Spore Print white.

Distribution

Reported from North West Himalayas (Kumar, 1987)

122. *Clitocybe tabescens* (Scop. ex Fr.) Bres.

Synonym

Armillaria tabescens (Scop. ex Fr.) Emel.; *Clitocybe monadelpha* (Scop.) Bres.

Habitat

Grows on and around the dead wood, may be in tufts.

Habit

Pileus upto 8 cm in diameter, convex or expanded, irregular at maturity, fibrillose and squamulose, brownish. Stipe upto 8 cm long, usually narrow, fibrillose. Spores (8-10×5-7 µm) light cream.

Distribution

Reported from Maharashtra (Sathe and Rahalkar, 1975).

123. *Clitopilus prunulus* (Scop. ex Fr.) Kummer

Common Name

Sweet bread mushroom, The Miller or Plum Clitopilus.

Habitat

Grows solitary or scattered on the ground, on grassland very common in both deciduous and coniferous forests.

Habit

Pileus 3-13 cm in diameter, convex to depressed then plain white or greyish, margin decurved often wavy, irregularly elevated, pale-grey. Flesh white, soft. Gills decurrent white becoming pink with age. Stipe 3-8 cm long, white, solid with striations. Spores (8.5-14×4.5-6.5 µm) ellipso-fusoid with distinct longitudinal ridges, polygonal in apical view, hyaline.

Distribution

Reported from Jammu and Kashmir (Watling and Gregory, 1980) and Kerala (Bhavani Devi, 1995).

124. *Collybia altargrettia* (Scynes) Pegler

Habitat

Grows singly or aggregated in frondose forest soils.

Habit

Sporophores small to medium with convex, purple coloured cap, long slender and cartilaginous stipe.

Distribution

Reported from Meghalaya (Verma *et al.*, 1995)

125. *Collybia aurea* (Beeli) Pegler

Habitat

Grows on dead wood.

Habit

A species, forming dense, tufted clusters, up to 6 cm tall, growing on dead stumps and logs. The bright yellow to orange colours and the clustered habit make this species easily recognizable.

Distribution

Reported from New Delhi (Saxena and Kapoor, 1988)

126. *Collybia butyracea* (Bull. ex Fr.) Kummer

Habitat

Grows in groups on humicolous soil and leaf litter of *Cedrus deodara* tree in coniferous forest.

Habit

Pileus upto 4 cm in diameter, convex to applanate, obtusely umbonate, surface moist, centre brown vinaceous, almost glabrous to minutely fibrillose, margin regular, non-striate, cuticle half peeling, flesh clay pink, upto 2.5 mm thick. Lamellae adnexed, crowded, unequal, white 2.5 mm in breadth, edge scattered. Spore deposit white, stipe upto 7 cm long, 4-10 mm broad, cylindric bulbons towards base, stuffed then hollow.

Distribution

Reported from Matiana, Himachal Pradesh by Upadhyay and Kaur (2003).

127. *Collybia confluens* (Pers. ex Fr.) Kummer

Synonym

Marasmius confluens (Pers. ex Fr.) Karst.

Common Name

Clustered tough-shank.

Habitat

Grows in dense clusters, often in rings in frondose wood or in pine forest leaf litters.

Habit

Sporophore leathery and plain white to light buff or pale clay with reddish purple tinge. Pileus 2-4.5 cm in diameter, convex becoming flat with finally striate margin, sometimes umbilicate, glabrous. Stipe central, upto 5.5 cm long, densely pubescent, thin, covered with a frosty bloom of hairs, very tough and fibrous, hollow when old. Spores (6.6-8.8×2.2-2.3 µm) globose to subglobose, smooth, white and inamyloid.

Distribution

Reported from Tamil Nadu (Natarajan and Manjula, 1982) and from Meghalaya (Verma *et al.*, 1995).

128. *Collybia dryophila* (Bull. ex Fr.) Kummer

Synonym

Marasmius dryophilus (Fr.) Karst.

Habitat

Grows in groups or in tufts in forests, on the ground in open places, groves or often amongst grasses.

Habit

Pileus 3-5 cm in diameter, hemispherical then convex and finally flattened with age, depressed at the centre, tan or reddish brown, surface smooth. Gills crowded adnexed or almost free, whitish, sometimes yellowish. Stipe 3-8 cm long, equal or slightly attenuated upward, cartilaginous, smooth, hollow and hairy at the base. Spores (6-7.6×3-4 μm) white. Flesh pale tan, soft, thick near the stipe, thinner towards the margin.

Distribution

Reported from Lachen, Himalayas (Butler and Bisby, 1960); Jammu and Kashmir (Watling and Gregory, 1980).

129. *Collybia maculata* (A. and S. ex Fr.) Kummer

Synonym

Rhodocollybia maculata.

Common Name

Spotted collybia, Rust spot fungus.

Habitat

Grows in groups amongst leaves.

Habit

Pileus 5-13 cm in diameter, obtusely convex to flattened with age, whitish with rusty red spots, margin inrolled when young sometimes wavy. Gills crowded, free, adnexed or short decurrent white. Stipe 10-18 cm long, uniformly thick or narrowed towards the base, white, firm and brittle, flbrous within, hollow with age, base of stipe penetrating the soil, without volva and annulus. Spores globose. Flesh whitish, thick at the stipe, thinner toward the margin of the pileus, fibrous and spongy.

Distribution

Reported from Lachen, Himalayas (Berkley, 1856; Butter and Bisby, 1960).

130. *Collybia radicata var. superbiens* (Berk) Sacc.

Synonym

Oudemansiella radicata var *superbiens* (Berk) Pegler and Young; *Agaricus radicatus* var. *superbiens* Berk; *Xerula radicata* var *superbiens* (Berk) Dorfelt.

Habitat

Grows on soil.

Habit

Pileus 1.5-3 inches, convex, dark brown when dry, slightly viscid, smooth, sometimes lobed. Stipe 5-6 inches high, 0.25 inch thick, attenuated upwards, minutely furfuraceous especially at the base, rooting deeply. Gills distant adnate, decurrent, yellowish at length, orange when dry.

Distribution

Reported from Khasi Hills, Assam (Berkeley, 1856).

131. *Cookenia sulcipes* (Berk) Kuntze.

Habitat

Reported to grow on unidentified dead stem.

Habit

Fruiting bodies (Apothecia) deep cupshaped to funnel shaped. The inner spore bearing surface of the apothecium, the hymenium bright colour, yellow to red, although the colour fade upon drying. The outer surface less brightly coloured. The excipulum, the tissue making of the wall of the apothesium thin and flexible. When hairs are present on apothesium, they are fesiculate and are made of bundles of cylindrical hyphae. Asci constricted abruptly below and form a blunt rounded base with a slim tail-like connection. All asci mature simultaneously not in series. They have paraphyses which anastomose and form a three dimensional network. Ascospores 20-40 µm long, ellipsoidal or slightly unequal and either smooth or ornamental with fine wrinkles.

Distribution

Reported from Jalpaiguri, W.B. (Kar and Pal, 1968).

132. *Cookenia tricholoma* (Mont.). Kuntze

Common Name

Pink Cup Fungi

Habitat

Grows on unidentified stump or fallen branches of angiosperms or even on fruits.

Habit

Apothecia deep–cup shaped to funnel shaped. The inner spore bearing surface of the apothecium, the hymenium brightly coloured, yellow to red which fade upon drying. The outer surface less brightly coloured. The excipulums, the tissue making up the walls of the apothecium thin and flexible. Hairs, it present on apothecium, are made up of bundles of cylindrical hyphae. Asci abruptly constricted below forming a blunt, rounded base with a slim tail like connection, asci mature simultaneously, paraphyses which anastomose and form a three dimensional network ascospores large (20–40 mm long), ellipsoidal or slightly unequal sizes, smooth or ornamented with fine wrinkles.

Distribution

Reported from Jalpaiguri, West Bengal (Kar and Pal, 1970).

133. *Coprinus atramentarius* (Bull. ex Fr.) Fr.

Synonym

Coprinopsis atramentaria.

Common Name

Ink cap (Local name-Mashi Koon)

Habitat

Sporophores grow scattered or in dense cluster near base of frondose trees, garden, fields and around on dead stumps, richly manured soil.

Habit

Caps dirty, usually smaller, not tender, centrally stipitate, fruiting body autodigesting. Pileus 2.0-7.0 cm in diameter, conical to ovate when young, broadly conical when old. Colour varying from different shades of grey, surface sometimes smooth or provided with small scales at centre, occasionally with striations and furrows, margin splitting into lobes. Gills crowded, free, white when young but black when old. Stipe central, cylindrical, 6-12×0.8-1.5 cm, smooth and white above annulus, light brownish and scaly below, hollow, annulate but without volva.

Distribution

Reported from Kerala (Bhavani Devi, 1995), entire North eastern Hill region (Verma *et al.*, 1995), U.P. (Vasudeva, 1962; Nita Behl, 1988), India (Kaul, 1971), Jammu and Kashmir (Kaul and Kachroo, 1974; Abraham *et al.*, 1981; Nita Behl, 1988), Allahabad (Saksena and Mehrotra, 1952), Solan, H.P. (Munjal *et al.*, 1974 and Thapa *et al.*, 1977).

134. *Coprinus brunneofebrillosus* Dennis.

Habitat

Grows in isolation on grassy soil.

Habit

Pileus upto 2.5 cm broad, applanate, revolute at maturity, grey or greyish brown, surface dry with few brown floccose velar squamules. Gills free, linear, subdistant, equal, narrow, black. Stipe up to 4.7 cm long, broad, cylindrical, slightly tapering upward, hollow, white, without annulus. Spores (9-12×6-7.5 µm) ellipsoid, smooth, thick walled, dark brown, apically truncated.

Distribution

Reported from Tamil Nadu (Natarajan and Raman, 1984) and Punjab (Atri and Kaur, 2004).

135. *Coprinus cinereus* (Schaef.) Cooke

Synonym

Coprinopsis cinerea.

Common Name

Inky Cap Mushroom.

Habitat

Grows on dung of Nilghai

Habit

Besides being edible, it is one of the two Basidiomycete with *Schizophyllum commune* that are commonly used in research as a model organism. Its genome has been sequenced entirely by the Broad Institute of M.I.T. (U.S.)

Distribution

Reported from Punjab (Ginai, 1936; Rea, 1922)

136. *Coprinus comatus* (Muller ex Fr.) S. F. Gray

Synonym

Agaricus comatus.

Common Name

Shaggy Cap; Lawyer's wig.

Habitat

Sporophores grow singly, scattered, or in clumps on grassy land, in lawns, in gardens, fields, roadsides and on refuse clumps.

Habit

Sporophores centrally stipitate, usually oblong, characterised by shaggy appearance of the pileus, almost white at maturity, fruiting body autodigesting. Pileus 6.0-10.0×2.5-5.0 cm, cylindrical or oblong when young but campanulate to expanded when fully grown, white when young but tan to purplish tan when old, surface covered with shaggy brown scales, pileus splitting at the margin. Gills crowded, free, white when young, then pink, finally black. Stipe centrally placed, tapering at the top, 3.5-8.0×.0.6-1.5 cm, whitish smooth or fibrillose, hollow, with a thin loose ring around the stipe, disappearing quickly, without volva, flesh white and fragile.

Distribution

Reported from Kerala (Bhavani Devi, 1995), North Eastern Hill (Verma *et al.*, 1995), Punjab (Chopra and Chopra, 1955; Kaushal and Grewal, 1992), West Bengal (Bose and Bose, 1940; Banerjee, 1947), Gujrat (Moses, 1948), Punjab, U.P. and Several other parts of India (Chopra and Chopra, 1955), Maharashtra and West Bengal (Butler and Birby, 1960), Jammu and Kashmir (Kaul and Kachroo, 1974; Abraham *et al.*,

1984), Maharashtra (Trivedi, 1972; Nita Bahl, 1988; Sathe and Deshpande, 1979) and from Delhi (Saxena *et al.*, 1969).

137. *Coprinus disseminatus* (Pers. ex Fr.) S. Fr Gray

Synonym

Agaricus disseminatus Pers ex Fr.; *Pasthyrella disseminatus* (Pers ex Fr.) Quel.; *Psathyrella etenodes* (Pars. Fr.) Petch., *Pseudocoprinus etenodes* (Pers. Fr.) Muchner.

Common Name

Tropping crumble caps, Fairies Bonnets.

Habitat

Sporophores grow in large numbers on and around rotten stumps and logs of trees or grows gregariously on the trunk base of the *Morus alba* tree.

Habit

Caps very small (1-2 cm) ovate, thin expanded, thin membranous, strongly grooved, grey or pale ochre, not autodigesting. Gills white, then grey. Stem thin, short, finely hairy. Spores 8-11×4-5 μm.

Distribution

Reported from Kerala (Bhavani Devi, 1995), Punjab (Atri and Kaur, 2004; Atri and Amanjeet Kaur, 2002) from India (Ray and Samajpati, 1979; Lange and Smith,1953) and from Jammu and Kashmir (Abraham *et al.*, 1980).

138. *Coprinus fimetarius* Linn.

Synonym

Coprinus cinereus (Schaeff) Gray; *C. macrorhizus* (Pers.) Rea; *C. fimetarius* var. *cinarius* (L.) Fr.

Habitat

Grows on cattle dung.

Habit

Pileus upto 5 cm broad, convex to campanulate, greyish white to greyish brown, surface sticky, striate towards the margin, soon autodigested. Gills free, narrow, greyish brown to black, soon autodigested. Stipe up to 16.7 cm long, brown, slightly tapering upwards with subbulbous base, white, fibrillose, without annulus. Spores (10.5-15×7.5-9 μm) ellipsoid, epiculate, thick walled, blackish brown.

Distribution

Reported from West Bengal (Bose, 1920), Maharashtra (Patil *et al.*, 1995; Sathe and Deshpande, 1979), Kerala (Bhavani Devi, 1995) and Punjab (Atri and Kaur, 2004).

139. *Coprinus hiascens* (Bull. ex Fr.) Fr.

Habitat

Sporophores grow gregariously on wooden log of *Melia azadirachta.*

Habit

Pileus upto 4.5 cm broad, convex then expanding with reflexed margin at maturity, surface brownish grey, grooved, grooves dark brownish grey. Gills adnexed to free, unequal, subdistant, narrow, not in series. Stipe up to 9 cm long, central, slightly tapering upward, hollow, white shining, glabrous, without annulus. Spores (7.5-10.5×4.5-6.7 μm) subovoid to ovate, dark brown, thick walled with prominent apical pore.

Distribution

Reported from Punjab (Atri and Kaur, 2004).

140. *Coprinus lagopus* (Fr.) Fr.

Synonym

Agaricus lagopus Fr.

Common Name

Bonfire ink Cap, Local name in Kerala–Mashi Koon.

Habitat

Sporophores grow solitary or in groups in the fields and on the paddy straw heaps; as a weed mushroom in the beds of *Volvariella volvacea*, on pine and beech test blocks and preservative treated service timber packing of cooling towers or on grassy humicolous soil.

Habit

Pileus upto 5 cm broad, thin and membranous, companulate when young, plano-convex at maturity soon revolute, pale grey, covered by white pilose fibrils, margin striate. Gills free, white to pale grey when young black at maturity. Stipe upto 15.0×0.5 cm, cylindrical, attenuate towards apex, hollow, white, covered with white cottony fibrils. Veil formed by abundant pilose fibrils on the pileus. Annulus absent. Context, thin, whitish, made up of thin walled hyphae, with clamp connections.

Distribution

Reported from India (Krishnamoorthy and Verma, 1974; Natarajan and Raman, 1984; Doshi and Sharma, 1997), Punjab (Garcha and Kalra, 1977; Atri and Kaur, 2004), Kerala (Bhavani Devi, 1995), Tamil Nadu (Natarajan and Raman, 1983a), Punjab (Atri and Amanjeet Kaur, 2002).

141. *Coprinus macrorhizus* Pers. Ex Rea.

Habitat

Sporophores grow in loose groups on decaying wheat straw.

Habit

Pileus upto 4.5 cm broad, conical to campanulate, greyish white with light brown centre, covered with white recurved fibrillose scales, margin splitting, digested at maturity. Gills free, crowded, white when young, black at maturity, autodigested at maturity. Stipe up to 17.5 cm in height, epigeal portion hollow, tapering upwards, white with minute fibrillose scales, annulus absent. Spores (10.5-15×6-9 μm) ellipsoid, smooth, thick walled, brownish black, apiculate.

Distribution

Reported from Punjab (Atri and Kaur, 2004).

142. *Coprinus micaceus* (Bull ex Fr.) Fr.

Synonym

Coprinellus micaceus

Common Name

Mica ink Cap, Glistering ink Cap.

Habitat

Grows on, fence, posts and trees, in dense tufts on rotten stumps and timber, and on the moist soil under shade, in dense clumps or more or less scattered on ground, sometimes at the base of the living trees or around stumps, rarely on logs in woods.

Habit

Sporophores tender and delicate, centrally stipitate. Pileus 2.0-6.0 cm in diameter, ovate when young and campanulate or expanded when old, light buff or yellowish brown, very delicate and tiny glistering scales on the surface of the pileus, prominent striations or furrows running from the edge. Gills crowded, at first white, then purplish brown and finally black, not wide, liquefying into an inky fluid. Stipe slender, 5.0-10.0×0.4-0.8 cm, white, silky, brittle, hollow, delicate fibrous ring present in young sporophore but disappear at maturity. Basidia polymorphic, separated by board sterile cells.

Distribution

Reported from West Bengal, Jammu and Kashmir, Maharashtra (Sathe and Rahalkar, 1978; Bose and Bose, 1940; Trivedi 1972; Watling and Gregory, 1980). Reported from Kerala (Bhavani Devi, 1995), Assam (Gogoi *et al.*, 2000), West Bengal (Banerjee, 1947; Butler and Bisby, 1960; Nita Bahl, 1988), Jammu and Kashmir (Kaul and Kachroo, 1974; Nita Bahl, 1988), Maharashtra (Trivedi, 1972; Nita Bahl, 1988; Sathe and Deshpande, 1979), Lucknow, U.P. (Ghosh *et al.*, 1974) and from Calcutta, W.B. (Banerjee, 1947).

143. *Coprinus micaceus* var. *microsporus* (Bull. ex Fr.) Atri *and* Kaur

Habitat

Sporophores grow in clusters on cattle dung and manured soil.

Habit

Pileus upto 4 cm broad, campanulate then expanded, white then pale to pale grey, surface moist, covered with fine granular velar squamules. Gills adnexed to free, crowded, unequal, pale then black at maturity. Stipe up to 8.5 cm long, white, cylindrical, hollow, slightly tapering upwards, covered with fine hairs, annulus absent. Spores (10.5-15×7.5-9 µm) ellipsoid in side view, smooth, thick walled, apically truncated by germ pore.

Distribution

Reported from Punjab (Atri and Kaur, 2004) and from Solan, H.P. (Thapa *et al.*, 1977).

144. *Coprinus niveus* (Pers. ex Fr.) Fr.

Common Name

Snowy ink-cap, snow-white Coprinus.

Habitat

Sporophores grow solitary or in groups on cow dung, horse dung, sometimes on wet or fertilized ground or on heaps of rotten straw.

Habit

Pileus 1.5-2.0 cm in diameter, oval then campanulate and finally with recurved margin, split and curled over on itself, surface covered with white, persistent and floccose scales. Gills white, then flesh coloured, finally blackish, adnexed, crowded. Stipe upto 8 cm long, central, almost equal, narrowing towards the top, straight, hollow, white and covered with floccose elements. Spores (13-16×10-12 µm) broadly elliptical, brown to blackish, smooth with germ pore.

Distribution

Reported from Punjab (Manju, 1933), West Bengal (Banerjee, 1947) and Kerala (Bhavani Devi, 1995).

145. *Coprinus patouillardii* (Quel.) Pat.

Habitat

Grows on dung under *Albizzia lebbek* tree.

Habit

Pileus upto 1.5 cm broad, convex companulate, greyish white with minute fibrillose squamules. Gills free, narrow, distant, unequal, grey. Stipe up to 4.3 cm long, hollow, almost equal in diameter, white, without annulus. Spores (7.5-10.5×5.2-6.7 µm) ellipsoids, smooth, double walled, apically truncated by a germ pore.

Distribution

Reported from Punjab (Atri and Kaur, 2004).

146. *Coprinus plicatilis* (Curtis ex Fr.) Fr.

Synonym

Agaricus plicatilis Curt ex Fr.

Common Name

Little Japanese Umb

Habitat

Sporophores solitary or in groups grows in grassland, gardens and on ground or scattered on humicolus soil under *Psidium guazava* and *Melia azadirachta* tree.

Habit

Pileus upto 3.0 cm broad, membranous, convex when young, planoconvex at maturity, greyish yellow, glabrous, radially plicate to the disc. Lamellae free, olive brown, becoming black. Stipe upto 4.0×0.3 cm, cylindrical with a bulbous base, hollow, Veil absent.

Distribution

Collected from different localities of Kerala (Bhavani Devi, 1995), Tamil Nadu Madras (Natrajan and Raman, 1983a), Punjab (Atri and Kaur, 2004), Jammu and Kashmir (Watling and Gregory, 1980), Rajasthan (Doshi and Sharma, 1997), Chandigarh, Punjab (Rawla *et al.*, 1982) and from Sirhind, Punjab (Atri and Amarjeet Kaur, 2002).

147. *Coprinus sterquilinus* (Fr.) Fr.

Habitat

Sporophores grow scattered or in groups on compost, manured ground or straw.

Habit

Pileus 4-6 cm in diameter, broad conical to flattened, white, light brown towards centre, surface silky but scaly with age, margin recurved with radial furrows. Gills narrow, free and black. Stipe 10-15 cm long, central, straight with inconspicuous volva. Spores large with eccentric germ pore.

Distribution

Reported from Srinagar, Jammu and Kashmir (Watling and Gregory, 1980) and from Solan, H.P. (Thapa *et al., 1977*).

148. *Cordyceps militaris* (L.) Link

Common Name

Keera Jhar

Habitat

It is an entomopathogenic mushroom. Its fruiting body originates from the head of larvae and pupae of insects. All species of *Cordyceps* are parasitic specially on insects, nematodes and sclerotia of *Claviceps* or hypogeous ascocarps of several species of *Elaphomyces*.

Habit

Fruiting body is 5–6 cm tall, nearly cylindrical or club–shaped, normally solitary and creamish white in colour. The fruiting body bears minute powdery mass on the top. The fruiting body is associated with the head of insect larvae and pupae. The associated body of the insect becomes mummified by the growth of the mycelium. Conidiophores cylindrical and conidia are barrel–shaped. Spores colourless or transparent, filiform, smooth, red–shaped or filiform, smooth, rod–shaped or elliptical, hyaline, 2–5 µm in size.

Distribution

Reported from Shimla, Himachal Pradesh (Sagar *et al.*, 2007) and from Maharashtra (Jagdale and Patil, 1983).

149. *Cordyceps ophioglossoides* (Ehrenlo. ex Fr.) Link.

Habitat

Grows on decomposed organic matter in the soil.

Habit

It grows as parasite on underground puffball. Fruiting body tough, club shaped, lacking clearly defined cap, the reddish brown to black top area and yellow to brownish or black lower area, the presence of yellow cord connecting the parasite and the puffball, and of course, in the presence of underground parasitized puffball which belong to the genus *Elaphomyces*.Fruiting body 2-8 cm long to 1 cm wide; club shaped with the top wider than the base, without a clearly defined cap but the upper portion reddish brown and smooth when young becoming blackish and roughened or pimply with maturity, lower portion smooth throughout development, yellow to brownish to blackish, rooting base attached to yellow cord that lead to the *Elaphomyces* fruiting body, flesh whitish and tough.

Distribution

Reported from Darjeeling (W.B.) by Kar and Gupta (1978).

150. *Cortinarius armillatus* (Fr.) Fries

Synonym

Hydrocybe armillatus.

Common Name

Red-banded Cortinarius

Habitat

Sporophores grow solitary or scattered on the soil, on the forest.

Habit

Cap 5-12 cm, convex, then expanded and flattened, dry, fibrillose, brick-red to tawny, orange when old. Gills whitish then cinnamon-brown. Stem tall, swollen, clavate, pale whitish brown with three or four reddish belts running obliquely round. Spores almond shaped, 9-12×5-6 µm.

Distribution

Reported from Kerala (Bhavani Devi, 1995)

151. *Cortinarius cinnabarinus* Fr.

Habitat

Sporophores grow on decaying coconut tree stumps.

Habit

Pileus upto 12 cm in diameter, convex with depressed centre, surface dry, ochraceous buff becoming reddish orange to brown, fibrillose, scales minute, margin entire, splits with maturity. Stipe upto 5 cm long, slightly tapering upwards, concolourous with the pileus surface, fibrous, striate and hollow. Sproes (7-10×5-7 µm) ellipsoid, inamyloid and thick walled.

Distribution

Reported from Kerala (Kaul and Kapur, 1988).

152. *Cortinarius collinitus* Fr.

Habitat

Grows scattered on leaf litter under Oak trees in mixed forest.

Habit

Pileus 5-10 cm in diameter; applanate to convex, surface moist, waxy, sticky, centre umber, margin camel coloured with violet tinge, smooth, margin irregular, striate, involute, cuticle fully peeling, flesh unchanging. Lamellae sinuate, decurrent, unequal, 5-6 sized, moderately crowded, fleshy, cigar brown, 1-1.3 cm in breadth, edge smooth. Spore deposit brown. Stipe central, 9 × 2 cm in size, equal in diameter, surface fibrous and straw coloured; fleshy, texture smooth, solid then hollow, flesh light gold yellow; annulus corticoid fugacious.

Distribution

Reported from Shimla H.P. (Upadhyay *et al.*, 2005) and from Chail (2001 m), H.P. (Upadhyay *et al.*, 2007).

153. *Cortinarius cyanopus* (Secr.) Fr.

Synonym

Cortinarius amoenelns.

Habitat

Sporophores grow in mixed conifer forest.

Habit

Pileus 5-8 cm in diameter, hemispherical to explanate, date-brown with bluish hue, pale tan towards the top, surface viscid, margin glabrous. Gills 60-80 mm wide, adnato-emarginate, violaceous, pale with age. Stipe upto 8 cm long, violaceous-white, solid, bulbous, bulb spongy, depressed, oblique. Spores (12-13×6.5-7.7 μm) reddish brown and ornamented.

Distribution

Reported from Pahalgam, Jammu and Kashmir (Watling and Gregory, 1980).

154. *Cortinarius purpurascens* (Fr.) Fr.

Synonym

Phlegmaceum purpurascens

Common Name

Bruising webcap.

Habitat

Grows on the ground in mixed forests.

Habit

Pileus 6-1 5 cm in diameter, convex then expanded then flattened, yellowish to dark umber brown, surface slightly viscid when moist. Gills sub-free to decurrent, blue when young and rust coloured at maturity. Stipe upto 10 cm long, central, bulbous, brown, surface scaly, veil bluish violet. Spores (9-11.5×5-6.5 μm) ellipsoidal and rusty brown.

Distribution

Reported from Himachal Pradesh (Sharma and Thakur, 1978).

155. *Cortinarius violaceus* (L. ex Fr.) Fr.

Synonym

Inoloma violaceum (Fr.) Wunsche

Common Name

Violet cortinarium.

Habitat

Sporophores grow in the forests, solitary or scattered.

Habit

Sporophores fleshy, decay easily, centrally stipitate. Pileus 5.0-15.0 cm in diameter, convex to flattened, violet, dry, covered with numerous persistent minute and erect scales, texture metallic shiny. Gills distinct, adnate when young, sinuate when old, sometimes with veins between adjascent gills at their junction. Cap dark violet when young but greyish brown when old. Stipe 8.0-12.0×1.0-1.5 cm, thick with a bulbous base, solid, dark violet with cobweb like rusty ochreaceous ring. Flesh violet.

Distribution

Reported from Myrong, Khasi Hills, Assam (Butler and Bisby, 1931 and 1960).

156. *Cratellarus verrucosus* Massee

Common Name

Khasi-Tit–Syiem.

Habitat

Sporophores grow in clusters on forest floor of Coniferous woods.

Habit

Fruiting bodies trumpet shaped, fleshy and orange yellow.

Distribution

Reported from upper Shillong in East Khasi Hills and Howai in Jaintia Hills districts of Meghalaya (Verma *et al.*, 1995).

157. *Craterellus cornucopioides* (L. ex Fr.) Pers.

Synonym

Peziza cornucopioides L.; *Elvella cornucopioides* Scop.; *Merulius cornucopioides* Pers.; *Cantharellus cornucopioides* Fr.

Common Name

Horn of Plenty, Trumpet of the dead.

Habitat

Sporophores grow solitary or in clusters amongst dead leaves of frondose woods, on ground, in woods and shady places, sometimes on clumps of bomboo.

Habit

Sporophores deeply funnel shaped, often perforate through to hollow stipe. Pileus 2.5-6.4 cm in diameter and 5.1-10.2 cm long tubiform, blackish brown with a few obscure fibrous tufts or scales, flexible, thin, fruiting body surface very uneven with folds and wrinkles, ash grey to pinkish brown or dark smoky brown. Stipe very short or nearly lacking.

Distribution

Reported from Meghalaya (Verma *et al.*, 1995); West Bengal (Banerjee, 1947; Basu, 1955).

158. *Crepidotus applanatus* (Pers.) Fr.

Habitat

Grows gregariously in overlapping clusters on dead hardwood stumps, logs and branches.

Habit

Cap 1-4 cm, shell shaped or petal shaped, somewhat flabby smooth or finely velvety (towards this point of attachment) in all stages of development, the margin often slightly lined, white, becoming brownish to pale cinnamon brown, hygrophanus. Gills close or crowded, whitish, turning brownish at maturity. Stipe absent. The cap may be nearly circular, creating the illusion of a rudimentary stipe where the mushroom attaches to the wood. Flesh soft, thin. Spores 4-6 µm, globose, finely punctuate to roughened. Pleurocystidia absent. Clamp connections present.

Distribution

Reported from Mussoorie, U.P. (Hennings, 1901).

159. *Crepidotus mollis* (Fries) Kummer

Common Name

Soft Slipper Toadstool

Habitat

Sporophores grow solitary or scattered, imbricated, grow in clumps on the decayed stumps and oil palm pericarp waste and decaying woods.

Habit

Pileus 1.0-7.0 cm in diameter, shell or kidney shaped, pale brownish or yellowish, soft, glabrous, composed of parallal hyphae. Stipe rudimentary and lateral. Gills crowded, decurrent to central point, whitish, then watery cinnamon, often spotted.

Distribution

Reported from Kerala (Bhavani Devi, 1995), Arunachal Pradesh (Mishra, 1999), Mahabaleshwar, South Western India (Sathe and Despande, 1980).

160. *Crepidotus variabilis* (Pers. ex Fr.) Kummer

Synonym

Cladopus variabilis (Pers. ex Fr.), *Crepidotus variable*

Habit*at*

Sporophores grow on twigs, fallen branches of deciduous trees, on dead woods and dry twigs.

Habit

Sporophore is tiny kidney shaped, 0.5 to 2 cm in diameter and often slightly loded, always sessile (no stalk). The cap is initially white, turning creamy ochre with age. Fruit body is nearly always laterally attached to its substrate–usually small twigs via its cap, rather than with stipe. The gills radiate from the point of attachment, moderately crowded. Spore print pinkish buff.

Distribution

Collected from differment localities of all districts of Kerala (Bhavani Devi, 1995).

161. *Cyathus limbatus* Tul.

Habitat

Grows in flower pot or on rotten Bamboo logs.

Habit

Peridium cupulate,7 -12 mm high, 7–10 mm broad at the maturity, attenuated downwards, base attached to the substratum by brown rhizomorphs, inner and outer surface longitudinally plicatugose at the apex. Peridioles ablate oblate,1.2 -2 mm in diameter, outer wall black, formed of thick hyphae, base attached to the cup by funiculus. Contacted part umbilicate between the peridiole and funiculus. Spores oval, 9–18 × 6–10 um, hyaline, inamyloid.Peridiole composed of two types of hyphae:one being brown, thick walled, aseptate, 3–5um thick; the outer one being hyaline to subhyaline, thin walled, aseptate, 2–4 um thick. Often with 5–6 × 3–4 um inflated cells among the hyphae.

Distribution

Reported from W. B. (Cook, 1880; Lloyd, 1906) and from Varanasi, U.P. (Khare, 1976).

162. *Cyathus stercoreus* (Schw.) de Toni

Synonym

Cyathus dimorphus

Common Name

Dung loving birds nest

Habitat

Sporophores grow on manure heaps or soil containing dung, gregarious on dead sticks and leaves, rotting gunny bags, rotten wheat straw and mud walls.

Habit

Fruit body tiny birds nest filled with eggs and is referred to as splash cups, because they are designated to use the force falling drops of water to dislodge and dispense their spores. The fruiting bodies, funnel- or barrel-shaped, 6–15 mm tall, 4–8 mm wide at the mouth, sometimes short-stalked, golden brown to blackish brown in age. The outside wall of the peridium, the ectoperidium covered with tufts of fungal hyphae that resembles shaggy, untidy hair. However, in older specimens this outer layer of hair (technically a *tomentum*) may be completely worn off. The internal wall of the cup, the endoperidium, is smooth and grey to bluish-black. The 'eggs' of the bird's nest–the peridiole blackish, 1–2 mm in diameter, and are typically about 20 in the cup. Peridioles often attached to the fruiting body by a funiculus, a structure of hyphae that is differentiated into three regions: the basal piece, which attaches it to the inner wall of the peridium, the middle piece, and an upper sheath, called the purse, connected to the lower surface of the peridiole. In the purse and middle piece is a coiled thread of interwoven hyphae called the funicular cord, attached at one end to the peridiole and at the other end to an entangled mass of hyphae called the hapteron.

Distribution

Reported from Gurdaspur (Punjab) and Dehradun by Ahmad (1940, 1942),

Ahmedabad (Rao, 1964), Chamba, H.P. (Lloyd, 1906), Varanasi, U.P. (Khare, 1976), Kolhapur (Parndekar, 1964) and from H.P. (Sohi *et al.*, 1964).

163. *Cystoderma amianthianum* (Fr.)

Habitat

Sporophores grow in mixed coniferous forest mycorhizal with *Cedrus deodara*

Habit

Cap ochre yellow to coca brown in colour, radially wrinkled, margin toothed, finally flat unchanging context, grannular floccose levanescent annulus. Stipe slender, ochre, scaly upto inconspicuous ring. Spores turn blue black in Melzer's iodine.

Distribution

Reported from North West Himalayas (Lakhanpal, 1986 and Kumar, 1987).

164. *Dacryopinax spathularia* (Schw.) Martin.

Synonym

Guepinia spathularia (Schw.) Fr.

Habitat

Grows on *Bambusa arundacea* and *Shorea robusta* or on log and decaying wood.

Habit

It is an edible jelly fungi. Fruit body orange, in colour, fan shaped and less than a centimeter tall.

Distribution

Reported from West Bengal (Banerjee, 1947; Currey 1874), Punjab (Ahmad, 1945) and Saharanpur, U.P. (Lloyd, 1898-1925).

165. *Daedalea quercina* L. ex Fr.

Common Name

Maze gill fungus.

Habitat

Sporophores grow solitary or scattered on dead timber.

Habit

Sporophores usually sessile, 3.0-15.0 cm in diameter, hoof shaped, grey to black with prominent concentric ridges on the upper surface. Context fibrous, tough, corky, wood-colour, thick at the base, gradually thinner towards the margin. Pores in the hymenial surface usually radially elongate, labyrinthinal, pale tan, pore tube 1.0 cm. or more long, thickness of wall equal or greater than the pore diameter.

Distribution

Reported from West Bengal (Bose, 1918; Banerjee, 1947).

166. *Daedalopsis confragosa* var. *tricolor* (Bolt. Ex Fr.) Schoet

This species can be highly variable in appearance, and the 'founding mycologists' who initially examined a large array of specimens from a variety of locales had the tendency to give each a new species name, resulting in a rather large list of synonyms.

Synonyms

Agaricus confragosus (Bolton) Murrill, *Agaricus tricolor* Bull., *Amauroderma confragosum* (Van der Byl) D.A. Reid, *Boletus confragosus* Bolton*Cellularia tricolor* (Bull.) Kuntze, *Daedalea bulliardii* Fr., *Daedalea confragosa* (Bolton) Pers., *Daedalea confragosa* f. bulliardii (Fr.) Domañski, Orlos and Skirg, *Daedalea confragosa* f. rubescens (Alb. and Schwein.) Domañski, *Daedalea confragosa* subsp. rubescens Alb. and Schwein., *Daedalea rubescens* Alb. and Schwein., *Daedalea sepiaria* var. tricolor (Bull.) Fr., *Daedalea tricolor* (Bull.) Fr., *Daedaleopsis confragosa* var. bulliardii (Fr.) Ljub., *Daedaleopsis confragosa* (Bolton) J. Schröt., *Daedaleopsis confragosa* var. rubescens (Alb. and Schwein.) Ljub., *Daedaleopsis confragosa* var. tricolor (Bull.) Bondartsev, *Daedaleopsis rubescens* (Alb. and Schwein.) Imazeki, *Daedaleopsis tricolor* (Bull.) Bondartsev and Singer, *Ischnoderma confragosum* (Bolton) Zmitr. [as '*confragosa*'], *Ischnoderma tricolor* (Bull.) Zmitr., *Lenzites confragosa* (Bolton) Pat., *Lenzites tricolor* (Bull.) Fr., *Lenzites tricolor* var. rubescens (Alb. and Schwein.) Teng, *Polyporus bulliardii* (Fr.) Pers., *Polyporus confragosus* Van der Byl, *Striglia confragosa* (Bolton) Kuntze, *Trametes bulliardii* (Fr.) Fr. [as '*bulliardi*'], *Trametes confragosa* (Bolton) Jørst., *Trametes confragosa* f. bulliardi (Fr.) Pilát, *Trametes confragosa* f. rubescens (Alb. and Schwein.) Pilát, *Trametes rubescens* (Alb. and Schwein.) Fr. *Trametes rubescens* var. tricolor (Bull.) Pilát, *Trametes tricolor* (Bull.) Lloyd.

Common Name

Thin-walled maze flat polypore, Blushing bracket.

Habitat

Grows on rotten wood of willow, oak and other kinds of tree.

Habit

Fruiting body sessile or with a reduced base, fan-shaped, plane, 1-5 x 1.5-8 cm wide, 0.2-2 cm thick, sometimes laterally connected, leathery to corky, surface velvety initially, turning glabrous, zonate, radiate rugose, russet brown to liver purple, fading to cinnamon brown or cinnamon, finally turning to greyish white; margin thin, acute, wavy. Context 0.1-0.2 cm thick, pale buff to brown. Tubes often variable in shape; white to brown, sometimes bruising pink; 1-6 mm broad, 0.5-1.5 mm wide, often forked, and anastomose behind, edge wavy, sometimes dentate. Spores: cylindrical, hyaline, smooth, nonamyloid, 7-9 x 1.5-3 µm. Spore print white.

Distribution

Reported from India (Anonymus,1956; Mundkar, 1938)

167. *Daldinia concentrica* (Bolton) Cesati de Notaris

Common Name

King alferd's cake, Cramp balls and Coal fungus, Carbon balls.

Habitat

Sporophores grow on dead and decaying coniferous wood and on logs of different dead angiospermic trees.

Habit

Fruit bodies lack stipe, with a distinct cap having global hymenium; spore print black and is ball–shaped with a hard friable shiny black body, 2–7 cm wide. The flesh of the fungus is purple, brown, or silvery-black inside, and is arranged in concentric layers. Each layer represents a season of reproduction. The asci are cylindrical and arranged inside the flask-shaped perithecium. When each ascus becomes engorged with fluid it extends outside the perithecium and releases spores.

Distribution

Reported from Bomdila (Arunachal Pradesh) by Bisht and Harsh (2001), Arunachal Pradesh, Kashmir and J and K (Padwick and Merh, 1943), Sagar, M.P. (Saksena and Vyas, 1962–64) and Nainital (Mitter and Tandon, 1932).

168. *Elvela crispa* (Scop.) Fries.

Habitat

Grows on soil covered with dead decaying organic matter in the forest areas.

Habit

Ascophores pileate, pileus mostly saddle shaped, sometimes reflexed, usually irregularly lobed reaching a diameter upto 5 cm. Hymenium white, becoming cream or yellowish with age, margin of the pileus free. Stipe slender, stout deeply fluted longitudinally throughout, white, slightly tapering above, upto 8 cm in length. Asci cylidric, paraphyses numerous, hyaline, enlarged above.

Distribution

Reported from Simla (Himachal Pradesh) and Kashmir (Sohi *et al.*, 1965).

169. *Elvela mitra* L.

Habitat

Grows on soil covered with dead decaying matter.

Habit

Ascophores irregularly saddle–shaped or usually lobed, upto 3.0 cm in diameter. Hymenium even or irregularly convolute, greyish, margin of the pileus free. Stipe slender, deeply fluted longitudinally, enlarged below and gradually tapering above, whitish to greyish in colour, upto 4.5 cm long. Asci cylindric. Paraphyses hyaline and enlarged above.

Distribution

Reported from Solan (Himanchal Pradesh) by Sohi *et al.* (1965).

170. *Entoloma microcarpa* (Fr.) Kummer

Synonym

Termitomyces microcarpus (B. et. Br.) Heim, *Agaricus microcarpur* B.et. Br, *Collybia microcapa* (B.et. Br) Lioahn.

Common Name

Urgi Chhatu

Habitat

Gregarious to caespitose in broad leaf forests near roots of bamboos stump under which being termite nests.

Habit

Basidiocarp mycenoid to collybioid. Pileus 1–3.8 cm broad, campanulate to convex, often which conspicuous umbo at the centre, fleshy, white to dark grey, dry, tomentose, margin reflexed, lacerate. Context white, thin. Lamellae white, 41–42 per cm. at the margin, unequal, free, edges even. Stipe central, cylindrical, white 0.8 to 6 cm long, 2–6 mm. thick, solid, fibrous, velotinate to longitudinally striate, base enlarged without pseudorhiza or an inconspicuous pseudorhliza below. Spore ellipsoid, 6–7×3.5–4 mm, smooth, hyaline to pale pink, inamyloid. Basidia clavate, 4–spored with sterigma. Cheilocystidia thick, clavate, pale yellow. Gill trama parallal.

Distribution

Reported from Midnapur District, West Bengal (Das *et al.*, 2002)

171. *Entoloma turci* (Bresadola) Moser

Synonym

Leptonia turci Bres; *Rhodophyllus turci* (Bres) Kuhner and Romagri.

Habitat

Sporophores grow scattered on soil among mosses.

Habit

Pileus upto 1.2 cm broad, convex, brownish, lighter towards margin, surface dry, centre covered with minute fibrillose brownish squamules, margin striate and incurved. Gills broadly adnate, shortly decurrent, distant, unequal, not in series. Stipe up to 2.2 cm long, central, cylindrical, solid, fragile, glabrous, concolourous, without annulus. Spores (9-12×7.5-9 µm) angular, thick walled, apiculate, inamyloid.

Distribution

Reported from Punjab (Kaur and Atri, 2002).

172. *Favolus brasiliensis* (Fr.) Fr.

Synonym

Favolus tenulcubu P. Boaur.

Habitat

Grows on dead wood in the forest, often imbricate and in large troops.

Habit

Pileus–7 cm in radius, 10 cm, wide, mesopodal and infundibuliform to pleuropodal and flabelliform (with slight pileate rim on the upperside of the stipe–apex), minutely spiculose–villous in the centre (mesopodal) or at the base (pleureoodal), smooth towards the margin, white, then cream to ochraceous; margin dentate fibriate. Stipe 412´2–4 mm, subcylindric, finely spiculose–villous with a decurrent network of pores, eventually often spiculose verrucose, concolorous, Flesh 4 mm thick, soft, watery–pubescent, rather deep ochraceous.

Distribution

Reported from Poona (Maharashtra) by Patil *et al.*, (1995) and from West Bengal (Bose,1946).

173. *Favolus spatulatus* (Jungh) Lev.

Synonym

Polyporus spatulatus (Jungh) Corner, *Aschersonia spathulata* (Jungh) Kuntze, *Hymeogramme spathulatus* (Jungh) Sacc and Cup, *Laschia spathulata* Jungh, *Tyromyces spathulatus* (Jungl) G. Cunn. (as. Spathulatus)

Habitat

Grows solitary or in groups on dead wood.

Habit

Fruiting body flabeliform with a small stipe, white light yellow tint when fresh, brownish and corky when dry.

Distribution

Reported from South Manipur, Jaintia Hill districts of Meghalaya and Aizawal district of Mizorum (Verma *et al.*, 1995).

174. *Fistulina hepatica* (Huds.) Fr.

Common Name

Beefsteak fungus or Ox-Tongue

Habitat

Grows solitary or in groups in a shelving fashion from dead tree trunks or stumps and rarely on living trees.

Habit

Sporophores 10.0-20 cm high, 8.0-15.0 cm in diameter, very soft, juicy, dark-red coloured, stipitate or not, semicircular or kidney shaped, upper surface when young covered with darker, minute elevations and when old covered with radial streaks, sticky when young, moist. Stipe lateral, very short and thick, sometimes stipe absent, but often long upto 15.0 cm. Context white with alternating light and dark red streaks, 1.0-2.5 cm, thick, soft and watery. Pore tubes in the hymenial surface at first short, then cylindrical, 0.3-0.7 cm long, yellowish or slightly pink.

Distribution

Reported from West Bengal (Chopra and Chopra, 1955).

175. *Flammulina velutipes* (Curt. ex Fr.) Karst

Synonym

Colybia velutipes (W. Curt. ex Fr.) Kummer

Common Name

Velvet Stem, Winter Fungus

Habitat

Sporophores grow in clumps on dead wood or on old stumps especially *Castanea* and on decaying woods either erect or prostrate. Basswood is a favorite host,

Habit

Sporophores smaller to medium sized, very viscid, velvet stemmed, centrally stipitate, several fruit bodies developing from a common and short rooting structure. Pileus 2.0-6.0 cm in diameter, convex to flattened, sometimes obtuse, orange to tawny, surface glabrous, viscid, margin inrolled. Gills subdistant, adnexed, white to yellowish white, broad, rounded near the stipe. Stipe central, 3.0-10.0×0.3-1.0 cm, pale yellow when young, stuffed or hollow, externally cartilaginous, lower half covered with obtuse reddish brown hairs, without volva and annulus. Flesh yellowish white, soft, thin.

Distribution

Reported from West Bengal (Banerjee, 1947; Butler and Bisby, 1960), Himachal Pradesh (Ghosh *et al.,* 1967), Jammu and Kashmir (Watling and Gregory, 1980), Arunchal Pradesh (Mishra, 1999). West Bengal, Sikkim, and Himachal Pradesh (Nita Bahl, 1988).

176. *Fomes fomentarius* (L. ex Fx) Kickx

Synonym

Polyporus fomentarius, Polyporus introstuppeus Berk and Cook

Common Name

Timber Fungus, Hoof Fungus, Timber Polypore, Ice Man Fungus.

Habitat

Grows on dead wood, on the branches of *Juglans regia, Betula* sp., *Celtis australis Cedrus deodara, Fraxinus excelsior, Picea morinda, Pinus excelsa, Quercus dilatata* and *Pyrus* sp. It infects the tree through broken bark and cause its rotting.

Habit

The large fruit bodies are shaped like horse hoof and vary in colour from a silvery grey to almost black, although normally they are brown, 5–45 cm across, 3–25 cm wide and 2–25 cm thick which attaches to the tree on which the fungus in growing, while typically like a hours hoof, it can be also more bracket like with an umbonate attachment to substrate. Species typically has broad concentric ridges with a blunt and rounded margin. Flesh hard and fibrous and a cinnamon brown colour. The upper surface tough and bumpy, hard and woody, usually light brown or grey. The margin whitish during the period of growth. The hard crust 1–2 mm thick and covers the tough–yellow brown trama, The underside with round pores of cream colour when new maturing to brown. The tubes 2–7 mm long and rusty brown. Spores lemon–yellow, oblong, ellipsoid in shape.

Distribution

Reported from Khasi Hills, Assam and W.B., North West India and Sonmarg, Kashmir (Lloyd, 1898–1925, 1904–1919;1915), from India (Anonymous, 1950), from H.P. (Thind and Rattan, 1971) and from Dehradun, U.P. (Bakshi and Reddy, 1972).

177. *Galerina mutabilis* (Schaeff ex Fr.) Orton

Synonym

Kuehneromyces mutabilis (Fr.) Singer and Smith; *Pholiota mutabilis* (Fr.) Kummer.

Common Name

Changing Pholiota or Two toned pholiota.

Habitat

Sporophores grow in clusters on the stump of *Betula utilis.*

Habit

Pileus upto 5 cm in diameter, convex then obtusely umbonate with age, date brown when moist, paler towards the centre when dry, water soaked in appearance. Gills crowded, pale to cinnamon coloured with maturity, adnato-decurrent, thin. Stipe slender, curved, brown, central, surface scaly upto ring. Spores smooth, brown, elliptical and without a germ pore.

Distribution

Reported from Sonamarg, Jammu and Kashmir (Watling and Gregory, 1980).

178. *Ganoderma applanatum (Pers)* *Pat*

Synonym

Boletus applanatus, Fomes applanatus, Fomes vegetus, Ganoderma aplani, Ganoderma lipsiense, Polyporus applanatus and *Polyporus vegetus.*

Common Name

Artists bracket, Artiots Conk or Flacher Lackporling

Habitat

Grows on trunk of *Albizia*, dead and fallen branches of *Mangifera indica* and other angiospermic plants and on logs of wood.

Habit

The spore bodies are up to 30-40 cm across, hard, woody-textured, white at first but soon turn dark red-brown. It is a wood-decaying fungus, using primarily dead heartwood, but also as a pathogen on live sapwood, particularly on older trees. It is a common cause of decay and death of Beech and Poplar, and less often of several other tree species, including Alder, Apple, Elm, horse-chestnut, Maple, Oak, Walnut, and Willow. A peculiarity of this fungus lies in its ability to be as a drawing medium for artists. When the surface is rubbed or scratched with a sharp implement, it changes from light to dark brown, producing visible lines and shading.

Distribution

Reported from Hamoti (Arunachal Pradesh) by Bisht and Harsh (2001), from Allahabad, U.P. (Singh *et al.*, 2001), from M.P. (Saxena, 1960; Verma, 1996), from Calcutta and its suburbs (Banerjee, 1947) and from Mussoorie Hills (Thind and Chatrath, 1960).

179. *Ganoderma lucidum* (Curtis) P. Karst

Habitat

Grows alone or in group on decaying hardwood logs and stumps (rarely on conifers).

Habit

Ganoderma lucidum is one of the most beautiful mushrooms in the world. When very young its varnished surface is Chinese red, bright yellow, and white. Later the white and yellow shades disappear, but the resulting varnished, reddish to reddish brown surface is still quite beautiful and distinctive. While *Ganoderma lucidum* is annual and does not actually grow more each year like some polypores, its fruiting body is quite tough and can last for months. Cap 2-20 cm; at first irregularly knobby or elongated, but by maturity more or less fan-shaped; with a shiny, varnished surface often roughly arranged into lumpy "zones"; red to reddish brown when mature; when young often with zones of bright yellow and white toward the margin. Pore Surface white, becoming dingy brownish in age; usually bruising brown; 4-7 tiny (nearly invisible to the naked eye) circular pores per mm; tubes 2 cm deep. Stipe sometimes absent, but more commonly present; 3-14 cm long; up to 3 cm thick; twisted; equal or irregular; varnished and coloured like the cap; often distinctively angled away from one side of the cap. Flesh brownish; fairly soft when young, but soon tough. Spore print brown.

Distribution

Reported from Senra (Arunachal Pradesh) by Bisht and Harsh (2001), from Bombay (Uppal *et al.*, 1935), from Mussoorie (Thind *et al.*, 1957), from Himachal Pradesh by Sagar *et al.*, (2007), from M.P. (Harsh *et al.*, 1933), from India (Anonymous,

1950), Calcutta, W.B. (Bannerjee, 1947), Nainital (Bhargava and Sehgal, 1954), from Mysore, Konkan, Nilgiri, Tamil Nadu (Hennings, 1901) and from Nicobar Island, Banglore (Venkatakrishnaiya, 1956).

180. *Geaster floriformis* Vitt.

Habitat

Grows on sandy soil in the forest areas.

Habit

Fructification small, 1.8-2.6 cm in height, 1.7-2.5 cm in diameter, submerged until mature. Peridium consists of two distinct parts, outer peridium thick, hard when dry, whitish smooth, splitting at maturity in 8-10 star-like unequal rays. Inner layer fleshy, whitish to brownish, cracked when dry, inner peridium persistent forming a difnite spore sac with a single apical mouth, pale in early stages, covered at first with fine granules, rays hygroscopic, covered at first with earth held on by the mycelium. Peristome only a puncture or slit with short radiating fissures with no well defined area around it.

Distribution

Reported from Himachal Pradesh, Gopalpur (6900 m) by Sohi *et al*. (1964).

181. *Geaster morganii* Lloyd

Habitat

Grows on the grounds in the forest areas.

Habit

Fructifiction large, broadly bulb shaped, pointed with a short or long point, 2-5 mm in length. Outer peridium splitting into 6-8 rays with long acuminate tips which become revolute under the convex arched base. The outer layer tending to crack and peel off in flakes. Inner fleshy layer dark brown to blackish, cracking inner peridium thin, subspherical forming a spore sac, minutely grannular, greyish to brown in colour. Peristome forming narrowly conical papilla with the sides crumbled all over or towards the top forming or pseudosulcate peristome with lacerated tip. Columella persistent and clavate.

Distribution

Reported from Himachal Pradesh, Maharastra, Mashobra (7500 m) by Sohi *et al*. (1964).

182. *Geastrum arenarius* Lloyd

Habitat

Sporophores grow along roadsides sometimes on sandy soil.

Habit

Sporophores 1.9-2.7 cm. high, 1.7-2.7 cm in diameter, small, submerged upto maturity, usually light grey, spherical on drying. Peridium 2-layered; outer peridium thick, splitting into several star like unequal lobes or rays with age, inrolled when dry, whitish in colour; inner peridium sub globose, with single short stalk, opening by an apical pore. Columella present. Capillitium threads attached to inner peridium, yellow to brown, fragmented, thick.

Distribution

Reported from Solan, Himachal Pradesh (Gupta *et al.*, 1974).

183. *Geastrum fimbriatum* Fr.

Synonyms

Geastrum novahollandicum Mull, *G. sessile* (Sowerby) Pouzar, *G. ruballum*

Common Name

Sessile Earth Star Fungus.

Habitat

Grows associated with roots of *Pinus wallichiana, P. roxburgii, Cedrus deodara*.

Habit

Sporophore with glebal hymenium, without distinct cap and is saprotrophic. This is puff–ball mushroom of which the outer tunic splits into a star like pattern. Once the sac of spores is exposed dispersal depends upon raindrops hitting the sack which causes puffs spores to shoot out. Spores print is brown.

Distribution

Reported from Dehradun, U.P. (Bakshi *et al.*, 1968)

184. *Geastrum saccatum* Fries.

Habitat

Grows on debris.

Habit

Sporophores attached by basal rhizomorph, ovoid when young, sub globose and 2-3 cm across at maturity. Exoperidium saccate, split to about middle into 6-10 pliable, equal rays. Fleshy layer brown, adnate; exterior smooth and free form debris, base concave with a prominent umbical scar. Endoperidium sessile, 1.5-2 cm in diameter, globose, glabrous, brown with a fibrillose peristome situated on a small, depressed, circular silky zone. Endoperidium partly enclosed by the saccate base of the exoperidium. Gleba dark brown, pseudocolumella present. Spores globose, 3.5-4 µm in diameter, verruculose. Capillitium of branched threads more or less the same diameter as spores.

Distribution

Reported from Ganeshkhind, Poona, Maharastra by Nair and Patil (1978).

185. *Geastrum triplex* Jungh.

Habitat

Grows on plant debris.

Habit

Fruit body reddish brown, subglobular with a pointed beak. Exoperidium thick and three layered, a mycelial outer layer, fibrillose middle and a fleshy inner layer. The outer mycelial layer sloughs off leaving some patchy remnants. The middle and inner layer splits longitudinally into 4-8 rays which event forming the star. Simultaneously the inner layer splits transversely in the middle and peals off partially from the fibrillose layer, the free lower halves curls upward forming a cup at the base of the subglobose, sessile inner peridium. The mouth is subdentate and somewhat raised with a light areole at the base. Columella prominent, persistent and elongate. Capillitial thread thicker than spores. Spores dark, globose, spinous and 4 µm in diameter.

Distribution

Reported from Ganeshkhind, Poona, Maharashtra by Nair and Patil (1978).

186. *Gomphus clavatus* (Pers. ex Fr.) Gray

Synonym

Neurophyllum clavatum (Fries) Pat.; *Cantharellus clavatus* Fr.; *Craterellus clavatus* Pers.

Common Name

Clustered chanterelle or Pig's ear

Habitat

Grows in clusters on ground under *Picea smithiana.*

Habit

Pileus 3-8 cm in diameter, ochre yellow above and violet or rosy below, usually astipitate, club shaped or turbinate, flattened or depressed at the apex, margin thin or lobed, longitudinal markings developed by the veins at maturity. Spores ochre yellow and elliptical.

Distribution

Reported from Gulmarg, Jammu and Kashmir (Watling and Gregory, 1980).

187. *Gomphus floccosus* (Schw.) Singer

Synonym

Turbinellus floccosus (Schwein) Earle.

Common Name

Wooly Chanterelle of Wooly Gomphus.

Habitat

Grows solitary or in linear groups on floors of coniferous or mixed forests.

Habit

Fructification infudibuliform with a distinct central stipe; yellow to deep–yellow, fleshy and slimy when fresh. Cap cylindrical becoming funnel–form upto 15 cm broad, margin plain to strongly uplifted, suface moist, nealy smooth. When young squamulose to coarsely scaly at maturity, yellowish–orange to reddish–orange fading in age. Fertile surface wrinkled or with blunt ridges and veins. stipe central to slightly eccentric, tapering downward, hollow to near the base. Veil absent.

Distribution

Reported from upper Shillong, East and West Khasi Hills and Jaintia Hills districts of Meghalaya and Manipur (Verma *et al.*, 1995).

188. *Gyroporus castaneus* (Bull. ex. Fr.) Quel.

Synonym

Boletus castaneus (Bull.) Fr.

Habitat

Grows solitary, scattered to gregarious on soil in mixed angiospermic forests with *Quercus incana*, *Rhododendron arboreum* and *Picea walichiana* or in open places.

Habit

Pileus 3.5-7 cm in diameter, hemispherical to convex, sometimes depressed reddish brown, surface velvety-then almost smooth, firm, brittle. Gills free shorter towards stipe. Stipe upto 7 cm long, equal, sometimes attenuated at the top, easily detachable from the cap, sinuate towards the top and base, often slightly curved. Spores (8-12×4.5-5.5 µm) ellipsoid to occasionally ovoid, cream coloured and thin walled. Flesh fragile, white, pinkish when exposed.

Distribution

Reported from Jammu and Kashmir (Watling and Gregory, 1980), Himachal Pradesh (Lakhanpal *et al.*, 1985; Lakhanpal, 1996).

189. *Gyroporus cyanescens* (Bull. Ex Fr.) Quel.

Synonym

Boletus cyanescens Bull. Ex. Fr.

Habitat

Grows on ground under deodar trees or on healthy soil below birch or spruce.

Habit

Pileus 5-14 cm in diameter, convex then flattened at maturity, dirty whitish to cream, pale straw or buff, texture rough, velvety, scaly, margin generally incurved, distinguished easily by lemon yellow pores. Stipe pale yellow, hollow, fibrillose to tomentose below smooth above, surface often cracking to form ring zones. Flesh firm white turning blue-green to indigo on cutting. Spores (8-13×4-6 μm) light yellow, ellipsoidal and smooth.

Distribution

Reported from Himachal Pradesh (Sharma *et al.*, 1978).

190. *Hebeloma fastibile* (Pers. ex. Fr.) Kummer

Habitat

Grows in coniferous forest especially *Pinus sylvestris*.

Habit

Stipe brown cream. Spore ellipsoidal, verrucose 764–960 × 468–639 μm. Basidia cylindrical, 2368–3756 × 725–903 μm. Cheilocistidia ventricose 3193–5573 × 547–1006 μm.

Distribution

Reported from Bombay (Maharashtra) by Sathe and Deshpande (1979) and Patil *et al.*, (1995).

191. *Helvella crispa* Fr.

Common Name

Saddle fungi, White Helvella.

Habitat

Grows singly or in groups of two, three or four under the shade of trees on damp place in pine forest and can be found rather readily because of their pure white colour.

Habit

Sporophores stalked pileus 4.6 cm across, irregularly convoluted, at first margin attached to stipe, later entirely free and reflexed. Stipe 4.0-8.0 cm long, whitish when young, yellowish with age, hollow, often swollen at the base and gradually attenuated towards the apex, deeply and unevenly longitudinally furrowed. Flesh light coloured, thin, tough. Hymenium white to cream, convoluted towards the centre, presence of the ridges on the upper end of stipe. Asci 8 spored, elongated to cylindrical 250-300×15-18 μm.

Distribution

Reported from Himachal Pradesh (Sohi *et al.*, 1965), Jammu and Kashmir (Kaul *et al.*, 1978; Ghosh and Pathak, 1962), India (Lloyd, 1904).

192. *Helvella elastica* Bull ex St. Amans.

Habitat

Grows on soil rich in humus under forest trees.

Habit

Cap 1–5 cm, folded over (with the top exposed) or loosely saddle shaped, with convex lobes that sometimes fuse by maturity, upper surface tan to grayish brown, under surface whitish or at least paler than the upper surface, smooth sometimes ingrown with stipe where contact occurs, the young margin folding down ward. Flesh thin brittle. Stipe 2–6 cm long to 1 cm thick, more or less even, cream colour, smooth. Spores with one central oil droplet and up to 5 small droplets at each end, 19.5–22.5×11.5–13.5 mm and elliptical.

Distribution

Reported from Senchal lake area, W.B. (Kar and Pal, 1970)

193. *Helvella lacunosa* Afz. ex Fr.

Synonym

Elvella mitra

Common Name

Slategrey helvella; Black helvella

Habitat

Grows scattered, sometimes solitary to gregarious on the ground in forests.

Habit

Sporophores usually smaller. Pileus irregularly saddle shaped, convoluted, 2.0-5.0 cm in diameter, dark brown with curved margin, attached to the stipe in some places. Stipe 5.0-10.0 × 1.0-2.0 cm, stout, ribbed, furrowed or chambered. Asci borne on sadle shaped fertile portion on long stalk. Ascospores elliptical, smooth with large internal oil drops.

Distribution

Reported from India (Kaul, 1971), H.P. (Sohi *et al.*, 1965), Gulmerg, Kashmir (Kaul *et al.*, 1978); Nainital, Ranikhet, Almora (Joshi *et al.*, 1982).

194. *Helvella leucopus* Pers.

Synonym

Helvella albipes Fuckel.

Habitat

Grows on damp soil, humus soil and commonly under the shade of *salix* trees.

Habit

Pileus 3-6 cm in diameters, irregular lobed, stipitate, margin free and reflexed. Stipe 3-7×4.5 cm, central whitish to buff, smooth, narrowed towards the apex, lacunose. Asci cylindrical, 8 spored. Spores (10-13×19-21 μm), smooth, ellipsoidal, uniseriate and slightly thick walled.

Distribution

Reported from Srinagar, Jammu and Kashmir (Kaul *et al.*, 1978).

195. *Hericium clathroides* (Pal ex. Fr.) Gray

Common Name

Icicle fungus

Habitat

Grows on dead wood of *Quercus incana*.

Habit

This fungus is branched to form a coral like formation. It ditters from the coral fungi in that if hangs off the host rather than growing upward from the ground or wood. Sporophore white in colour. The spores formed on the outer surface of its many branches as they do not have gills or spongy tube layer. It can grow upto several ponds in weight. Cap shelf like in shape, broadly attached to wood, upper surface smooth to very slightly velvety orange brown in colour with bands of darker and lighter shades, the margin being the palest. The margin of the cap is usually quite wavy. Spines dull and pale yellow, less than 1 cm deep, crowded.

Distribution

Reported from Chamba, H.P. (Thind and Khare, 1975)

196. *Hericium coralloides* (Scop. Ex Fr.) S. F. Gray

Synonym

Hydnum coralloides Scop.

Habitat

Sporophores appearing in tufts on rotten logs or branches.

Habit

Fruiting body negatively geotrophic spines gathered together in an irregularly shaped mass, 5.0-10.0 cm or more in diameter, pure white when young but tan with age, fruit body branched, branches numerous arising from a common stalk forming a long, ascending coralloid clump; spine crowded, small, numerous, delicate, 0.3-0.6 cm long distributed over the under surface.

Distribution

Reported from West Bengal (Chopra and Chopra, 1955) and from North Himalaya (Thind and Khera, 1975).

197. *Hericium erinaceus* (Bull ex Fr.) Pers.

Synonym

Hydnum erinaceus Bull.

Common Name

Lions Mane

Habitat

Sporophores develop on trunks of dead trees, sometimes or wounds in living trees.

Habit

Sporophores 5.0-14.2 cm long and 3.2-13.3 cm in diameter, base stipitate, sometimes not stipitate, 1.8-3.8 cm thick, often much larger, whitish to creamy white, forming unbranched, tubercular mass, often roundish or somewhat heart shaped, fleshy, upper surface sometimes sparsely fibrillose; spines branched, short, long spines upto 4 cm., straight or curved or flexuous, pendent in straight parallel lines; flesh soft or tough.

Distribution

Reported from Sikkim (Butler and Bisby, 1960) and from Shimla, H.P. (Thind and Khera, 1975).

198. *Heterobasidium annosum* (Fr.) Br.

Synonym

Fomes annosus (Fr.) Cke., *Polyporus annousus* Fries., *Polyporus irregularis* Underw.

Common Name

Root Fomes

Habitat

Grows solitary or imbricate on stumps and logs among coniferous trees or among hard wood

Habit

Sporophores sessile, sometimes applanate, 5.0-10.0 cm in diameter, often larger, greyish brown when young but dark brown when old, occasionally blackish, tough and corky when fresh, hard after drying. Fruit body forming a multistriatous tube layer, upper surface with red brown crust, concentrically grooved, texture velvety to smooth, margin thin. Context white to isabelline, corky, 0.1-0.5 cm thick. Hymenial surface white when fresh, light brown when dry. Pore tube 0.2-1.0 cm long, white, generally in one layer, sometimes in few layers, distinctly visible, round to angular, 2-3 (4) per mm.

Distribution

Reported from U.P., Meghalaya and Assam (Buttler and Bisby, 1960; Nita Bahl 1988), Himachal Pradesh (Vasudev, 1962; Nita Bahl, 1988), Himalayas (Bakshi, 1971).

199. *Hirneola auricula-judae* (Bull. pr St. Amans) Berk.

Synonym

Auricularia-auricula judes (L.) Schroet.; *Auricularia aurienla* (L. ex Hooker) Underwood.

Common Name

Wood ear or Ear of the tree or Jew's ear.

Habitat

Grows solitary or gregarious or in dense tufts on wood and logs or on tree trunks.

Habit

Fruiting body jelly like or gelatinous, sessile to substipitate, shallow cup shaped or flattened or shaped like an ear, 3.0-7.0×0.8-1.2 mm thick, yellow brown to reddish brown when moist, surface convoluted, flexible, becoming horny on drying and brittle. Hymenium about 150.0 mm thick. Basidia cylindrical, septate 50.0-60.0×5.0-60 mm.

Distribution

Reported from West Bengal (Banerjee, 1947; Nita Bahl, 1988), Sikkim, Himalayas, Khandala, Bombay (Nita Bahl, 1988), Jammu and Kashmir (Butler and Bisby, 1960; Vasudeva, 1960), Kerala (Bhavani Devi, 1995), Sikkim (Banerjee, 1946).

200. *Hirneola polytricha* Mont.

Synonym

Auricularia polytricha (Mont) Sacc.

Common Name

Jelly fungus. , Jew's Ear, Judas's ear, Cloud ear Fungus.

Habitat

Grows solitary, lignicolous on dead branches of *Ficus* species.

Habit

Fruit body 2.0-3.0 cm when young and 8.0-10.0 cm when old, cup or ear shaped, red-brown when fresh, grey or tan on drying, rubbery gelatinous when fresh, brittle, cartilaginous on drying, surface velvety, eight distinct hyphal zones in transverse section, hymenium 80-150 mm wide, smooth, papillate. Basidia cylindrical, transversely septate.

Distribution

Reported from Arunchal Pradesh (Mishra 1999); Punjab, Uttaranchal (Butler and Bisby, 1931).

201. *Hohenbuehelia petaloides* (Bull ex Fr.) Schulz.

Synonym

Pleurotus petaloides (Bull. ex Fr.) Quel.

Habitat

Grows on dead wood or on tree trunk.

Habit

Sporophores with gelatinous zone in pileus, resembling flower petal. Pileus 2.0-10.0 cm long, 1.0-5.0 cm in diameter, wadge shaped to spathulate, irregularly petaloid or cochoid, white to pale reddish brown or brown, margin involute when young but expanded later, when wet, margin occasionally covered with fine striations, upper surface nearly plane or more or less depressed, densely hairy or smooth near the junction of the pileus and the stipe. Gills soft, fleshy, crowded, decurrent, white or yellowish, narrow. Stipe usually eccentric, short, 1.0-3.0 cm high. Thick-walled sterile cells present in hymenium.

Distribution

Reported from U.P. (Ghosh *et al.,* 1967); Tamil Nadu (Natarajan and Raman, 1981).

202. *Hydnum imbricatum* Linn.

Synonym

Sarcodon imbricatum

Common Name

Hawk's wing; Shingled Hedgehog.

Habitat

Grows in groups in conifer and mixed broadleaf conifer woods.

Habit

Cap 6.0-30.0 cm or more, convex then flat, often slightly umbilicate, eventually funnel shaped, floccose, tersellated and squamose with large, grey-brown scales, persistent or slightly caducous. Teeth decurrent, ash-white then brown 1.0-1.2 cm long. Stipe 2.5-7.5×2.5-5.0 cm, short, thick, smooth, whitish or cap-coloured. Flesh whitish then light grey-brown, thick, consistent, sometimes zoned. Odour slightly iodized, sometimes horse like, flavour astringent or slightly bitter.

Distribution

Reported from Assam (Bhattacharya and Barua, 1953).

203. *Hydnum repandum* L. ex. Fr.

Synonym

Dentinum repandum (L. ex. Fr.) S. F. Gray

Common Name

Wood Hedgehog; Hedgehog mushroom.

Habitat

Grows solitary or in clusters on ground in woods and open places or in grass.

Habit

Sporophores 1.0-2.0 to 10.0-12.0 cm high 3-10 cm in diameter, stipitate, usually eccentric, convex or plane, white or buff to dull brown, fragile, surface smooth, margin wavy, teeth distinct. Stipe 3-6 cm long usually eccentric, rarely lateral, uniform or club shaped, whitish or light coloured, solid then hollow with age. Spores (6.5-9×5.5-7 µm) ellipsoid, smooth, whitish in mass. Flesh white, 0.5-1.0 cm thick near the stipe, spongy. Basidia clavate.

Distribution

Reported from Uttaranchal (Chopra and Chopra, 1955; Vasudeva, 1960; Butler and Bisby, 1960; Sohi *et al.*, 1964), Himachal Pradesh (Thind, 1961; Sohi *et al.*, 1965; Nita Bahl, 1988).

204. *Hygrocybe calypraeformis* (Berkeley and Broome) Fayod.

Synonym

Hygrophorus calypraeformis (Berkeley) Broome

Common Name

Ballerina hygrophorus, Pink Wax Cap

Habitat

Grows scattered on grassland.

Habit

Sporophores caespitose. Cap 2.5 cm, margin expanding, often splitting into 3-4 large wings, smooth, dry, slightly fibrillose, clear pale rose-pink. Gills pale rose, then whitish, waxy, distinct adnexed, very narrow at stem, stem tall, graceful fragile, striate, fibrillose, white to pale pinkish. Spores white, elliptic, 7-8×4-5 mm.

Distribution

Reported from Kerala (Bhavani Devi, 1995).

205. *Hygrocybe cornica* (Schaeff : Fries) Kumm

Habitat

Grows scattered or in small groups on soil in mixed coniferous forests

Habit

Pileus slightly viscid, deep orange to greyish orange, conical, gills waxy. Stipe greyish orange to bright yellow. Pileus context yellow, changing to bluish green and then black on exposure.

Distribution

Reported from North West Himalayas (Kumar, 1987).

206. *Hygrocybe miniata* (Fr.) Kummer

Synonym

Hygrophorus miniatus (Fr.) Fr., *Hygrophorus congelatus.*

Common Name

Miniature waxy Cap.

Habitat

Grows solitary or in clusters on rotten wood or on old tree stumps and sometimes on bare soil.

Habit

Sporophores usually centrally stipitate, small. Pileus 1-4 cm in diameter, convex then flattened with age, bright red, often fading to orange, surface glabrous often with minute scales, margin wavy and cracked. Gills blunt, waxy, thick, adnate to sinuate, 0.3 to 0.6 mm wide, yellow sometime tinged with red. Stipe 2-7 cm long, central, slender, concolourous, sometimes light coloured, solid then hollow when old, without ring and volva. Spores (7.6 µm long) elliptical and white. Flesh pale and thin.

Distribution

Reported from Lachen, Sikkim (Butler and Bisby, 1960).

207. *Hygrocybe pratensis* (Pers. ex Fr.) Donk.

Synonym

Hygrophorus pratensis (Pers.) Fr.

Common Name

Butter Mushroom, Meadow wax cap.

Habitat

Grows solitary or in clusters, sometimes appearing in ring, in pastures, on the ground in open forests and fields.

Habit

Sporophores smaller, centrally stipitate. Pileus 3.0-9.0 cm in diameter, hemispherical, then convox to flattened, sometimes with broad and high umbo, buff, yellow or tawny, often white, not viscid, surface glabrous, sometimes cracking in dry weather, margin thin. Gills blunt, waxy, thick, distant, decurrent, whitish or yellowish, 0.4-1.0 cm wide. Stipe 3.0-7.0 cm long, narrowed downward, whitish, smooth, outer firm, inner spongy, without ring or volva. Flesh whitish.

Distribution

Reported from Madhya Pradesh (Trivedi, 1972).

208. *Hygrocybe psittacina* (Schaeff ex Fr) Kummer

Synonym

Hygrophorus psittacinus

Common Name

Parrot Mushroom

Habitat

Grows Solitary to scattered to gregarious in damp soil, moss, humus; most common under redwoods

Habit

Pileus 1.5–4 cm broad, convex when young, broadly convex to plane in age; colour highly variable, bright green to dark green to olive green when young, changing to some shade of pink, yellow, or orange in age; surface glabrous, gelatinous to viscid; flesh thin, waxy; gill adnate to subdecurrent, sometimes seceding; at first greenish, then changing colour like the cap. Stipe 4-9 cm long, 3-5 mm broad at apex, equal or tapering, hollow; surface glabrous, viscid; greenish when young, changing to yellow, orange or pink, although apex may remain green. Spore print white.

Distribution

Reported from North West Himalayas (Watling and Gregory, 1980) and from Orissa (Dhancholia and Sinha, 1988).

209. *Hygrocybe punicea* Fr.

Synonym

Hygrophorus puniceus

Common Name

Scarlet wax gill.

Habitat

Grows scattered on ground, grassy fields, in woods under broad leaf trees, conifers.

Habit

Cap 5.0-11.0 cm scarlet-red tending to turn pale with age starting from the center, campanulate, obtuse, margin normally involute, lobate viscid. Gills yellow, often red at base, ascending, apparently free, ventricose, broad, thick and distant. Stipe 7.0-11.0×1.0-2.5 cm, cap coloured or bright yellow, base invariably white, fusiform, often curved, fibrillose, striate, squarulose at top, solid, quickly becoming hollow, flesh cap coloured, initially white, slightly watery, waxy, slightly fibrous in the stipe. No particular order of flavor.

Distribution

Reported from Kerala (Bhavani Devi, 1995).

210. *Hygrophoropsis aurantiacus* (Wulf. Fr.) Martin-Sow.

Synonym

Cantherellus auriantiacus (Wulf.); Fr. *Clitocybe auriantiaca* (Wulf. ex. Fr.) Studer.

Common Name

False Chanterella

Habitat

Grows solitary or in groups of two or three on ground or on rotten logs, woods etc.

Habit

Sphorophores orange coloured, casually centrally stipitate. Pileus 2-7 cm in diameter, convex, later becoming plain, funnel shaped or depressed, yellowish orange, velvety, fleshy, soft, thick, surface minutely tomentose, margin repand or undulate or plain. Gills crowded, decurrent, frequently forked, bright orange, sometimes paler, narrow. Stipe 3-5 cm long, equal, central, clay coloured or ochre yellow, soft and solid. Spores (6.3-7.6×4-4.6 μm) subelliptical, dextrinoid and white. Flesh yellowish or whitish, soft.

Distribution

Reported from India (Bose and Bose, 1940).

211. *Hygrophorus cerasius* Fr

Synonym

Agaricus cerasius Berk, *A. agathosmus* Fr *Hygrophorus agathosmus* var. *aureofloccosus* (Bres) A Pearson and Dennis. *H. agathsomus*, *F. aureofloccosus* Bres.

Common Name

Gray almond waxy cap or Almond Woodwax.

Habitat

Grows scattered on grassy grounds.

Habit

The cap 4–8 cm in diameter, convex with edge rolled inwards becoming flat with age, center slightly depressed or slightly elevated, dull ashy grey and when moist the cap surface sticky, surface smooth with minute soft hair along the edges. Flesh soft, whitish, watery gray. Gills have adnate attachment to the stipe but at maturity the gills start to extend down to the length of stipe, gill white but become grayish with age. Stipe 4–8 cm long by 0.6–1.4 cm thick, whitish, becoming pale ashy with age, surface covered with tiny fibrils and a fine whitish powder when young later become smooth.

Distribution

Reported from Trivandrum, Kerala (Bhavani Devi, 1995).

212. *Hygrophorus chrysodon* (Batsch ex Fr.) Fr.

Habitat

Grows on the ground in mixed conifer woodland.

Habit

Pileus 4-7 cm in diameter, convex then expanded with age, white, surface viscid, numerous golden or light yellow squamules scattered all over the fruit body, margin involute in the early stage. Gills distant, decurrent, white or yellowish. Stipe equal, white, soft and spongy. Spores white, oval to elliptical.

Distribution

Reported from various places in Jammu and Kashmir valley (Watling and Gregory, 1980).

213. *Hygrophorus citrinus* Rea

Synonym

Hygrocybe citrina.

Habitat

Grows solitory or subgregarious on the ground in grass–land and mossy places

Habit

Pileus–5–20 mm, obtusely conical, soon expanded to applanate, hygrophanous, when moist lemon yellow or chrome–yellow, often yellow–orange in places, occassionally entirely yellow–orange when young, translucently striate up to centre, strongly glutinous, on drying, cream coloured to straw yellow. Lamellae, L=13–21, l=1–2 (–3), sub–distant, broadly adnate to short–decurrent with tooth, first arcuate then segmentiform, thickish, upto 2 mm broad, whitish, then sulphur, lemon–yellow or chrome yellow. Stipe 15–45 × 1.2 mm cylindrical lemon–yellow or chrome–yellow often with orange yellow apex. Sometimes orange–yellow with orange–red apex when young, strongly viscid when moist. Spore print 'white'.

Distribution

Reported from Jorhat (Assam) by Gogoi *et al.*, (2000)

214. *Hygrophorus eburneus* (Bull. ex Fr.) Fr.

Common Name

Ivory waxy Cap.

Habitat

Grows on ground, usually under Oak trees.

Habit

Pileus upto 14 cm in diameter, plain, fleshy, shining white, brownish towards the centre with age, smooth, viscid, margin incurved when young. Stipe solid, stout, squamulose and attenuated towards the base. Spores (6-8×3-5 µm) hyaline and ellipsoid.

Distribution

Reported from Solan, Himachal Pradesh (Sharma *et al.*, 1978).

215. *Hygrophorus marzuolus* (Fr. : Fr.) Bresadola

Habitat

Grows solitary or gregarious on the ground, beneath litter of leaves and moss in coniferous and broadleaf woods, especially in mountainous areas.

Habit

Pileus 3.0-10.0 cm, fairly dark ash grey, sometimes with ochreous shading, then spackled blackish grey, convex, becoming flat and depressed, irregularly humped with margin expanded, wavy-lobate, cuticle initially moist, drying out rapidly. Gills crowded, large, short, distant, then become thinner, arcuate, decurrent, white tending to turn grey or blackish. Stipe 4-8×1.2-3 cm, squat, full, cylindrical, straight or curved or tapered at base, white at top, otherwise silvery grey, solid. Flesh white, thick, faintly grey beneath cuticle.

Distribution

Reported from Kerala (Bhavani Devi, 1995)

216. *Hygrophorus psittacinus* (Schaleff ex Fr.) Fr.

Common Name

Parrot wax cap or Parrot toad stool.

Habitat

Grows scattered or in groups on the ground in the community of *Picea smithiana.*

Habit

Pileus upto 4 cm in diameter, campanulate then expanded and finally umbilicate when mature but column disappears on drying, surface sticky when moist. Gills adnate, pale green to red or yellow. Stipe upto 7 cm long, thick, greenish at the top and reddish to yellowish at the base spores amyloid and smooth.

Distribution

Reported from Gulmarg, Jammu and Kashmir (Watling and Gregory, 1980).

217. *Hygrophorus pustulatus* (Pers.ex Fr.) Fr.

Synonym

Limacium pustulatus (Pers. ex. Fr.) Kummer.

Habitat

Grows gregarious in the community of *Picea smithiana*.

Habit

Pileus 2-4.5 cm in diameter, convex then plain and arched with maturity, ash coloured with the dark brownish centre, surface viscid and margin involute. Gills decurrent, white and narrow. Stipe 6-9 cm long, equal, white, viscid with dark grey dots present at the top. Spores (7-9×4-5 µm) whitish, smooth, inamyloid and ellipsoidal.

Distribution

Reported from Gulmarg, Jammu and Kashmir (Watling and Gregory, 1980).

218. *Hypholoma capnoides* (Fr. ex Fr.) Kummer.

Synonym

Naematoloma capnoides (Fr Karsten)

Common Name

Smoky-gilled wood lover

Habitat

Grows in dense tuft on old coniferous stump.

Habit

Pileus 2-6 cm in diameter, ochre yellow. Gills whitish yellow or bluish grey, sinuate. Stipe pallid, whitish at the top. Spores (7.7-8.8×4.5 µm) purple-black and oval.

Distribution

Reported from Darjeeling, West Bengal; Shimla, Himachal Pradesh (Berkeley, 1854) and Jammu and Kashmir (Watling and Gregory, 1980).

219. *Hypholoma fasciculare* (Huds Fr.) Kummer

Common Name

Sulphur Tuft/Clustered Woodlover

Habitat

Grows in clumps on stumps, logs or diseased deciduous trees.

Habit

Pileus upto 8 cm in diameter, convex to slightly umbonate, smooth, bright sulphur yellow with tint of orange at the centre, margin with veil remains. Gills sinuate, sometimes adnate, yellow then greenish and finally dark purple brown. Stipe long, fibrous, pale yellow, brownish at base with a dark ring zone above. Spore purple brown, 5-7×3-5 µm.

Distribution

Reported from Kerala (Bhavani Devi, 1995) and from W. B., Simla (H.P.) by Berkeley (1856).

220. *Hypholoma sublateritium* (Fr.) Quel.

Common Name

Brick top

Habitat

Grows scattered or in dense tufts on dead wood.

Habit

Pileus 3-11 cm in diameter, convex then expanded with age, orange or brick red, fading towards margin, smooth. Gills crowded, adnate, narrow, yellow then greyish black, then dark brown. Stipe solid, often curved, attenuated downwards, smooth, yellow at the top and reddish brown downwards, sometimes with flat scales. Spores purple to olivaceous, elliptical to oval.

Distribution

Reported from West Bengal (Berkeley, 1856).

221. *Irpiciporus lacteus* (Frios.) Murr.

Synonym

Polyporus tulipiferae (Schw.) Overh.; *Boletus tulipiterae* Schw.

Habitat

Grows on hard woods and also on the dead wood of conifers.

Habit

Sporophores effuso-reflexed or resupinate, thin and leathery, upper surface whitish but yellowish after drying, hairy with distinct zonations, margin thin and entire, inwardly curved. Flesh 0.5-2 cm thick, white to yellow. Hymenial surface white or yellow, pore tubes 0.1-0.3 cm long, pores large, angular or sinuous, irregular when immature, sometimes concentrically arranged. Spores (4.5-6×2-3 μm) cylindrical or ellipsoidal, smooth and hyaline.

Distribution

Reported from Uttaranchal and Himachal Pradesh and Punjab (Vasudeva, 1962); Kashmir (Gardezi, *et al.*, 2003).

222. *Laccaria amethystina* (Bull.) Murr.

Habitat

Sporophores grow on soil in forest.

Habit

Pileus upto 4 cm in diameter, plano-convex, vinaceous to deep violet, smooth.

Stipe upto 5 cm long, cylindrical, hollow, concolorous with the pileus. Spores (7-11.2×7-10 μm) globose to subglobose, echinulate, hyaline, inamyloid, thick walled.

Distribution

Reported from Uttaranchal (Rawla and Arya, 1983; Adhikary, 1992), Maharastra (Sathe and Deshpande, 1979).

223. *Laccaria laccata* (Scop. Fr.) Cooke

Synonym

Clitocybe laccata Scop

Common Name

Waxy Laccaria or Deseiver.

Habitat

Grows solitary, sometimes scattered or in clumps on the ground or on rotten wood in fields, forests and other waste places, or on humiculus soil among mosses under the trees of *Quercus incana* and *Rhododendron arboreum.*

Habit

Pileus 1.5-4.0 cm in diameter, convex then flatten when old, sometimes shallow, funnel shaped, salmon coloured or purple when fresh, pale when moist, light coloured when dry, surface smooth or with minute scales, thin, watery appearance. Gills decurrent, adnate or notched, often intervenose, pink or red, thick, surface waxy. Stipe 3.0-8.0 cm long, whitish or light pink, fibrous tough.

Distribution

Reported from Uttaranchal and Sikkim (Hennings, 1901; Nita Bahl, 1988), Meghalaya and Manipur (Verma *et al.*, 1995), South India (Natrajan, 1977), Himachal Pradesh (Saini and Atri, 1993), Meghalaya (Shajahan *et al.*, 1988; Sarkar and Ram Dayal, 1983).

224. *Lacrymaria velutina* (Pers. ex Fr.) Pat.

Synonym

Hypholoma velutinum (Pers.) Fr.; *Psathyrella velutina* (Pers. ex Fr.) Singer.

Common Name

Weeping widow

Habitat

Grows solitary or in groups, sometimes scattered on the moist ground in forests, on road and path sides in gardens and fields.

Habit

Pileus 3.0-10.0 cm, in diameter, convex to expanded, usually yellowish brown but darker at the center, surface pubescent, later scaly. Gills crowded, adnate or sinuate, light yellow turning into purple brown, edge whitish and with drops of liquid. Stipe 2.0-8.0×0.4-1.0 cm, pale brown, scaly upto the attachment of veil, upper part white, fragments of veil remaining on the pileus margin. Flesh thick and watery. Basidiospores ornamented.

Distribution

Reported from West Bengal (Butler and Bisby, 1960), Maharashtra (Trivedi, 1972).

225. *Lactarius camphoratus* (Bull ex Fr) Fr.

Common Name

Curry Milk cap

Habitat

Grows scattered or gregarious under Oak and mixed coniferous wood on ground.

Habit

Fruiting body 5–15 cm. Cap 2.5–5 cm across, convex, then with a depression and often with a small umbo, red-brown, bay or dark brick, sometimes with a violet tinge, surface smooth and matt, not sticky, margin slightly inrolled at first, often furrowed. Stipe 30–50 × 4–7 mm, cylindrical or narrowing downwards coloured as the cap or deeper. Flesh pale rusty brown. Gills decurrent, closely spaced, narrow, pale reddish brown. Milk rather watery but with whitish clouds. Smell weakly of bugs when fresh, but a strong curry-like scent develops on drying. Spore print creamy.

Distribution

Reported from North West Himalayas (Bhatt and Lakhanpal, 1988b, 1994).

226. *Lactarius controversus* (Fr. ex Fr.) Fr.

Common Name

Milk Cap.

Habitat

Grows in groups at the border of the ditch in willow or mulberry plantation.

Habit

Pileus upto 16 cm in diameter, infundibuliform, light tan coloured, viscid, dotted or zoned with blood red blotches, margin inrolled when immature, then raised. Gills very closed, white to pink, short, decurrent at maturity. Stipe upto 4 cm long, central, narrowed downwards with lateral root like out growths. Spores spherical, rough amyloid.

Distribution

Reported from Jammu and Kashmir (Watling and Gregory, 1980; Abraham *et al.*, 1981).

227. *Lactarius corrugis* Peck.

Habitat

Grows scattered on humicolous soil in the Gymnospermous forest composed of *Pinus wallichiana* and *Cedrus deodara*.

Habit

Sporophores upto 8 cm in height, pileus upto 5.5 cm broad, expended with a slightly depressed centre, margin occasionally splitting, involute, surface dry, feebly corrugated, brownish orange with darker tone, latex milky, scanty, mild, unchanging, flesh upto 7 mm thick, white, on exposure turn brown. Lamellae adnate, close, unequal, highly branched, moderately broad, pale yellow, of late dark brown where bruised, edges fimbriate. Stipe upto 7 × 1 cm, cylindrical, concolorous with pileus, smooth, solid, exudes unchanging milky latex where injured.

Distribution

Reported from Jammu and Kashmir, Bactone (1500 m) by Saini and Atri (1993), from Himachal Pradesh by Bhatt and Lakhanpal (1990) and from Garhwal, Uttar Pradesh by Bhatt and Bhatt (1999).

228. *Lactarius deliciosus* (L. ex Fr.) S. F. Gray

Common Name

Saffron milk cap, Delicious Lactarid.

Habitat

Grows scattered on the ground in damp woods.

Habit

Sporophores centrally stipitate, 3.0-10.0 cm high, fruiting body producing milk like or coloured fluid (orange) when broken. Pileus 5.0-12.0 cm in diameter, convex and centrally depressed, then expanded and funnel shaped, orange coloured, usually with alternating light and dark orange concentric zones, margin incurved when young. Gills crowded, adnate or short decurrent, sometimes forked, indistinct vein between the gills where the gill attached with the pileus, yellowish orange, sometimes with darker spots, green when old, often hairy at the base, solid when young, hollow at maturity, ring and volva absent. Flesh white shaded with orange, greenish near the stipe.

Distribution

Reported from Sikkim (Butler and Bisby, 1931, Chopra and Chopra, 1955 and Berkeley, 1956); Himachal Pradesh (Saini and Atri, 1982).

229. *Lactarius deterrimus* Groger

Common Name

False saffron milk cap or Bitterer Milching.

Habitat

Grows on the ground amongst pine duff.

Habit

Pileus upto 15 cm in diameter, convex but expanded at maturity with a slight depression in the middle, orange or apricot coloured, central surface moist, margin incurved, smooth and entire. Stipe upto 6 cm long, orange, surface some what striated, often moist. Spores (7.5-10×7-8 µm) white to cream, broadly ellipsoidal, apiculate with warted or reticulate ornamentation.

Distribution

Reported from Gulmarg, Jammu and Kashmir (Abraham *et al.*, 1980) and Tanmarg, Jammu and Kashmir (Watling and Gregory, 1980).

230. *Lactarius eccentrica*

Common Name

Sarai Pehare (Sal Mushroom)

Habitat

Grows solitary to scattered on soil in ectomycorrhizal relationship with sal tree (*Shorea robusta*).

Habit

Fructification upto 4 cm in height. Pileus upto 3 cm in diameter, depressed at the centre, surface glabrous, white to cream, milky latex oozes when bruised, margin soft, uneven, firm. Gills adnate to decurrent, distant, thick. Stipe upto 2 cm long, clavate, tapers downward, short, solid, eccentric, glabrous, white. Spores (6.5-7×3.5-6 µm) subglobose to globose, whitish, apiculate, amyloid.

Distribution

Reported from Madhya Pradesh (Rahi *et al.*, 2003).

231. *Lactarius piperatus* (Scop.) Fr.

Common Name

Peppery milk-cap.

Habitat

Grows in mycorhizic association with *Quercus incana.*

Habit

Pileus varies from 6–16 cm across and convex with a widely funnel-shaped centre. The cap creamy-white in colour, glabrous and not glossy; its surface may become cracked when dry. The stipe white in colour, smooth, 3–7 cm long by 2–3 cm thick and cylindrical, sometimes tapering towards the base. There is a thick layer of firm white flesh, and the decurrent gills are particularly crowded and narrow, sharing the white colouration of the stipe but becoming creamy with age. As with other species of *Lactarius*, there is abundant milk (latex), which is white, and dries olive-green. It has a white spore print with elongate, elliptic or amyloid spores which are ornamented.

Distribution

Reported from North West Himalayas (Bhatt and Lakhanpal, 1990) and from Jammu and Kashmir (Saini and Atri, 1993).

232. *Lactarius princeps* Berk

Habitat

Grows singly or in groups in pine, mixed or frondose woods particularly of Oak.

Habit

Sporophore stipitate, fleshy with a brick–red, flat cap, depresed at the centre and a stout spongy to hollow stipe, yields a serious white latex when cut or broken.

Distribution

Reported from East and West Khasi Hills of Meghalaya, Jaintia Hills, Manipur, Mesorum–Aizwal, Arunachal Sefari and Bomdila (Verma *et al.*, 1995).

233. *Lactarius sanguifluus* (Paulet ex Fr.) Fr.

Common Name

Lal Chhatri (H.P.) and Khunyo

Habitat

Grows scattered in mixed conifer forests usually under the fern *Onychium contiguum.*

Habit

Pileus 6-15 cm in diameter, convex, plane to subinfundibuliform, viscid, brittle, glabrous with carrot coloured and paler zones, margin incurved, regular and smooth. The fruiting body exudes a blood red to purple red latex which turns greenish on exposure. Stipe 5-8 cm long, central, cylindrical, dull than pileus, stuffed or hollow. Basidiospores (7-9×6-7.5 μm), ellipsoid to subglobose, pale yellow in mass, ridges and warts form reticulum, amyloid.

Distribution

Reported from Himachal Pradesh (Lakhanpal *et al.*, 1987); Uttaranchal (Bhatt *et al.*, 2000)

234. *Lactarius scrobiculatus* (Scop: Fr.) Fr

Habitat

Grows isolated and scattered in pine forest on soils, occasionally beneath willows.

Habit

Sporophores light yellow becoming yellow ochraceous, with age, cap convex with or central depression, 6-20 cm in diameter, margin involute, wooly with numerous not very long hairs, slimy when mosit. Gills yellowish, lighter than cap, crowded, slightly decurrent. Stipe. 3.5-6.0×2-3.5 cm, hollow, tapering towards the base, whitish with many yellow or yellowish red pits. Whitish flesh, yellowing when cut because of latex, tending to blacken with time, quick to rot. Slight odor of geranium. Latex white unchanging.

Distribution

Reported from Meghalaya (Verma *et al.*, 1995); Jammu and Kashmir (Abraham *et al.*, 1980); Uttaranchal (Saini and Atri, 1982).

235. *Lactarius subdulcis* (Pers. ex Fr.) S. F. Gray.

Common Name

Mild milk cap or Beech milk cap.

Habitat

Grows solitary or scattered in groups in fields, forests, wet places or in base soil or on humicolus soil under *Nyrica esculenta, Quercus leucotrichophora* and *Rhododendron arboreum*; under hardwoods.

Habit

Sporophores centrally stipitale, fruiting body producing milk like liquid. Pileus 2.0-5.0 cm in diameter, at first convex later becoming flattered or subinfundibuliform slightly umbonate or not, tan to reddish brown, smooth, thin, even, usually without zones, moist or dry, margin not inrolled, sometimes wavy or flexuous. Gills crowded, distinctly formed, adnate or short decurrent, sometimes forked, whitish sometimes with reddish tinge, narrow. Stipe equal or slightly tapering upward, 4.0-7.0 cm long, 0.3-0.8 cm thick, colour lighter than pileus, usually glabrous, sometimes hairy at the base, stuffed or hollow at maturity, ring and volva absent. Flesh pinkish or reddish grey.

Distribution

Reported from India (Butler and Bisby, 1960); West Bengal (Berkeley, 1851); Uttaranchal (Bhatt *et al.*, 2000).

236. *Lactarius subisabellinus* var. *subisabellinus* Murrill.

Habitat

Grows scattered on humicolous soil under *Quercus incana*.

Habit

Pileus upto 3.5 cm broad, convex, flattened depressed at maturity, margin regular, incurved. Latex dilute, white. Gills decurrent, close, narrow, wavy or serrate. Stipe 6-8.5 cm long, concolourous, central, smooth, hollow. Spores (6.5-7.5 × 5.6-6.8 µm) globose to subglobose, warty, apiculate and amyloid.

Distribution

Reported from Himachal Pradesh (Atri *et al.*, 1993).

237. *Lactarius subpurpureus* Peck

Common Name

Purplish Lacterius.

Habitat

Grows alone, scattered or gregarious on moist grassy places or on shaded or exposed slopes of conifer forests (mycorrhizal with conifers) or in association with *Picea smithiana* and *Quercus senecarpifolia.*

Habit

Pileus upto 7.5 cm in diameter, convex then plain or infundibuliform, pale blood red turning greenish with age, fleshy, surface glabrous, slightly viscid when wet. Stipe upto 7.5 cm long, blood red, even or slightly tapering at the top, surface glabrous or often hairy downwards bearing red spots. Spores (8.4-12×6.9 µm) ellipsoid, apiculate, amyloid, ornamented.

Distribution

Reported from Jammu and Kashmir (Watling and Gregory, 1980; Atri, 1981; Saini and Atri 1990), Uttar Pradesh (Saini and Atri, 1990; Atri and Saini, 1986), Uttaranchal (Bhatt *et al.*, 2000).

238. *Lacterius vellereus* Fr.

Habitat

Grows on Fir wood

Habit

Like other fungi in the *Lactarius* genus, it has crumbly rather than fibrous flesh and when this is broken, the fungus exudes a milky liquid. The mature caps are white to cream, funnel shaped and upto 25 cm in diameter. It has firm flesh, and a stipe which is shorter than the fruit body and is wide. The gills are fairly distant (quite far apart), decurrent and narrow and have brown specks from the drying milk. The spore print white.

Distribution

Reported from Sikkim (Berkeley, 1856).

239. *Lactarius volemus* (Fr.) Fr.

Common Name

Orange brown lactarius.

Habitat

Grows scattered or in groups on the ground in Gymnospermous forests or open places on humicolous soil.

Habit

Sphorophores upto 14.5 cm in height. Pileus 5-12 cm in diameter, convex to flattened, depressed at the centre, golden brown or orange brown, surface glabrous, dry, occasionally wrinkled, margin irregular, spilitting and maturity. Gills crowded, adnate or subdecurrent, white or light brown. Stipe upto 10 cm long, central, cylindrical, solid, surface glabrous, sometimes pruinose. Spores whitish and globose. Latex milky.

Distribution

Reported by Garcha (1980) from India and by Saini and Atri (1982) from Himachal Pradesh (Shimla).

240. *Lactarius yazooensis* Hesler and Smith.

Habitat

Grows scattered on humicolous soil under the trees of *Quercus incana*.

Habit

Fructification upto 4.4 cm in height. Pileus upto 6.6 cm broad, expanded with deep centre, finally infundibuliform, margin irregular, surface moist, stomate, pellucid and naked along the margin, pale yellow, more towards orange side, zonate, with zones of lighter and darker colours alternating, latex milky, unchanging on exposure. Lamellae subdecurrent close, unequal, not in series, branched, moderately broad, colourless with the pileus unchanging where bruised but exude acrid milky latex, edges smooth. Spore deposit yellowish white. Stipe upto 3.8 cm long, 0.6 cm broad, central, slightly broad at the base, short, thin, concolorous with the pileus.

Distribution

Collected from Mussoorie, on way to Mossy fall (1800 m) by Atri and Saini (1988).

241. *Laetiporus sulphureus* (Bull. ex Fr.) Murr.

Synonym

Polyporus sulphureus (Bull.) Fr.; *Boletus sulphureus* Bull.

Common Name

Sulphur shelf Mushroom; Sulphur Polypore.

Habitat

Grows on rotten woods, dead stumps or trunks of coniferous trees and also found on living trees, usually gregarious.

Habit

Sporophores sessile or sub-stipitate. Pileus 30-40 cm in diameter, fun or tongue shaped with the margin undulate, lobed, orange-yellow or tawny, margin sulphur yellow, fading to dull and dirty with age, surface smooth, sometimes with concentric furrows, Stipe very short or absent as such many overlapping brackets arising from a common base on the host. Pores elliptic or circular. Tubes short and sulphur yellow.

Besidia club shaped, 4-spored. Basidiospores elliptical, 5-7×4-5 µm thin walled, whitish.

Distribution

Reported from West Bengal (Bose and Bose, 1940), Jammu and Kashmir (Ahmad, 1941), Meghalaya, Manipur (Verma *et al.*, 1995), Arunchal Pradesh (Mishra, 1999), Sikkim, Jammu and Kashmir (Berkley, 1856), Meghalaya (Bose, 1946), U.P. (Puri, 1955), M.P. (Saxena, 1960), Assam, Sikkim, Jammu and Kashmir, H.P. and U.P. (Nita Bahl, 1988).

242. *Langermannia gigantea* (Pers.) Rost Kovius

Common Name

Giant Puff Ball.

Habitat

Grows solitary, rarely gregarious, common in the pastures, gardens.

Habit

Pileus 20.0-45.0 cm in diameter, round or globose or slightly flattened on the top, whitish, then yellowish to olive brown, surface smooth or may be covered with fine scales, cracking. Flesh or gleba, firm and white, then yellowish and finally olive brown and pulverunt. Stipe absent or only present as a small cone of tissue, capillitium yellow and finally dull olivaceous.

Distribution

Reported from West Bengal (Bose and Bose, 1940), Jammu and Kashmir (Ahmad, 1942), Arunchal Pradesh (Mishra, 1999).

243. *Lasiosphaera gigantea* (Batsch ex Fr.) Rost.

Synonym

Langermannia gigantea (Pers.) Rost., *Calvatia gigantea* (Bastsch) Fr., *Lycoperdon giganteum* Batsch ex Pers.

Common Name

Giant Puff Ball

Habitat

Grows on the ground in grassy areas.

Habit

Sporophores large, sessile, white becoming yellowish or smoky when old. Peridium 20-40 cm in dimeter, spherical or depressed surface smooth or covered with fine scales. Flesh white when young, brownish when spores ripen subgleba absent, replaced by capillitium threads and spores. Capillitium yellow or dull olivaceous. Spores (4 µm in dimaeter) globose, greenish yellow to dull olivaceous.

Distribution

Reported from West Bengal (Bose and Bose, 1940) and Jammu and Kashmir (Ahmad, 1941).

244. *Leccinum eximium* (Peck) Sing.

Habitat

Grows on the ground.

Habit

Pileus upto 6 cm in diameter, convexo-plane, brown, tomentose, margin incurved, whitish. Stipe upto 12 cm long, tomentose, central, cylindrical, solid with bulbous base. Spores (7.5-13×5.5-7 μm) elliptic-oblong, smooth, hyaline, inamyloid.

Distribution

Reported from Himachal Pradesh (Rawla *et al.*, 1983).

245. *Leccinum holopus* (Rostk.) Watling

Synonym

Boletus niveus Fr.

Habitat

Grows on the ground in the mixed forest.

Habit

Pileus upto 9 cm in diameter, convex greyish white, surface pitted, viscid when moist. Stipe upto 10 cm long, thick, slender, brown, solid, covered with dark brown scales. Spores (11-17×4.5-6 μm) elliptical, brownish, smooth.

Distribution

Reported from Himachal Pradesh (Sharma *et al.*, 1978).

246. *Leccinum oxydabile* (Singer) Singer

Synonym

Krambholtzia oxydabilis Singer.

Habitat

Grows in groups of two three or solitary and scattered as ectomycorrhizal with *Betula* in mixed canopy of *Rhododendron* and *Betula.*

Habit

Pileus 2-9 cm broad, globose to parabolic when young, expanded to convex when mature, reddish brown to light brown to yellow brown with age, surface dry, viscid when moist, smooth to finely tomentose, margin regular, context 8-12 mm and smooth. Tubes 15-20 mm deep, depressed around stipe, pores roundish, yellowish brown. Stipe 6-12.5 cm long, gradually tapering upwards, whitish to yellowish brown with age, central, dark brown scabers at the apex, grading into darker scales below. Spores (16-19×5-6 μm) elliptical, yellowish and inamyloid.

Distribution

Reported from Himachal Pradesh (Lakhanpal, 1996) and Arunachal Pradesh (Mishra, 1999).

247. *Leccinum rugosiceps* (Peck.) Singer.

Synonym

Boletus rugosiceps (Peck.) Bull.

Habitat

Grows scattered, rarely gregarious, underneath *Quercus* sp., Long lasting among all *Leccinum* species.

Habit

Pileus 4-9 cm broad, convex becoming almost plane at maturity, surface glabrous, areolate and pitted with age, medium yellowish brown turning dark yellowish brown to orange yellow at maturity, margin slightly extended beyond the tubes. Context 1-2 cm deep, deeply depressed tubes yellowish white, unchanging, pores small, one per mm. Stipe 5-9 cm long, 10-20 mm broad at the base, unequal, narrowed at apex, solid, surface sticky to touch, light to medium brown, changing to blackish with age. Spore mass deep yellowish brown. Pileus cutis of rounded to ellipsoid cells, thin walled, hyaline in KOH.

Distribution

Reported from Shimla, Himachal Pradesh by Lakhanpal (1996).

248. *Leccinum scabrum* (Bull. ex Fr.) Gray.

Synonym

Boletus scaber Bull. ex Fr.

Common Name

Birch rough stalks or Brown birch-bolete.

Habitat

Grows usually on soil.

Habit

Pileus upto 10 cm in diameter, convex then flattened, brownish black, surface glabrous, dry, viscid when moist, margin not hanging. Stipe upto 15 cm long, grooved, obclavate, whitish or grey, hollow without bulbous base. Spores (15.5-18.5×5-7 μm) ellipsoidal, brown, smooth, amyloid, guttulate.

Distribution

Reported from Himachal Pradesh (Sharma *et al.*, 1978; Rawla *et al.*, 1983).

249. *Lentinellus cochleatus* (Pers. ex Fr.) Karst.

Synonym

Lentinus cochleatus (Pers ex Fr.) Fr.

Habitat

Grows in clusters on stumps and on decaying timbers of deciduous trees.

Habit

Sporophore funnel shaped, brown or grey to beign in colour with stipe much longer than cap diameter, attaining a length less than 5 cm. Pileus 3-7 cm in diameter, funnel shaped, tan to reddish brown, surface smooth or with short erect scales, margin irregular or lobed, sometimes radially cracked or furrowed. Gills coarse, thick, tough, crowded, decurrent, whitish or pinkish, strongly notched at the margin. Stipe 20–5×8–15 mm central, often eccentric, cylindrical, narrowed slightly towards the base, solid, several united at the base. Spores (less than 8 µm) whitish in mass, amyloid, spherical and minutely ornamented. Spore print white.

Distribution

Reported from Maharasthra by Trivedi (1972) and Sathe and Deshpande (1979).

250. *Lentinula edodes* (Berk.) Pegler

Synonym

Lentinus shiitake (Schroet.) Singer; *Cortinellus edodes* (Berk.) Ito and Imai.; *Lentinus edodes* (Berk.) Singer.

Common Name

Shiitake mushroom of Japan.

Habitat

Grows in clusters on dead twigs and branches of *Castanopsis sp.* and *Quercus sp.* in mixed forest.

Habit

Sporophores centrally stipitate with convex subumbonate cap. Pileus upto 11 cm in diameter, convex then depressed, brown, scales darker in the centre, often cracked to deeply fissured, scales triangular or areolate, consisting of filamentous interwoven hypae, a true epicutis of pileal surface not formed. Gills crowded, edge denticulate, adnate, whitish to brownish or greyish with age. Stipe 3-4 cm long, usually eccentric, straight to slightly curved, solid, pale, reddish brown or dark brown scales present at the joints of cortina. Spores cylindrical or elliptical, smooth, thin walled and inamyloid. Volva absent.

Distribution

Reported from Jammu and Kashmir (Dhar, 1976; Bakshi and Puri, 1978), Meghalaya, Nagaland and Manipur (Verma and Singh, 1981; Verma *et al.*, 1995).

251. *Lentinula lateritia* (Berk.) Pegler

Habitat

Like *L. edode* it also grows in clusters on dead twigs and branches of *Castanopsis* sp. and *Quercus* sp. in mixed forests but comparatively more common.

Habit

Sporophores smaller than those of *L. edodes* but easily mixes with the latter. Cracks or fissures on the cap are almost absent.

Distribution

Reported from Meghalaya and Manipur occurring together with *L. edodes* which is less common of two (Verma *et al.*, 1995).

252. *Lentinus badius* Berk.

Synonym

Panus badius Berk., *Agaricus verrucarius* Berk., *Lentinus inquinans* Berk., *Lentinus brevipes* Cooke., *Pleurotus verrucarius* (Berk.) Sacc., *Panus brevipes* (Cooke) Singer, *Panus inquinans* (Berk.) Singer.

Habitat

Grows on grassland.

Habit

Sporophores soft, corky, centrally stipitate. Pileus infundibuliform, finely pubescent and covered by soft piles, hymenophore lamillate, basidiospores hyaline, cylindric, smooth and thin walled.

Distribution

Reported from West Bengal and Sikkim (Pegler, 1975).

253. *Lentinus cladopus* Lev.

Habitat

Grows singly or in groups on dead and decaying wood and roots of angiospermic plants eg. *Grevia mierocos* and on the soil.

Habit

Sporophores white to creamish white, stipitate. Pileus upto 10 cm in diameter, infundibuliform, cap with depressed centre and with scales, margin squarrose, sometimes recurved. Gills crowded, unequal, white or cream, decurrent, thin. Stipe cylindrical, solid, branched but arising from the common stalk. Spores (4.2-5.6×2.5-3.5 μm) ellipsoid, thin walled, hyaline and non-amyloid.

Distribution

Reported from Tamil Nadu (Natrajan, 1975, 1978), Meghalaya and Manipur (Verma *et al.*, 1995).

254. *Lentinus connatus* Berk.

Synonym

Lentinus revelatus Berk.; *Panus ochraceus* Manee.

Common Name

Bamboo Mushroom

Habitat

Grows singly or in groups in bamboo forest.

Habit

Sporophore stipitate, white to cream when fresh, turning to yellowish brown on storage. Pileus 3-12 cm in diameter, slightly depressed at center infudibuliform, surface smooth, sometimes with minute scales, margin irregular and splitted. Stipe 2-7 cm long, central to eccentric, cylindrical, minutely tomentose and tough. Spores (4-6×14-17 µm) cylindrical and white or whitish.

Distribution

Reported from Meghalaya, Manipur and Arunachal Pradesh (Singh *et al.*, 1992).

255. *Lentinus edodes* (Berk) Singer

Common Name

Shiitake

Habitat

Grows single but most often in densely caespitose clusters on logs and stumps of conifers, occasionally hardwoods (members of family Fagaceae).

Habit

Cap 5–12 cm broad, convex, nearly flat in age, viscid when young but soon dry, white to buff at first, soon breaking up into small scales which are cinnomon to wood–brown and raised in age, margin incurved at first, straight in age. Flesh tough, white except for stalk base, which is yellow to rusty brown. Gills adnexed, close and edges toothed, white to buff in age, rusty brown stains when bruised. Stalk 3–10 cm long, 10-15 mm thick, equal white and minutely brown in age. Veil membranous, superior ring, but in age if may weather away. Spores long elliptical, smooth, thin walled, pure white to buff spore print.

Distribution

Reported from Himachal Pradesh, Shimla (Sagar *et al.*, 2007).

256. *Lentinus glabratus* Mont.

Habitat

Grows on wood.

Habit

Sporophores stipitate, pale or whitish, fleshy, coriaceous, sometimes rigid. Pileus hemispherical to umbilicate, glabrous, concentrically rugulose, margin nearly involute, surface rough, tawny. Gills more or less crowded, decurrent, concolorous, narrow, edge dentatelacerate. Stipe eccentric, solid, glabrous, base attenuated and shield like. Flesh tough but fleshy. Hymenophoral trama not entirely irregular, some times regular to subregular in young fruit-bodies; hyphae non-amyloid, thin walled, thick walled with maturity, subparallel, interwoven. Basidia 4-spored. Basidiospores non-amyoid. Spore print white or cream.

Distribution

Reported from Raj Mahal Hills, Jharkhand (Currey, 1874).

257. *Lentinus lepideus* (Fr. ex Fr.) Fr.

Habitat

Grows solitary or in group of 2 or 3 or scattered on the logs of coniferous trees, living deodar or roots.

Habit

Sporophores centrally stipitate or not so, tough or poliant, becoming hard at maturity with a definite colour. Pileus 4.0-12.0 cm broad, convex when young but flattened when old, sometimes with slight depression in center or undulate, white or light-brown, course brown scales on the surface of pileus, fleshy, compact, hard when dry. Margin with conspicuous cracks often extending to the flesh. Gills subdistant, sinuate or decurrent, white or yellow, broad, dentate at the edges. Flesh white, tough. Stipe central or eccentric, short 2.0-7.0 cm or long, 1.0-2.0 cm thick at the top and pointed at the base, hard and solid, rough with recurved scale, occasionally with evanescent ring on the stipe, sometimes few basal scales representing a volva.

Distribution

Reported from Himalayas (Smith, 1949; Vasudeva, 1962), Himachal Pradesh (Bakshi *et al.*, 1955).

258. *Lentinus lecomtei* Fr.

Synonym

Panus rudis Fr., *Panus locomtei*.

Habitat

Grows on wood

Habit

When it grows on the upper side of logs the pileus is sometimes regular and funnel shaped (Cyathiform), but it is often irregular and produced on one side, especially if it grows on the side of the substratum. In most cases, however, there is a funnel–shaped depression above the attachment of the stipe. The pileus is tough, reddish or reddish brown or leather colour, hairy or sometimes strigose, the margin incurved. The stipe is usually short, hairy or in age it may become more or less smooth. The gills are narrow, crowded. The spores small, ovate to elliptical 5–6×2–3 mm.

Distribution

Reported from Tongio, Sikkim, East Nepal and Gulmarg, Kashmir (Berkeley, 1856 and Petch, 1916).

259. *Lentinus polychorus* Lev.

Synonym

Lentinus praerigidus Berk.; *Lentinus eximis* Berk.

Habitat

Grows on living mango trees or on dead stumps during rainy season.

Habit

Pileus upto 15 cm in diameter, convex to infundibuliform, lobed and grooved, white, surface covered with squarrose scales, margin entire and downward. Gills crowded, white, decurrent. Stipe 4×2 cm, cylindrical, central, concolorous, scaly. Spores (5-6×1.5-2.5 μm) ellipsoidal, colourless, thin walled and smooth.

Distribution

Reported from Bihar (Hooker, vide Pegler, 1975) and Tamil Nadu (Natarajan and Manjula, 1978), Bihar and Tamil Nadu (Singer, 1961).

260. *Lentinus praerigidus* Berk

Habitat

It grows on dead wood, *Shorea robusta* or on *Trminalia paniculata* and *Vateria indica*.

Habit

Sporophore with cone like pileus, the upper part of which is hollow. It is fine and velvety on the upper surface. It is 5–10 cm across. The stipe is 1–2 cm long by 5–10 mm wide. It is cream to brown in colour.

Distribution

Reported from Sone River in Bihar (Berkeley, 1856 and Lloyd, 1904–1919), Calcutta, W.B. (Banerjee, 1947) and from India (Anonymous, 1950).

261. *Lentinus sajor-kaju* (Fr.) Fr.

Habitat

Grows singly or in clusters on dead and decaying wood.

Habit

Pileus upto 10 cm in diameter, broad and tough. Sporophores stipitate, infundebuliform, pale grey to greyish yellow when young, yellowish at maturity. Stipe 4-6 cm long, central with fungacious annulus. Generative hyphae thin walled, frequently branched with clamp connection. Spores (7-8.4×2.5-4.5 μm) cylindrical, hyaline, smooth walled, thin.

Distribution

Reported from Arunachal Pradesh, Mizoram, Meghalaya and Manipur (Verma *et al.*, 1995), Tamil Nadu (Natarajan, 1978)

262. *Lentinus squarossulus* Mont.

Synonym

Pleurotus squorrosulus (Mont.) Sing.

Common Name

Scaly Oyster (local name in Kerala- Mamkoon, Marakoon)

Habitat

Grows singly or in groups on rotting wood, dead wood stumps, branches of trees.

Habit

Pileus upto 15 cm in diameter, convex to infundibuliform, sometimes flabelliform of dimidiate, surface white, lobed with grooves, squamose scales present all over the surface, margin decurved, entire. Lamellae decurrent, white, crowded, edge entire, lamellulae present. Stipe usually central, sometimes eccentric, cylindric, solid, surface with appressed scales, concolorous with pileus, upto 4 cm long and 2 cm wide, context upto 4 mm thick at the center, thinner near the margin, fibrous.

Distribution

Reported from Meghalaya, Manipur, Mizoram, Tripura and Nagaland (Verma *et al.*, 1995); Kerala (Bhavani Devi, 1995), Tamil Nadu (Natarajan and Manjula, 1978).

263. *Lentinus strigosus* (Schwein.) Fr.

Synonym

Panus rudis Fr.

Common Name

Ruddy Panus

Habitat

Grows solitary but mostly in clusters on deciduous logs and stumps or on dead and rotting wood.

Habit

Sporophore reddish with fine hairs. Cap 2.5–7.5 cm wide, Stipe (when present)

1–2 cm long and 0.3–1 cm thick. Cap pinkish tan to reddish with violet tints when young, tan with age, dry, densely hairy and velvety. Gills decurrent, white to tan. Stipe, if present, pinkish brown to tan, densely hairy, stubby, lateral to off centre. Veil absent. Spore print white.

Distribution

Reported from Arunachal Pradesh, Nagaland and Manipur (Verma *et al.*, 1995).

264. *Lentinus tigrinus* (Bull. ex Fr.) Fr.

Synonym

Panus tigrinus (Bull ex Fr.) Sing.

Habitat

Grows solitary or in clusters on dead and decaying wood, on old hard wood stumps of Poplar and Willow.

Habit

Sporophores yellowish and stipitate soft when young but hard at maturity. Pileus 2.5-15 cm in diameter. Convex with central depression, expanded convex or float at maturity, dry, firm, leathery, fibrillose, dull white to tan with age, margin incurved, split radially with age. Stipe upto 7 cm long, eccentric, rarely central, solid, without ring and volva. Spores (6-7×2-3 μm) ellipsoid or oblong, white, smooth, thin walled, apiculate non–amyloid and germpore absent.

Distribution

Repoted from Jammu and Kashmir (Watling and Gregory, 1980); Meghalaya and Manipur (Verma *et al.*, 1995); Kerala (Bhavani Devi, 1995) and from U.P. (Hennings, 1901).

265. *Lentinus tuber-regium* (Fr.) Fr.

Synonym

Pleurotus tuber-regium (Fr.) Singer.

Habitat

Grows solitary or in tufts. Sclerotia subterranean.

Habit

Pileus 10-20 cm in diameter, infundibuliform or cyathiform, depressed in the centre, white then light brown with age, glabrous, margin incurved, thick and sometimes incised. Gills decurrent, ash grey, serrate, thick, yellow with sharp edge. Stipe central, with concolorous surface. Sclerotium spherical to ovoid, upto 30 cm thick, surface dark brown, white within. Spores hyaline and ellipsoidal.

Distribution

Reported from Lucknow, Uttar Pradesh (Pathak and Gupta, 1982).

266. *Lentinus velutinus* Fr.

Habitat

Grows attached to burried wood.

Habit

Pileus covered with soft, erect hairs and closed velvety pile formed of single hyphae and have curious thick walled arising directly from tramal elements which have pierced the hymenium. Gills not anastomosing at the base, whole sporophore at first mauve, soon turning brown. Gill broad.Stipe long, central and stout, typically attached to the buried wood.

Distribution

Reported from Bengal (Lloyd, 1904 -1919; Currey, 1874).

267. *Lenzites betulina* (Fr.) Fr.

Habitat

Grows scattered to clustered in overlapping shelves on hardwood logs, but not limited to birch as the species name would suggest.

Habit

Fruiting body annual or short-lived perennial, 2-10 cm broad, 0.5-2.0 cm thick, at first resupinate, then forming sessile, fan-shaped, tiered brackets or occasionally rosettes emanating from a common base; surface tomentose, concentrically-zoned, often multi-coloured, cream, pale-buff, dingy yellow-brown, or greyish-brown, in age sometimes green from encrusted algae; Flesh thin, pliant, becoming tough, cork-like, white, unchanging; Gills white, radiating from attachment point, broad, tough, cream-colored. Spores 4-6 × 1.5-2.0 μm, smooth, cylindrical to elongate bean-shaped, nonamyloid; Spore print white.

Distribution

Reported from Namnai (Arunachal Pradesh) by Bisht and Harsh (2001), Manipur (Verma *et al.*, 1995), W. B. (Anoymous, 1950), U.P. (Bagchee *et al.*, 1954), M.P. (Saxena, 1960).

268. *Lenzites elegans* (Fr) Pat.

Synonym

Daedalea elegans Fr., *Polyporus aesculi* Fr., *Lenzites pallida* B., *Trametes incana* B., *Daedalea milliauni* Beel.

Habitat

Grows on hardwood logs. Imbricate on rotten wood of broad leaf tree.

Habit

Basidiocarp sessile. Pileus 2–6 cm long, 2.2–3.8 cm broad, 13.20 mm thick near the base, flabellate to dimitiate; initially appresed tomentulose, then glabrescent, zonate and radially rugose, yellow with brown tints, gradually becoming greyish black with KOH, margin thin, only 2–5 mm thick, slowly changing to greyish black with KOH. Hyphal system trimitic : generative hyphae thin walled 2–4 µm thick colourless or yellowish, without clamp–connections.

Distribution

Reported from Namsai (Arunachal Pradesh) by Bisht and Harsh (2001).

269. *Lepiota americana* Peck

Habitat

Grows solitary or gregarious on the grassy land, sometimes on old stumps or under hedges.

Habit

Pileus 2-10 cm in diameter, ovate, convex or expanded with an umbo, surface white but the umbonal region, scales reddish or reddish brown. Gills crowded, white, free and close to the stipe. Stipe white to brownish red with age, hollow and more or less thickened downward. Spores (7.5-10×5-7.5 µm) sub-elliptical, smooth and without germ pore.

Distribution

Reported from Maharashtra (Sathe and Rahalkar, 1978; Sathe and Deshpande, 1979 and 1982) and Kerala (Bhavani Devi, 1995).

270. *Lepiota anthomyces* var. *macrospora* (Sacc.) Atri

Habitat

Grows in open grassland.

Habit

Pileus upto 2.5 cm broad, convex, campanulate at maturity, reddish white, surface dry with superficial scales throughout, margin irregular, splitting. Gills free, unequal, close, arranged in series. Stipe upto 4 cm long, clavate, white then light brown, solid, surface with cottony fibrils, annulate. Spores (5-7×3-4.3 µm) ellipsoid, smooth, double walled, apiculate, amyloid.

Distribution

Reported from Punjab (Atri *et al.*, 1996).

271. *Lepiota badhami* (Berk. and Br.) Locquin

Synonym

Lepiolophyllum badhami (Berk. and Br.) Locquin.

Habitat

Grows on the ground.

Habit

Pileus usually 3-10 cm in diameter. Sporophore similar to *Lepiota americana* but characterised by less scaly pileus, spindle shaped stipe and non-striate margin of the pileus.

Distribution

Reported from Thane, Maharashtra (Butler and Bisby, 1931) and Chennai, Tamil Nadu (Ramakrishnan *et al.*, 1952).

272. *Lepiota clypeolaria* (Bull. ex Fr.) Kummer

Common Name

Shield mushroom; shield lepiota.

Habitat

Grows solitary or scattered on the ground.

Habit

Pileus upto 8 cm in diameter, campanulate then flattened with an umbo, surface whitish or yellowish with yellow or brown scales except umbonal region, margin with velar remains. Gills crowded, white, free, soft. Stipe up to 10 cm long, slender, slightly attenuated upwards, fragile, hollow when old, annulus white, thin and delicate. Spores oblong or fusiform, white, smooth, pseudoamyloid, without germ pore.

Distribution

Reported from Uttar Pradesh (Hennings, 1901) and Maharashtra (Sathe and Rahalkar, 1978).

273. *Lepiota cristata* (A. and S.) Fr.

Habitat

Grows solitary on the debris.

Habit

Pileus upto 5.6 cm broad, flattened, pale orange to brown, scaly, margin regular. Gills free, close, broad, yellowish white. Stipe upto 6.6 cm long with swollen base, concolorous, annulate, sheathed above. Spores (6-8.4×3.8-5.3 µm) amygdaliform, apical pore absent.

Distribution

Reported from Uttar Pradesh (Hennings, 1901), Himachal Pradesh (Sharma *et al.*, 1978); Punjab (Atri *et al.*, 1996); Maharastra (Sathe and Deshpande, 1979).

274. *Lepiota erythrogamma* (Berk. and Br.) Sacc.

Habitat

Grows on the ground.

Habit

Pileus 2-8 cm in diameter, convex then flat, reddish purple but darker near the centre, radially fibrillose or minutely squamulose, fleshy. Gills more or less crowded, free, white with denticulate. Stipe 4-7 cm long, central, cylindrical, enlarged at the base, white, surface smooth and shining, annulated. Spores hyaline, thin walled and pseudoamyloid.

Distribution

Reported from West Bengal (Ray and Samajpati, 1980).

275. *Lepiota leprica* (Berk. and Br.) Sacc.

Habitat

Grows on the ground.

Habit

Pileus 3-9 cm in diameter, convex then hemispherical and finally expanded with obtuse centre, pale white with reddish brown at the centre, suface with reddish brown squamules. Gills crowded, white, free with smooth margin. Stipe 4-11.5 cm long, central, cylindrical, uniform, annulated, squamulose below annulus. Spores (7.8-10.4×3.9-6.5 µm) hyaline, thin walled, smooth, ellipsoidal and pseudoamyloid.

Distribution

Reported from West Bengal (Ray and Samajpati, 1980).

276. *Lepiota mimica* Massee

Habitat

Grows on the ground.

Habit

Pileus 3-5 cm in diameter, companulate to expanded, whitish or bluish grey, fibrillose with fimbriate margin. Gills crowded, white, free but close to stipe. Stipe 5-6 cm long, attenuated upwards, fibrillose, pale, annulated and more or less hollow. Spores (8×5 µm) ellipsoidal.

Distribution

Reported from West Bengal (Saccardo, 1925).

277. *Lepiota oregonensis* var. *macrospora* (Smith) Atri.

Habitat

Grows scattered on grassy soil under *Pongamia glabra.*

Habit

Pileus 3-4.5 cm broad, convex-campanulate, white with brown squamules crowded around the centre, surface dry, margin entire, splitting at maturity. Gills free, crowded, unequal, white. Stipe upto 5 cm in length broadening to a bulbous base, white, solid. Spores (4.5-6×3-3.7 µm) ellipsoid, smooth.

Distribution

Reported from Punjab (Atri *et al.*, 2000).

278. *Lepiota rhacodes* (Vitt.) Quel.

Habitat

Grows on flat grassy grounds amongst weeds in conifer wood predominantly of *Picea smithiana* and *Abies pindrow*, gregarious or in groups.

Habit

Pileus 70-150 (-200) mm broad, ovate or conical, becoming expanded, conical with age, whitish when young soon becoming dark-brownish or tawny, fleshy, firm, scaly squamulose, forming definite ionations, scales either recurved or reflexed, flushed, cigar brown or sometimes darker; margin incurved, apendiculate-flocculose with brownish velar material, not striate; flesh 15-30 mm thick, white soon pinkish or clay-pinkish, becoming brownish with age. Lamellae free, crowded, broadened at the middle, white, soon pink, finally pinkish brown, fleshy, distinctly formed, easily separateble from the flesh, 8-12 mm broad, 4 lengths with entire margin. Stipe 80-200 × 25-50 (-65) mm, central, clavate towards the base, hollow, smooth, ivory white when young soon pinkish brown or darker with age. Spore print white or ivory.

Distribution

Reported from Pahalgam, Kashmir (2100 m) by Abraham *et al*. (1984).

279. *Lepiota roseoalba* P. Henn.

Habitat

Grows scattered on humicolous soil.

Habit

Pileus upto 3.5 cm broad, convex to campanulate, surface dry, pastel red, margin splitting at maturity. Gills free, distant, broad, unequal, not in series. Stipe upto 4 cm in length, white, very slightly tapering below, annulate, sheathed above. Spores (6-9.2×3.8-5.3 µm) smooth, double walled, hyaline, guttulate, apical pore absent.

Distribution

Reported from Himachal Pradesh (Atri *et al.*, 1996) and from Calcutta (Pegler, 1969).

280. *Lepiota serena* var. *macrocorpa* Atri

Habitat

Grows in groups on humicolous soil among grasses.

Habit

Pileus 7-9.6 cm broad, convex when young and flattened with broad umbo at maturity, surface dry, yellowish white to pale yellow, scaly, scales greyish orange.

Gills free, subdistant, unequal, not in series. Stipe 10-13.5 cm in length, concolourous, smooth, annulate. Spores (6-8.4×4.6-5.3 µm) ellipsoidal, smooth, double walled, apiculate, guttulate, amyloid.

Distribution

Reported from Punjab (Atri *et al.*, 1996).

281. *Lepiota sistrata* (Fr.) Quel.

Synonym

Leipota seminuda (Lasch) Kummer

Habitat

Grows on the ground.

Habit

Pileus 2-3.5 cm in diameter, campanulate then expanded, pinkish or yellowish white but darker toward the centre, surface mealy with shining particles. Gills crowded, white and free. Stipe uniform in width, cylindrical, central, white or flesh coloured, annulated, annulus fibrillose. Spores white, smooth, thick walled, pseudoamyloid, without germ pore.

Distribution

Reported from Uttar Pradesh (Hennings, 1901).

282. *Lepista irina* (Fr.) Bigelow

Synonym

Tricholoma irina Dickinson and Lucas

Habitat

Grows scattered on leaf litter under *Eucalyptus* tree

Habit

Fructification 4–9 cm in height. Pileus 5–10 cm in diameter, depressed, surface white to orange white or light flesh coloured, dry; scales appressed fibrillose; margin irregular, splitting at maturity; flesh light–pinkish coloured, unchanging, 0.5 to 1.0 cm thick. Lamellae almost decurrent, subdistant, unequal with lamellulae of five lengths, 0.5–0.9 cm broad, yellowish brown; gill edges wavey; spore deposit light orange to greyish orange. Stipe 3.5–8.0 cm long, 1.0 to 1.7 cm broad, concolorous with the pileus, fleshy, equal in diameter throughout, solid, fibrillose, stipe cuticle shreded into scales. Veil absent. Clamp connection present.

Distribution

Reported from Patiala, Punjab (Kaur *et al.*, 2008)

283. *Lepista nuda* (Bull. ex Fr.) Cooke

Synonym

Tricholoma nudum (Bull. ex Fr.) Kummer; *Rhodopaxillus nudus* (Bull. ex Fr.) Maire.

Common Name

Wood blewit, Naked mushroom or Naked Tricholoma.

Habitat

Grows under coniferous trees, gregarious.

Habit

Pileus upto 10 cm in diameter, convex then expanded, depressed at the centre, bluish or violet then reddish brown with age, surface smooth, margin moist and sometimes incurved. Gills crowded, narrow, adnate or sinuate. Stipe upto 10 cm long, violet, solid, fleshy and fibrillose. Spores light pink, minutely echinulate and ellipsoidal.

Distribution

Reported from Pune, Maharashtra (Sathe and Rahalkar, 1975; Sathe and Deshpande, 1979).

284. *Leptopodia atra* (KÖnig ex Fr.) Boudier

Synonym

Helvella atra; Elvella atra.

Habitat

Grows solitary or scattered on ground.

Habit

Sporophores usually smaller in size. Pileus very irregular, with margin lobed, usually saddle shaped, surface smooth, black, under surface darkr brown. Stipe 3.0-5.0×0.3-0.4 cm, slender, black, but whitish and enlarged at the base, surface dark brown. Stipe 3.0-5.0×0.3-0.4 cm, slender, black, but whitish and enlarged at the base, surface even. Asci borne on saddle shaped fertile portion on long stalk. Ascospores elliptical, smooth, with large internal oil drops, 17.0-18.011.0-12.0 µm, Paraphyses cylindrical.

Distribution

Reported from H.P. (Sohi *et al.*, 1965a), Jammu and Kashmir (Kaul, 1971).

285. *Leptopodia elastica* (Bull ex St. Amans) Boud.

Synonym

Hevella elastica Bull.

Habitat

Grows on humus under forest trees.

Habit

Sporophores saddle-shaped, stipitate. Pileus 1.0-2.2 cm in diameter, cream, yellow or tan, irregularly lobed. Stipe 4.5-9.0 cm long, 0.4-0.7 cm thick at base and 0.3 cm thick at top, slender, cylindrical, hollow light coloured, usually, smooth but few lacunae present towards the base. Asci 8-spored, cylindrical, 248.0-290.0 × 16.5-23.1 µm. Ascospores broadly elliptical, warted, uniseriate. Paraphyses branched, swollen at the top. Flelsh usually tough.

Distribution

Reported from Darjeeling, West Bengal (Kar and Maity, 1970).

286. *Leucoagaricus excoriatus* (Schaeff ex Fr.) Singer

Synonym

Lepiota excoriata (Schaeff. ex Fr.) Kummer

Habitat

Grows in large number in pastures.

Habit

Pileus upto 10 cm in diameter, globose then expanded, central region darker, other light fawn, fleshy with scaly surface. Gills crowded, white, free and soft. Stipe short, white, fibrillose, slightly bulbous towards the bottom, annulated. Spores white, elliptical with conspicuous germ pore.

Distribution

Reported from Darjeeling, West Bengal (Ramakrishnan and Subramanian, 1952).

287. *Leucoagaricus gongylophorus* (Moller.) Singer

Common Name

Leaf Cutting Ant Fungus

Habit and Habitat

This is a symbiotic mushroom.It grows in symbiotic association with nestropical leaf cutting ants of the genera *Acromyremax* and *Atta lira.* Attine ants live in an intensely studied tripartite mutualism with this mushroom which provide food to the ants. This mushroom can be successfully cultivated by the leaf cutting ant.

Distribution

Reported from Poona, Maharashtra (Patil *et al.*, 1995)

288. *Leucoagaricus holosericeus* (Fr.) Moser.

Synonym

Lepiota holoseriea (Fr.) Gillet

Habitat

Grows singly on ground in open fields among grasses.

Habit

Pileus convex, white and shiny like silk, fleshy, waxy on drying, fibrillose, sometimes smooth. Stipe solid, central, long bulbous, fibrillose with persistent annulus. Spores (7×5 µm) ornamented with germ pore.

Distribution

Reported from Uttar Pradesh (Hennings, 1901), Maharashtra (Sathe and Rahalkar, 1978; Patil *et al*., 1995) and East Khasi, Meghalaya (Verma *et al*., 1995).

289. *Leucoagaricus hortensis* (Murr.) Pegler.

Habitat

Grows solitary to scattered on soil.

Habit

Pileus 1.5-8 cm broad, conico-companulate, expanding convex, lowly umbonate, surface nought at the disc, disrupting into nought appressed to recurved squamules on a whitish ground, entire surface brushing red, margin straight, becoming ribose. Lamellae free, white upto 5 mm wide, crowded with lamellulae of different lengths. Stipe 3.5-7 cm × 4-8 mm, central, cyclindric or with a clavate bulbose base, surface white, vinaceous red on bruising, glabrous. Annulus superior, white membranous. Hyphae occasionally with clamp-connection. Lamellae edge sterile, cheilo-cystidia cylindric to cylindro-clavate, crowded thin walled. Hymenophoral trama regular of thin walled hyphae.

Distribution

Reported from Kerala by Vrinda *et al*. (1999).

290. *Leucoagaricus naucinus* (Fr.) Sing.

Synonym

Lepiota naucinoides P.K.; *Lepiota leucothites* (Vitt.) Qrton.; *Lepiota naucina* Fr.; *Annularia laevis* Krombh.

Common Name

Smooth lepiota

Habitat

Grows scattered or in groups on the ground in pastures and fields, sometimes in open woods.

Habit

Sporophores centrally stipitate. Fruiting body lacking powdery veil. Pileus 5.1-10.2 cm in diameter, globose when young, convex to flattened with age, white or smoky white, surface usually smooth, retaining shape, rarely cracked into minute scales. Gills crowded, free at first white turning to pink brown or smoky brown with age, edge fringed. Stipe cylindrical, equal, 5.1-7.6 cm or more long, shightly enlarged at the base and tapering upward, smooth, hollow or stuffed with cottony threads, annulus present but volva absent, ring distinct, white, movable, sometimes disappearing. Flesh white, firm, thick.

Distribution

Reported from Maharashtra (Patil *et al.*, 1995), Kerala (Bhavani Devi, 1995), M.P. (Trivedi, 1972).

291. *Leucocoprinus cepaestipes* (Sow. ex Fr.) Patouillard

Synonym

Lepiota cepaestipes (Sow.) Quel.; *Agaricus cepaestipes.*

Habitat

Grows caespitose or in partial fairy rings in soil, freshly manured ground or sometimes in decomposed vegetable matters or on logs, rotten woods.

Habit

Sporophores centrally stipitate. Fruiting body with powdery veil and collapsing at maturity. Pileus ovate when young, later expanded or broadly conical with an umbo, white but umbo and squamules brownish, thin, surface dry, margin plicale-striate. Gills crowded, free, white, narrow, thin. Stipe slender, 5.0-8.0×0.2-0.4 cm, enlarged at the base, white, smooth, covered upto ring with a mealy white mycelium, stuffed or hollow, ring thin, disappearing latter, Flesh white, very soft, thin.

Distribution

Reported from West Bengal (Bose, 1920; Banerjee, 1947), Gujarat (Moses, 1948), U.P., Maharashtra (Butler and Bisby, 1960), Maharashtra (Patil *et al.*, 1995), Kerala (Bhavani Devi, 1995), West Bengal, Gujarat, U.P., Maharashtra (Nita Bahl, 1988, Sathe and Deshpande, 1979).

292. *Leucopaxillus giganteus* (Sow. ex Fr.) Singer

Synonym

Clitocybe gigantea (Sow. ex. Fr.) Quel.

Habitat

Grows commonly under *Pinus wallichiana.*

Habit

Pileus upto 20 cm or more in diameter, initially flat then depressed, milky white to cream, margin inrolled. Gills well formed, crowded and decurrent. Stipe 4-7 cm long, whitish and tough. Spores ovoid to ellipsoidal with broadly rounded base and apex, thin walled and smooth.

Distribution

Reported from Jammu and Kashmir (Watling and Gregory, 1980).

293. *Limacella guttata* (Pers ex Fr.) Kanrad and Maublane

Synonym

Lepiota lenticularis (Larch) Gillet; *Lepiota lenticularis* (Larch) Maire.

Habitat

Grows in the forest of *Picea* or in open places.

Habit

Pileus convex then hemispherical then expanded with a central, shallow umbo, white with grey squamules, viscid, margin involute, thin, striated, sometimes fissured. Gills free, whitish, thick and comb like. Stipe central, cylindrical, whitish, well-developed, solid, thickened at the base, annulate. Spores subglobose to globose, hyaline, inamyloid, smooth, thin walled.

Distribution

Reported from West Bengal (Roy and Samajpati, 1978).

294. *Lycoperdon atropurpureum* Vitt.

Habitat

Grows scattered on soil covered with dead decaying organic matter.

Habit

Fructification 2.4-5 cm high and 2.2-3.6 cm broad, usually depressed globose or nearly pyriform, sometimes obovate to turbinate with large distinct stalk like base, greyish tan to buff or leather coloured, cortex covered with very delicate hair like spicules and grannules. These are persistent in fresh collections but absent in old. Gleba deep smoky purple at maturity. Subgleba occupying the stem like base which is usually more than half of the whole fructifiction. Capillitium composed of long, brown, unpitted, branched threads, attenuated at the ends.

Distribution

Reported from Gopalpur, Himachal Pradesh (6900 m) by Sohi *et al*. (1964).

295. *Lycoperdon curtisii* Berk.

Synonym

Lycoperdon wrightii Berk and Curt.

Habitat

Sporophores grow usually gregarious, sometimes closely crowded, forming a tuft on grassy land, pastures in cultivated fields.

Habit

Sporophores usually small with distinct apical pore, peridium 1.2-5.1 cm wide, globose, usually without stem, white or whitish. Adorned with deciduous, stellate, crowded spines or pyramidal warts, upper surface smooth and velvety, dark coloured, under surface often plicate. Capillitium dark, olive in colour, columella present.

Distribution

Reported from Uttaranchal (Ahmad, 1942).

296. *Lycoperdon gemmatum* Batsch

Habitat

Gross on soil.

Habit

Stipe prolonged and tapering from above suggesting the specific name, pear–shaped, colour dingy white, surface covered with deciduous warts; substance, young state white, spore dust brown, height two to three inches.

Distribution

Reported from Gopalpur, H.P. (Sohi *et al.*, 1964)

297. *Loycoperdon giganteum* Per

Synonym

Calvatia gigantea

Common Name

Putca Chhatu, The Giant Puffball

Habitat

Grows on sandy soil

Habit

The giant puffball reaches a foot (30 cm) or more in diameter and is difficult to mistake for any other fungus. The true puffball do not consist of a visible stalk. Young puffballs in the edible state have undifferentiated white flesh within. It has been estimated that large specimen of this fungus, when mature produces around 7×10^{12} spores. If collected before spores have formed while the flesh is still white, it may be cooked as slices fried in butter, with a strong earthy, mushroom flavor. It can often be used in recipes that would ordinarily call for eggplant. It can often be used in recipes that would ordinarily call for eggplant. It does not store well in freezer, the entire freezer rapidly acquires a strong mushroom smell.

Distribution

Reported from Midnapur District, West Bengal (Das *et al.*, 2002)

298. *Lycoperdon marginatum* Vitt.

Habitat

Grows on ground or sandy soil

Habit

Fruit body 1.5 cm high, globose sometimes with a small rooting base; white becoming brownish; with an outer covering of pointed wart or spins, which fall away in irregular sheets, exposing the olive–brown inner wall or endosperm. Spore mass olive to gray–brown. Sterile base well developed, chambers about 1 mm across, capilitial threads 3–6 mm wide, spore globose, minutely ornamented, olive–brown, 3.5–4.2×3.5–4.2 mm.

Distribution

Reported from Sahranpur and Dehradun, U.P. (Ahmad, 1942)

299. *Lycoperdon molle* Pers

Synonym

L. muscorum Morg.

Habitat

Grows among moss, especially *Polytrichnm* in coniferous forest and among woods in forest.

Habit

Fruiting body rounded or pear shaped with a stem like sterile base, 3-4 cm. high, 2-5 cm across, white to light brownish grey in colour. Outer skin covered with short, separate, erect, hair-like spines, these become flattened towards base. The spine become worn away in mature specimens leaving the pale-brown shiny inner skin. Gleba at first white, turning yellow, then olive, finally purplish-brown or dark olive- brown.

Distribution

Reported from Cherthala, Alleppey, Kerala by Bhavani Devi (1995).

300. *Lycoperdon muscorum* Morgan

Habitat

Grows scattered under shade.

Habit

Sporophores stalked, 5.7 cm high or less, 5.0 cm in diameter or less, fruiting body with distinct apical pore. Peridium turbinate, globose to depressed, inner peridium olive brown, occasionally granulated. Gleba usually brown with age, subgleba present. Capillitium yellow at first, brown with age.

Distribution

Reported from Himachal Pradesh (Gupta *et al.*, 1974).

301. *Lycoperdon perlatum* Pers.

Synonym

Lycoperdon gemmetum Batsch.; *Clavatia cranniformis.*

Common Name

Common Puff ball.

Habitat

Grows solitary or gregarious or scattered on the ground, in open places or in forests, sometimes on rotten woods or on moist soil after ploughing.

Habit

Sporophores smaller, 3.0-7.0 cm high and 2.0-6.0 cm in diameter, white when young, dark-grey or greyish brown at maturity, with distinct apical pore, fruiting body ellipsoid, globose or portel shaped and not star-shaped. Peridium spherical or with rounded top, usually with as stem like base. Stipe larger than the rounded top, stipe stout, white when young, then brown with age, warts falling off at maturity leaving scar at the top of sporophore, capitlitium greenish yellow with an olive tinge, later pale brown. Columella present.

Distribution

Reported from Himalayas (Ahmad, 1942), Himachal Pradesh, West Bengal, Sikkim (Butler and Bisby, 1960), Himachal Pradesh (Sohi *et al.*, 1964), Assam (Gogoi *et al.*, 2000), Arunchal Pradesh (Mishra, 1999), Himachal Pradesh, Punjab, Madhya Pradesh, West Bengal, Sikkim (Nita Bahl, 1988).

302. *Lycoperdon pusillum* Pers.

Habitat

Grows in group on the finely tilled soil after light and intermittent rainfall. It is generally found in moist and cool grassy areas.

Habit

It is smallest of our common puffballs. Fruiting body small in size, spherical in shape, smooth white surface and absence of a sterile base are good field characters. Fruiting body 1 to 2 cm in diameter, round, peridium smooth, white, gleba white when young, becoming brown with age, sterile base lacking. Spores brown, round with minute spines; often with a pedicel attached; 3.5-4.5 μm.

Distribution

Reported from Assam (Gogoi *et al.*, 2000), Plains of Punjab (Ahmad 1940, Chaal, 1963), Poona (Nair and Patil, 1978), Ahmadabad, Gujarat and Ludhiana, Punjab (Rao, 1964).

303. *Lycoperdon pyriforme* Schaeff ex Pers.

Synonym

Lycoperdon emodense Berk.

Common Name

Stump Puffball

Habitat

Grows singly or in clusters on burried wood or woody stumps, decaying logs, rotten wood in forests or in groves, open fields or on ground. It is very widely spread puffball.

Habit

Fruiting bodies often grow in dame clusters. Sporophores smaller in size 3-4×2-4.5 cm, almost sessile or with a very short stipe, white and coarse strands of mycelium at the base, fruiting body with distinct apical-pore. Peridium pear shaped, whitish or brownish, with sub–persistent nearly uniform warts and scales, slender deciduous spinules scattered in between. Peridium cracking during wet weather. Capillitium greenish yellow turning to dull olivaceous with age. Columella present.

Distribution

Reported from Manipur and Meghalaya (Verma *et al.*, 1995), Himachal Pradesh (Ahmad, 1942), Jammu and Kashmir and Sikkim (Butler and Bisby, 1960), Himachal Pradesh, Jammu and Kashmir and Sikkim (Nita Bahl, 1988).

304. *Lycoperdon rimulatum* Peck

Habitat

Grows scattered on humicolous soil.

Habit

Fruit body depressed globose but plicate below, attenuated to a slender pointed root, reddish brown when old, 2.5-3.0×2.5-3.0 cm, with distinct apical pore at maturity. Peridium with definite apical mouth, greyish brown. Gleba light yellow to grey, turning to purplish grey and eventually chocolate brown, subgleba present small. Capillitium simple or moderately branched, ferruginous brown, attenuated at the ends.

Distribution

Reported from Himachal Pradesh (Gupta *et al.*, 1974).

305. *Lycoperdon umbrinum* Pers.

Habitat

Grows solitary on the ground.

Habit

Sporophores 4.5 cm high and 3.2 cm in diameter, subglobose, with a rooting base, apical pore present at maturity. Exoperidium furfuraceous, fugacious and granular. Endoperidium brownish and papery. Gleba purple-brown with light coloured subgleba. Capillitium aseptate, freely branched, smooth and brown. Spores (4.5-5.5 μm) globose and verrucose.

Distribution

Reported from Kulu, Himachal Pradesh (Ahmad, 1942) and Sonamarg, Jammu and Kashmir (Bilgrami *et al.*, 1979).

306. *Lyophylum fumosum* (Pers ex Fr.) P.D. Orton

Synonym

Tricholoma conglobatum (Vitt.) Ricken.

Common Name

Shemigi mushroom

Habitat

Grows on the wood of *Acacia*, Willow and Poplar.

Habit

Pileus fleshy, fuscous, whitish or blackish, smooth, margin thin, inflexed. Gills free, rounded, whitish. Stipe short, solid, swollen at the middle, subtomentose. Spores (6-8 µm) globose or angulate, hyaline, nonamyloid.

Distribution

Reported from Jammu and Kashmir (Dhar, 1976).

307. *Macrocybe lobayensis* (Helm) Pegler and Lodge.

Synonym

Tricholoma lobayense Heim.

Habitat

Grows scattered on ground among grass.

Habit

Pileus upto 21 cm in diameter, convex to applanate, surface dull white, smooth and glabrous, dry, margin lobate, inrolled, non-striate. Stipe upto 13 cm long, central, cylindrical, solid, finally fistulose, dull white, smooth. Spores (5.5-6.6×4-4.5 µm) ovoid to ellipsoid, hyaline, thin walled, containing single large refractive guttule.

Distribution

Reported from Kerala (Vrinda and Pradeep, 2006).

308. *Macrolepiota mastoidea* (Fr.) Singer

Synonym

Lepiota mastoidea (Fr.) Kummer.

Habitat

Grows singly on in group in open fields near cowdung compost or on soils rich in organic content

Habit

Sporophores with conical to arched cap, centrally stipitate. Pileus 3.0-6.0 cm in diameter, companulate when young, then convex and later becoming umbonate, white, surface with greyish or brownish appressed scales. Gills free, crowded, white, attached to the stem by a broad collar. Stipe 7.0-10.0 cm long, cylindrical tapering upward, base bulbous, white with minute scales, ring movable with brownish margin, volva absent, clamp connection present.

Distribution

Reported from Mizoram and Manipur (Verma *et al.*, 1995), West Bengal (Bose, 1921; Bose and Bose, 1940; Banerji, 1947, Chopra and Chopra, 1955; Butler and Bisby, 1960), Dehradun, U.P. (Ghosh and Pathak, 1965), West Bengal and U.P. (Nita Bahl, 1988), U.P. (Pathak and Gupta, 1982) and from H.P. (Sohi *et al.*, 1964).

309. *Macrolepiota procera* (Scop. ex Fr.) Sing.

Synonym

Lepiota procera (Scop ex Fr.) S.F. Gray., *Agaricus procerus* Scop ex Fr.

Common Name

Parasol mushroom, Tall Lepiota

Habitat

Grows solitary or scattered, sometimes in clusters on soil, in pastures, lawns, woods and in garden.

Habit

Sporophores 12.0-20.0 cm or more high. Pileus 7.6-15.2 cm in diameter, ovate then companulate and finally expanded with an umbo, greyish brown or reddish brown, surface cracking into large, brown scales concentrically arranged except at umbo, white small scales, present between the large ones. Gills free crowded but not forked, white when young but pinkish at maturity broader in middle and narrow near the stipe. Stipe slender, cylindrical with prominent bulbous base, 15.2-25.4 cm long. Colour paler than the pileus, covered with minute darker scales, hollow; Annulus, tough, white movable. Flesh white, remaining unchanged when bruised, soft, thick, clamp connection present.

Distribution

Reported from Kerala (Bhavani Devi, 1995), West Bengal (Banerji, 1947; Nita Behl, 1988), U.P. (Chopra and Chopra, 1955; Butler and Bisby, 1960; Ghosh and Pathak, 1965; Nita Behl, 1988), Himachal Pradesh (Sohi *et al.*, 1964), Maharashtra (Patil *et al.*, 1995; Sathe and Deshpande, 1979).

310. *Macrolepiota puellaris* (Fr.) Mapud G.

Habitat

Grows on soil in fields.

Habit

Sporophores usually smaller in size, centrally stipitate. Pileus 5.0-6.0 cm in diameter, white, surface adorned with coarse scales. Gills free. Stipe separating readily from the pileus, ring present, volva absent; flesh does not turn red on cutting. Besidiospores oval, 10.0-13.0×5.0-6.0 μm.

Distribution

Reported from Uttar Pradesh (Ghosh and Pathak, 1965).

311. *Macrolepiota rachodes* (Vitt.) Singer.

Synonym

Lepiotophyllum rachodes (Vitt.) Locquin; *Lepiota rachodes* Vitt.

Common Name

Shaggy parasol, Wood parasol.

Habitat

Grows solitary, sometimes in dense clusters, often forming fairy ring on soil in fields, in pastures or in open forests.

Habit

Sporophores centrally stipitate. Pileus 8.0-15.0 cm in diameter, globose, at first later coarse flat, scales formed by the rupture of the cuticle, scale concentrically arranged, edge of the pileus not red. Gills crowded, remote from the stipe, whitish or pinkish, broad, edge finely toothed. Stipe 7.0-18.0 cm. long, firm, thick, tapering at the apex, surface glabrous, base enlarged, annulus not movable, edge not red, Flesh white, soft, thick, fibrous, turning red on cutting.

Distribution

Reported from Kerala (Bhavani Devi, 1995), U.P. (Butler and Bisby, 1960), Lucknow, U.P. (Ghosh and Pathak, 1965), Maharashtra (Patil *et al.*, 1995), H.P. (Sohi *et al.*, 1964).

312. *Macroscyphus macropus* Pers. ex Gray

Synonym

Helvella macropus (Pers. ex Fr.) Karst.

Habitat

Grows solitary or gregarious on damp soil or on humus under the shade of *Populus* and *Salix* trees in spring.

Habit

Sporophores saddle shaped and stalked. Pileus 3-6 cm in diameter, light coloured, granulose or minutely fibrillose. Stipe 3-5 cm long, central, stout, having longitudinal ridges, whitish when young, pale or dark brown with maturity. Ascospores (20.5-23×12-14.2 μm) elliptical to fusiform, uniseriate, smooth.

Distribution

Reported from Jammu and Kashmir (Kaul *et al.*, 1978) and Uttaranchal (Joshi *et al.*, 1982).

313. *Marasmius alliaceus* (Jacq. ex Fr.) Fr.

Common Name

Garlic Mushroom

Habitat

Grows solitary or scattered on leaf litter and dried twigs of deciduous forests.

Habit

Cap 2-4 cm, convex then soon expanding, with slightly rondially grooved margin, pale greyish to clay-brown. Gills whitish, adnexed, rather distinct. Stipe tall, slender, rooting, smooth and polished, black, spores white, ovate, 9-12×6-7 μm.

Distribution

Reported from Kerala (Bhavani Devi, 1995).

314. *Marasmius androsaceus* (L. ex. Fr.) Fr.

Synonym

Androsaceous androsaceus (L. ex. Fr) Rea

Common Name

Horse hair fungus

Habitat

Grows solitary in large numbers on leaf litter.

Habit

Pileus 0.5–1 cm across, convex with the center depressed, membranous and radially wrinkled, clay pink with red brown centers. Stipe 20–60 × 1 mm black hair like, stiff and tough. Flesh thin, white in cap, dark in stipe. Gills distant, clay pink. Spore print white.

Distribution

Reported from Allepy district of Kerala (Bhavani Devi, 1995) and from Mahabaleshwar (Maharashtra) by Sathe and Deshpande (1979) and Patil *et al.*, (1995).

315. *Marasmius grandisetulosus* Sing

Habitat

Grows in group on the decoying leaves among grasses.

Habit

Carpophore upto 7.5 cm in height. Pileus upto 1.5 cm. broad, campanulate to convex with acute brown umbo, often papillate at the center, surface dry, light brown at the center, margin irregular, cuticle not peeling, flesh yellowish white, very thin, unchanging on exposure. Lamellae free to finely adnexed, distant, narrow, 0.2 cm deep, pale yellow when young, yellowish white at maturity, unchanging where bruised, gill edges serrate. Stipe upto 7.8 cm in length, 0.1 cm in width, equal in diameter through out, concolourous with the gills above and pileus below, stiff hair like, smooth, hollow, unchanging where bruised. Pleaurocystidia clavate to ventricose, thick walled, cheilocystidia pyriform with 5–8 pointed apical setules, hyaline to pale brown with thin walls. Clamp connections present.

Distribution

Reported from Patiala, Punjab by Atri *et al.* (1996). and from Kerala by Natarajan and Manjula (1982).

316. *Marasmius oreades* (Bolt. ex Fr.)Fr.

Synonym

Scorteus oreades (Bolt; Fr.) Earle.

Common Name

Fairy ring mushroom.

Habitat

Grows in "fairy rings" or in a semicircle, occasionally scattered in grassy lands, by roadsides in pastures and weed free soil.

Habit

Sporophores pale coloured. Pileus 1.5-4.0 cm in diameter, convex when young, expanded to subumbonate when old, pale tan or white, fleshy, smooth, tough, margin with striations, when moist, reflexed when old or dry. Gills free distant with veins, white or pale tan, thick, broad, round near the stips. Stipe 4.0-7.0×0.3-0.5 cm thick, white or pale tan, hairy, tough, minutely powdered at apex and with strigose mycelium at base, corticated, solid. Flesh whitish, tough, thick.

Distribution

Reported from Gujarat (Moses, 1948), Assam (Gogoi, *et al.*, 2000), Maharashtra (Patil *et al.*, 1995), Kerala (Bhavani Devi, 1995).

317. *Marasmius purpureus* Pegler

Habitat

Grows in groups on humicolous soil.

Habit

Pileus upto 1 cm broad, convex whth acute umbo, deep red with purple tone, surface dry, margin regular. Gills free, subdistant, narrow, unequal, not arranged in series. Stipe up to 2 cm long, equal in diameter throughout with slightly bulbous base, smooth. Spores (3-4.6×2.3-3.8 µm) subglobose to avoid, smooth, double walled, pigmented, apiculate, hyaline.

Distribution

Reported from Punjab (Atri *et al.*, 1996).

318. *Melanogaster ambiguus* (Ritted.) Tue.

Habitat

Grows in groups on ground, a few inches below the soil surface.

Habit

Sporophores usually globose or subglobose, blackish, covered by dark brown fibrils which disappear on drying. Gleba jet black mixed with white tramal plates. Sporés (10.5×16.5 µm) dark brown, elliptical and pedicellate.

Distribution

Reported from Himachal Pradesh (Ahmad, 1941; Bakshi and Puri, 1978; Sohi *et al.*, 1964).

319. *Melanogaster durissimus* Cooke

Habitat

Sporophores grow subterranean or sometimes exposed at maturity.

Habit

Sporophores usually without stalk. Gleba containing polyhedric or rounded units (Basidial nests) formed by sheets of tramal tissue, basidia form an irregular hymenium which surrounds the hymenial cavities, basidia slendar, 2-8 spored, spores formed within fruiting body, basidiospores brown, ellipsoidal, frequently coated with a gelatinous layer

Distribution

Reported from Himachal Pradesh (Chopra and Chopra, 1955), U.P. (Cooke, 1879).

320. *Melanoleuca alboflavida* (PK) Murrill

Habitat

Grows singly or in group on the ground on leaf litter in deciduous or mixed forests and in open fields.

Habit

Pileus dull white to white, areolate, context whites unchanging on bruising. Cap 3–10 cm across, broadly convex, becoming flatter then often depresed with a low broad umbo remaining darker, smooth, dry to moist. Gills sinuate, crowded, narrow, whitish. Stipe 30–100 × 4–10 mm, slender with a small bulb at the base, whitish, cartilaginous, longitudinally lined with minute hairs. Flesh solid, white. Spore deposit white.

Distribution

Reported from North West Himalayas (Bhatt and Lakhanpal, 1990).

321. *Melanoleuca melaleuca* (Fr.) Murr.

Synonym

Tricholoma melaleucom; Melanoleuca vulgaris Patouillard

Habitat

Grows usually in groups of a few to several on grassy lands, in lawns and forests, solitary on ground under *Cedrus deodara*.

Habit

Pileus 2.0-7.0 cm in diameter, convex to flattened and slightly umbonate, grey or greyish brown, viscid when moist, shining when dry. Margin striate. Gills crowded, sinuate or nearly free, white when young, pale brown when old. Stipe cylindrical, tapering towards the top, 2.0-9.0 cm long, light brown, striate or twisted, base slightly bulbous. Flesh white, soft, easily peeled; thick-walled cystidia with apical crystals.

Distribution

Reported from Jammu and Kashmir (Murril, 1924), Uttar Pradesh (Dhancholia *et al.*, 1991).

322. *Microporus xanthopus* (Fr.) Kuntze.

Common Name

Yellow footed Tinypore

Habitat

Grows on hardwood logs and branches in the broader ranges rainforest ragion.

Habit

Fruit bodies small pored on the back with yellow foot because it is yellow where it holds the log. They can often grow in groups and measure upto 12 cm across.

Distribution

Reported from Namsai (Arunchal Pradesh) by Bisht and Harsh (2001).

323. *Morchella angusticeps* Pk.

Common Name

Black Morel.

Habitat

Grows in soil. It occurs widely in coniferons forests of middle elevation. It grows in cluster within a few feet of each other.

Habit

Sporophore usually smaller in size. Pileus small, pointed more than *Morchella conica,* rarely thicker than stipe, sometimes variable in size and shape, pitted by broad shallow depressions bearing hymenium and usually separated by sterile ridges.

Ridges on the pileus of large specimens may range from grey to black. Stipe long and narrow, cylindrical, hollow, occasionally curved, stipe narrowed downward in some cases. Asci 8-spored. Ascospores lacking internal oil drop but having numerous small external grannules adhering at each end.

Distribution

Reported from Jammu and Kashmir (Butler and Bisby, 1931; Batra and Batra, 1963; Waraitch, 1976), Himachal Pradesh (Mukerjee and Juneja, 1974; Munjal and Sharma, 1975) and from H.P. (Sohi *et al.*, 1965).

324. *Morchella conica* Pers.

Common Name

Conic morel

Habitat

Grows solitary or scattersd on the sandy loam rich organic substances (leaf-litter) in thick forest.

Habit

Sporophore 7.5-12.5 cm in height, buff to yellow when young, darker when old. Pileus 3.0-5.0 cm in diameter at the base 5.0-8.0 cm long, conical in shape with a pointed top, ridges extended longitudinally and run parallal from the base to the top, pits arranged in rows, usually longer than broad, sometimes irregular. Stipe cylindrical 4.0-8.0×1.0-3.0 cm, hollow, fragile. Asci borne on stalked receptacle. Ascospores lacking internal oil drop.

Distribution

Reported from Uttaranchal (Butler and Bisby, 1960; Ghosh and Pathak, 1965; Theissen 1911; Nita Bahl, 1988), Himachal Pradesh (Sohi *et al.*, 1965; Munjal and Sharma, 1975), Madhya Pradesh (Rai *et al.*, 1999), Jammu and Kashmir (Waraitch, 1976; Kaul *et al.*, 1978), Punjab (Atri and Kaur, 2005).

325. *Morchella crassipes* (Vent.) Pers. ex Fr.

Habitat

Grows on the ground in the coniferous forests or under shrubs, scattered or gregarious.

Habit

Pileus conical but elongated, apex obtuse or subacentric, ridges longitudinal or anastomosing irregularly, pitted, pits ochraceous or brownish near the margin. Stipe upto 10 cm long, cylindrical with distinctly swollen base, dirty white to yellowish, hairy. Spores (21-26×12-15.5 μm) ellipsoidal, yellowish in mass, smooth.

Distribution

Reported from Jammu and Kashmir (Waraitch, 1976).

326. *Morchella deliciosa* Fr.

Common Name

Grey Morel or White Morel

Habitat

Grows on the ground in grassy places.

Habit

Sporophores 4.0-8.0 cm high. Pileus long and narrow giving a cylindrical appearance with a blunt top, 2-3 times long as it is broad, length greater than the stipe, middle often narrowed, pits usually narrow, elongated. Stipe usually short. Asci borne on stalked receptacle. Ascospore lacking internal oil drop but having numerous small external grannules adhering to each end.

Distribution

Reported from Punjab (Sydow and Butler, 1911; Ginai, 1936; Butler and Bisby, 1960; Kaul *et al.*, 1978; Nita Bahl, 1988), Jammu and Kashmir, Uttaranchal (Butler and Bisby, 1960; Kaul *et al.*, 1978; Nita Bahl, 1988), Himachal Pradesh (Munjal and Sharma, 1975; Nita Bahl, 1988).

327. *Morchella esculenta* Pers. ex St. Amans

Common Name

Sponge Mushroom, Yellow Morel.

Habitat

Grows usually solitary on the ground, under trees in open woods, grassy lands, orchards, roadsides, often in sandy or clay soil.

Habit

Sporophores upto 20 cm high. Pileus 3.0-5.0 cm in diameter, usually longer than broad with a rounded top, first greyish, later becoming yellowish brown, narrow ridges around the pits and depression giving a honey comb like appearance, pits, round or irregular, 0.5-1.0 cm wide, 0.5-1.0 cm deep. Stipe cylindrical or slightly enlarged at the base, 4.0-8.0 cm long, pale white, fragile, usually hollow. Ascospores broadly elliptical, hyaline, 8 in number.

Distribution

Reported from Punjab (Baden Powell, 1863 vide Bose and Bose, 1940), India (Cooke, 1876; Watt, 1880), Jammu and Kashmir, Himachal Pradesh (Chopra and Chopra, 1955), Uttaranchal (Butler and Bisby, 1960; Sohi *et al.*, 1965a), Punjab, Kashmir, Chamba, Kumaon Hills, the Himalayas, H.P. (Nita Bahl, 1988), Jammu and Kashmir (Waraich, 1976), Kodai Kanal Hills (Western Ghats), Tamil Nadu; South India (Kaviyarasan *et al.*, 2006).

328. *Morchella hybrida* (Sow) Pers.

Habitat

Grows scattered on the ground.

Habit

Pileus 2-3 cm long, conical or bell shaped, tan, margin free from the stipe with longitudinal branched ridges. Stipe white, 4-10 cm long, 1-2 cm broad, occasionally enlarged downward, hollow and brittle. Ascospores lacking internal oil drop but having numerous small external grannules to each end.

Distribution

Reported from H.P. (Sohi *et al.*, 1965a), Jammu and Kashmir (Cooke, 1870; Kaul, 1975 vide Purkayastha and Chandra, 1985).

329. *Morchella vulgaris* (Pers.) Bond.

Common Name

Common Morel.

Habitat

Grows on the ground.

Habit

Sporophores sometimes reaching a height of 12 cm. Pileus ovoid, greyish brown when young, ochraceous or light coloured with age, pits and ridges irregular. Stipe usually hollow. Ascospores 8 in number lacking internal oil drops.

Distribution

Reported from Assam (Bhattacharya and Baruah, 1953).

330. *Mycena galericulata* (Scop ex Fr.) S. F. Gray

Synonym

Mycena rugosa (Fries) Quelet

Common Name

Bonne Mycena

Habitat

Usually grows singly or in clusters, few to many in a group on dead logs, stumps, branches or on decaying wood in forests.

Habit

Pileus 1.0-4.0 cm in diameter, conical to bell shaped with a prominent umbo, brown near the center, light tan towards the margin, smooth, striations extending from margin to the umbo. Gills not crowded, adnate, sinuate or short decurrent, intervenose, white at first, light reddish brown with age, edge often slightly wavy. Stipe slender 4.0-12.0 cm long, 0.1-0.3 cm thick brown or whitis, upward. Sometimes twisted, smooth but hairy at the base, firm, tough, hollow, often several fruit bodies joined at the base without ring and volva. Flesh pale yellowish brown 0.2 to 0.3 cm thick, tough.

Distribution

Reported from Sikkim (Butler and Bisby, 1960; Berkeley, 1856); Kerala (Bhavani Devi, 1995).

331. *Mycena galopus* (Pers and Fries) Kummer

Common Name

Milky Drop Mycena

Habitat

Grows in groups on the fallen leaves, pinus needles, twigs etc. of mixed forests.

Habit

Pileus upto 3 cm in diameter, ovoid to conic when young, becoming companulate with margin slightly upturned, translucent and striate, surfaces minutely powdery, blackish in center, becoming white towards margin. Flesh greyish white, lacks distinct odor. Gills adnate, white or slightly greyish. Stipe upto 10 cm long, slender, pale grey at apex, becoming almost black at base, exuding white latex when broken. Spores white, ellipsoid; smooth; sometimes slightly amyloid, 9-13×3-6.5 µm.

Distribution

Reported from Kerala (Bhavani Devi, 1995).

332. *Mycena haematopus* Pers.

Synonym

Agaricus haematopus Fries.; *Galactopus haematopus* Earle.

Common Name

Bleading Mycena

Habitat

Sporophores grows clustered to closely in caespitose, gregarious on decaying wood and logs of broad leaved trees.

Habit

Cap 1-3 cm bell shaped, greyish brown, sometimes with white powdery bloom, striate at margin. Gills white then flesh tinted to deeper reddish brown, adnate. Stem slender rigid, fragile, colour as cap darkening below to reddish brown. When broken yields red-brown latex. Spores white, elliptic, 10×6 µm.

Distribution

Reported from Kerala (Bhavani Devi, 1995; Kaul and Kapur, 1983).

333. *Mycena pura* (Pers ex Fr.) Kummer

Synonym

Agaricus purus Pers. ex Fr.

Common Name

Clean Mycena

Habitat

Grows solitary, scattered or clustered in pine wood or in mixed conifer forest.

Habit

Pileus 2-3 cm in diameter, conical, then campanulate to convex, then expanded and umbonate, variously coloured (violet or pink shades). Epicutis poorly formed. Gills adnate to sinuate, occasionally free, broad, connected to veins. Stipe cylindrical, smooth, hollow, tough, pinkish violet. Spores oval, white, amyloid and smooth.

Distribution

Reported from Sikkim (Berkeley, 1856), Jammu and Kashmir (Watling and Gregory, 1980).

334. *Mycenastrum corium* (Guereslent) Desv.

Habitat

Grows on ground.

Habit

The fruit body open with earth star–like rays resembling *Scleroderma*. Fruit body measure 5–15 cm broad, rounded to compressed with irregular bumps and shallow depressions; the base slightly grooved to folded, attached to the substrate via tuft of mycelium, sometimes breaking free at maturity, outer peridium white, matter tomentose, sloughing away in patches, revealing a dark–brown to dull–purple brown, thick, leathery inner peridium, the latter weathering to light brown or buff–brown in age. Inner peridium dehiscing via cracks and fissures or recurving earthstar like rays, or forming a large apical crater; sterile base absent. Gleba at first white becoming olive brown, finally dark brown.

Distribution

Reported from Kulu Hills and Dalhausie Road, H.P. (Ahmad, 1941; Cunningham, 1942)

335. *Mycoleptodonoides aitchisonii* (Berk.) Maas. G.

Habitat

Grows in Oak Woodlands and margins of grasslands.

Habit

Pileus upto 8 cm in diameter, imbricate, pale or whitish, minutely tomentose or glabrous, margin inflex and torn into lobes. Stipe central, 1.2-1.8 cm long. Spines decurrent colour fuscous, thin, sharp, upto 0.8 cm long.

Distribution

Reported from Gulmarg, Jammu and Kashmir (Berkeley, 1876).

336. *Otidea leporina* (Fr.) Fuckel

Synonym

Pezia leporina Batsch ex Fries.

Habitat

Grows solitary, gregarious or crowded or scattered on humicolous soil in coniferous forest.

Habit

Sporophores 2-3.5 cm in diameter, cupulate or ear shaped, stipitate, surface yellowish or yellowish brown with small stipe. Hymenium yellowish brown, concave and smooth. Spores hyaline, cylindrical, apex usually rounded, sometimes subtruncate.

Distribution

Reported from Mussoorie (Thind and Batra, 1957) and Shimla, Himachal Pradesh (Thind and Waraitch, 1964).

337. *Oudemansiella canarii* (Jungl) Hoehnel

Habitat

Grows always on dead wood in frondose wood.

Habit

Sporophores large, stipitate, umbrella like, finally flattened, white, fuming light brown with age, felty patches from the rupture of a universal veil and the stipe has an ephemeral annulus. Any rain soon remove this velar remnants. The subdistant, pearl–white lamellae have fimbriate edges indicating the presence of copious cheilocystidia.

Distribution

Reported from foot hills of Khasi Hills of Meghalaya (Verma *et al.*, 1995).

338. *Oudemansiella indica* Sathe and Deshpande

Habitat

Grows solitary on ground.

Habit

Pileus upto 5 cm in diameter, convex then expanded, depressed in centre, greyish brown, darker at margin, margin entire, incurved. Gills adnexed, unequal, cream coloured, fleshy, distant. Stipe upto 16 cm long, central, concolourous with pileus, exannulate, hollow, fibrose. Spores (13.8-17.7 × 1.0-13.8 μm) subglobose, hyaline, apiculate, inamyloid, smooth.

Distribution

Reported from Maharashtra (Sathe and Deshpande, 1980).

339. *Oudemansiella mucida* (Fries) Hoehnel

Synonym

Armillaria mucida (Fries) Kummer; *Mucidula mucida* (Fries) Fries Patonillard.

Common Name

Porcelain fungus.

Habitat

Sporophores grow singly or scattered on twigs or decayed coconut husk.

Habit

Unmistakable glistening white to greyish caps measuring 3-6 cm, convex then flattened, extremely slimy. Gills white, broad, distant, adnexed. Stem slender, tough, whitish to grey, fibrillose. Spores white, ovate, 13-18×12-16 µm.

Distribution

Reported from Maharashtra (Patil *et al.*, 1995), Kerala (Bhavani Devi, 1995).

340. *Oudemansiella radicata* (Relh. ex Fr.) Singer

Synonym

Collybia radicata Relh ex Fr.

Common Name

Rooting Shank.

Habitat

Sporophores grow in groups. Sometimes solitary on the ground in forests or in groves around trees trunks or hardwood stumps or in association with Oak trees.

Habit

Pileus 3.0-10.0 cm in diameter, sometimes broader, companulate on convex to flattened, sometimes with an umbo which is usually darker in colour, pileus usually whitish or pale tan, occasionally grey to brown, fleshy, surface smooth,often with radially arranged wrinkles or furrows, viscid when moist, sometimes margin slightly raised. Gills rather distant, adnexed, broad white or yellowish, thick. Stipe central, 5.0-20.0 cm long above ground, 4.0-8.0 cm thick, whitish or coloured like the pileus but paler, firm, surface glabrous or sometimes striate or grooved, base of the stipe enlarged and extending in the soil into a long slender root like process. Flesh white and thin.

Distribution

Reported from North West Himalayas (Kumar, 1987), Tamil Nadu (Natarajan and Raman, 1981), Karnal, Haryana (Patil *et al.*, 1995), Lucknow, U.P. (Pathak and

Gupta, 1982), Maharashtra (Sathe and Deshpande, 1979), Khashi Hills, Assam (Butler and Bisby, 1960), Assam, Kurseong, West Bengal (Ghosh *et al.*, 1967).

341. *Panaeolina foenisecii* (Pers. ex Fr.) Marie

Synonym

Panaeolus foenisecii (Fr.) Kuhner.; *Psathyrella foenisecii* (Fr.) A.H.; *Agaricus foenisecii* Pers.; *Prunuluss foenisecii* (Fr) Gray.; *Psilocybe foenisecii* (Fr.) Quelet.; *Drosophila foenisecii* (Fr.) Quelet.; *Cortinarius foenisecii* Schroter.; *Psathyra foenisecii* (Fr.) Bertrand.

Habitat

Grows solitary on ground among grasses.

Habit

Pileus upto 1.0 cm broad, convex, umbonate, surface smooth viscid, greyish orange, margin regular, without velar remnants. Lamellae adnate, upto 5 mm wide, yellowish brown finally turning to darkbrown. Stipe upto 6.0×0.3 cm, cylindrical, surface, smooth, yellowish white. Spores 11.2-14.0×7.0-9.8 µm broadly ellipsoid, distinctly verrucose brown in 10 per cent KOH and water, not discoloured in conc. H_2SO_4. Hyphae without clamp connections.

Distribution

Reported from Tamil Nadu (Natarajan and Raman, 1983).

342. *Panaeolus cyanescens* Berk. and Br.

Habitat

Grows on dung heaps.

Habit

Pileus companulate or conical, slightly fleshy, surface viscid, shining when dry, cuticular cells pyriform. Gills adnate or adnexed, greyish then black. Stipe equal, sometimes slightly tapering at the top, solid, later hollow. Spores lemon shaped, blackish, smooth walled, germ pore present.

Distribution

Reported from West Bengal (Bose, 1920).

343. *Panus conchatus* (Bull : Fries) Fries

Synonym

Lentinus torulosus, Lentinus conchatus.

Habitat

Grows solitary, scattered or occasionally in aespitose cluster on hard good logs and stumps

Habit

Cap 3–9 cm broad, plano convex becoming centrally depressed at maturity, margin wavy to lobed, inrolled when young, incurved at maturity, surface dry, minutely hairy, nearly smooth, or with small flattened scales in age, colour when fresh violet to lilac–brown, soon fading to vinaceous brown or tan, flesh white, tough. Gills decurrent, close, narrow, pallid when young, becoming buff to tan, frequently tinge violaceous. Stipe 2–4 cm long 2.5–1.5 cm thick, solid, tough, pliant, tapering to narrowed base, attachment variable, central, eccentric or lateral; surface pubescent, sometimes weathering smooth in age, concolorous with the cap. Veil absent. Spores 5–7×2.5–3 mm elliptical, smooth, spore print white.

Distribution

Reported from Darjeeling, W.B. (Berkeley, 1856).

344. *Panus tigrinus* (Bull ex Fr.) Singer

Synonym

Lentinus tigrinus (Bull ex Fr.) Fr.

Habitat

Sporophores grow solitary or in groups, on wood.

Habit

Fruiting body lacking gelatinized zone, thick walled cells absent. Pileus 2.0-5.0 cm in diameter, convex when young becoming expanded with a narrow depression in the center when old, tough and leathery, surface adorned with dark brown or biack hairy scales, margin split or wavy in old specimens. Gills crowded, decurrent, white, edge dentate. Stipe white or tan, smooth or sometimes minutely scaly, annulus found in young specimens but disappearing with age. Flesh white, tough. Hymenial tissues poorly differentiated.

Distribution

Reported from India (Butter and Bisby, 1960).

345. *Paxillus involutus* (Batsch ex Fr.) Fr.

Habitat

Grows singly or in groups on damp soil.

Habit

Pileus upto 15 cm in diameter, compact, planoconvex then depressed, shiny yellowish or tawny, smooth, margin strongly involute. Stipe upto 12.8 cm long, central, solid, paler than the pileus. Spores (8-16×6 μm) ellipsoid or oblong, rust coloured.

Distribution

Reported from Kerala (Kaul and Kapur, 1982; Bhavani Devi 1995 and from India by Verma and Upadhyay, 2000).

346. *Paxillus panuoides* (Fr. ex Fr.) Fr.

Synonym

Tapinella panuoides (Fr.) Gilbert.

Habitat

Sporophores grow on *Picea smithiana*.

Habit

Pileus 2-11 cm in diameter, semicircular, sometimes conchoid or resupinate, dull yellow, surface glabrous at maturity. Gills crowded, decurrent, brownish, thin. Stipe very short or absent. Spores yellowish, smooth, thin walled, without germ pore.

Distribution

Reported from India by Bagchee and Singh (1960).

347. *Paxina acetabulum* (L. ex Fr.) Kuntze

Synonym

Helvella acetabula (L ex St. Amans) Quel.

Common Name

Vinegar Cup.

Habitat

Grows on soil under coniferous forest or amongst leaf litter in wood usually on calcarious soil.

Habit

Cap 4–6 cm across, deeply cap shaped, inner surface dark chestnut brown, outer paler and finely downy. Stipe 10–40×20–30 mm, whitish, deeply furrowed and strongly ribbed along the underside of the cup, more or less hollow. Asci 400×20 mm. Spores broadly elliptical, 18–22×12–14 mm, white, cream or yellowish.

Distribution

Reported from Musroorie, U.P. (Thind and Batra, 1957)

348. *Peniophora gigantea* (Fr.) Massee

Habitat

Grows on wood of *Picea morinda* or *Abies pindrow.*

Habit

Sporophores expanded sometimes resupinate, hyaline or white, waxy, cartilaginous, papery when dry, strignose. Hymenium subhyaline, slightly roughened. Spores (4-5×3 μm) hyaline and ellipsoidal.

Distribution

Reported from Uttar Pradesh and Himachal Pradesh (Bagchee and Bakshi, 1954).

349. *Peziza badia* Pers. ex Merat

Habitat

Grows on ground or on charcoal beds in oak forests, either crowded or scattered.

Habit

Sporophores 3-7 cm in diameter, sessile, cup-shaped, reddish to dark brown, margin wavy, incurved, thick and brittle. Outer surface pale brown, rough with reddish pustules. Hymenium dark brown, smooth, concave. Spores (16-20×8-10 μm) ellipsoid, broadly rounded at the ends with a long oil drop within, sub hyaline or pale brown, verrucose with incomplete ridges.

Distribution

Reported from Uttaranchal (Thind and Batra, 1957; Thind and Sethi, 1957) and from India (Dastur,1946; Subramanian,1952).

350. *Peziza succosa* Berk.

Synonym

Aleuria succosa Gill, *Galactinia succosa* (Berk.) Sacc.

Habitat

Sporophores grow on soil under mixed forest of pine and oak.

Habit

Sporophores 3-4 cm in diameter, cup-shaped, later expanded, margin entire to wavy, fleshy and brittle, flesh exudes yellow liquid when broken. Outer surface milky white and smooth. Hymenium smoky brown, smooth and concave. Spores (16-20×8-12 μm) broadly ellipsoidal, hyaline, verrucose with small warts and ridges, uni-or biguttulate.

Distribution

Reported from Uttaranchal (Thind *et al.*, 1957) and from Mussoorie (Thind and Batra, 1957).

351. *Peziza vesiculosa* Bull ex St Amans.

Synonym

Pustularia vesiculosa (Bull ex Fr.) Fuckel.

Habitat

Grows on soil.

Habit

Fruiting body usually gregarious or densely caespitose (clustered), forming a mass of deep cups of irregular shapes; clumps may attain a diameter of upto 15 cm; brown on inner surface; much paler externally, becoming whitish with age, external surface marked by small wart like processes. A very stipe like base often, but not always, present. Spores elliptic, 20-24×12-14 μm smooth.

Distribution

Reported from U.P. and Punjab (Saxena, 1960; Thind and Waraitch, 1983).

352. *Phallus* impudens L.ex. Pers.

Common Name

Common Stinkhorn, sometimes called Witch's egg.

Habitat

Grows in broadleaved and Coniferous woodlands and parks where it is associated with rotten wood. It also occurs in garden and cementries.

Habit

Immature fruiting body is whitish to pinkish, egg shaped and typically 4-6 cm.× 3-5 cm on thè outside is a thick whitish volva also known as the peridium covering the olive coloured gelatinous gleba which contains spore and stinks to attract flies for spore dispersal, within this layer is a green layer which becomes the head of expanded fruit body and inside this there is a white structure called the receptaculum (The stalk when expanded) that is hard but has an airy structure like a sponge. The mature fruit body is 10-25 cm tall and 4-5 cm in diameter topped with a conical cap 2-4 cm high that is covered with a greenish brown slime termed the gleba which is removed at maturity exposing the base yellowish pitted and ridge (reticulate) surface. This has a passing resemblance to the common morel *Morehella esculenta*.

Distribution

Reported from Mahabaleshwar, Maharashtra by Nair and Patil (1978) and Bhide *et al*. (1987).

353. *Phellorina inquinans* Berk.

Habitat

Grows solitary or in groups on the ground, in sandy soil, sometimes on compost heap.

Habit

Sporophores grows underground covered by an universal veil, head of the plant body elevated on a stalk (stipe), dehisces by inrreular cracks at the apex. Preridium 5 cm in diameter, usually capitate, more or less globose and depressed, woody, sometimes flaccid and deciduous. Endoperidium develops from stipe and covers the gleba. Stipe 5 cm long conical, solid and fibrous.

Distribution

Reported from Punjab (Ahmad, 1941), Himanchal Pradesh (Jandaik, 1976a), Delhi, Haryana, Rajasthan (Kannaiyan and Ramasany, 1980; Sharma and Doshi, 1996).

354. *Phellorina strobilina* (Kalchbr.) Kalchbr. and Cke.

Habitat

Sporophores grow solitary or in groups in sandy soil.

Habit

Sporophores 4.5-15 cm in high. Peridium 2.5-6.0 cm high and 0.8-4 cm in diameter. Stipe 2.8 cm long and 0.8-4.0 cm wide. Exoperidium with large, thick scales. Gleba with persistent fascicles of basidia, pale yellow or brown. Spores (5-7.5 µm in diameter) globose or ellipsoidal, sometimes with short hyaline pedicel, covered with warts giving a reticulate appearance.

Distribution

Reported from Punjab (Ahmad, 1941); Punjab, Haryana, Delhi and Rajasthan (Kannaiyan and Ramasamy, 1980).

355. *Phlebonus portentosus* (Berk and Br) Boedjin

Synonym

Boletus portentosus Berk and Br.; *Phaeogyroporus portentosus* Berk and Br. Mc. Nabb; *Polyporus olivaceofuscus* Berk and Br. Mc. Nabb.

Habitat

Sporophores grow solitary to scattered on ground.

Habit

Pileus 8-24 cm diam, fleshy, convex becoming plano-convex, often with shallow depression at the centre, surface olive brown to sepia brown at the center, paler elsewhere. Pileal surface turns violet in NH_4OH. Pileus slimy when wet, otherwise dry, with a non-separable cuticle, smooth and glabrous, margin involute projecting beyond the hymenophore, undulate. Hymenophore tabulate, adnexed to adnate, tubes lemon yellow, upto 0.8 mm diameter, bluing when injured. Stipe 6-17×4-10 cm, central, robust, clavate with a swollen base, solid; surface concolorous with pileus, discolouring "henna" when bruised. Context upto 5 cm wide at the center, spongy, butter yellow, bluing on bruising or cutting when fresh and young. Odour pleasant.

Distribution

Reported from Western Ghats region of Kerala (Vrinda *et al.*, 2000).

356. *Pholiota aurivella* (Batsch ex Fr.) Kummer

Synonym

Pholiota cerifera (Karst) Karst.

Common Name

Golden pholiota.

Habitat

Sporophores grow on the stumps of trees.

Habit

Pileus 5-12 cm in diameter, sticky, campanulate, then convex, finally glabrous, ferruginous at the centre, margin floccose, involute. Gills crowded, light yellow then ferruginous, adnate, broad. Stipe central, equal, curved, rusty brown, fading at the top, annulus not persistent, fibrillose. Spores oblong, smoky brown, smooth.

Distribution

Reported from Darjeeling, West Bengal and Sikkim (Berkley, 1856).

357. *Pholiota destruens* (Brond.) Gillet.

Habitat

Sporophores grow solitary or rarely in groups on the old stumps of Poplar.

Habit

Pileus 6-10 cm in diameter, convex then expanded, young fruit body compact and hard, brown with white scales, surface dry, margin incurved. Gills crowded, brown adnexed, broad. Stipe with bulbous base, solid, scaly and annulated. Spores oval and brown.

Distribution

Reported from Jammu and Kashmir (Watling and Gregory, 1980; Abraham *et al.*, 1984).

358. *Pholiota nameko* (T. Ito) S. Ito *et* Imai

Synonym

Collybia nameko T. Ito; *Pholiota glutinosa* Kawam; *Kuehneromyces nameko* (T. Ito) S. Ito.

Common Name

Nameko or Nami-Sugi-take (in Japan).

Habitat

Sporophores grow on wood.

Habit

Pileus upto 15 cm in diameter, hemispherical, then convex and finally plain, brown, lighter near the margin, glabrous, covered with mucilage, margin entire. Gills adnate, variable length, white or yellow, brown at maturity. Stipe central, equal, solid or hollow, annulate, annulus fugeceous. Spores (5-6×2.5-3 μm) ellipsoidal, sometimes oblong, cinnamon, smooth, with apparent germ pore.

Distribution

Reported from Himachal Pradesh (Dhar, 1976).

359. *Pholiota praecox* Pers (Fr.)

Synonym

Agrocybe praecox.

Habitat

Grows amongst grasses.

Habit

Cap 1.25-2.25 cm across, smooth and convex, with margin incurved for a long time, dotted with few remnants of veil. Gills crowded, broad, adnexed with decurrent tooth; spore powder brown. Stipe tall, hollow at length, somewhat woolly and fibrillose below the ring. Smell and taste mealy.

Distribution

Reported from Kerala by Bhavani Devi (1995) and from Saharanpur (Hennings,1901).

360. *Pholiota squarrosa* (Fr.) Kummer

Synonym

Agaricus squarrosus Fr.; *Dryophila squarrosa* (Fr.) Quel.; *Hypodendrum floccosum* (Schaef.) Overh.

Common Name

Shaggy Pholiota; Scaly Cluster fungus.

Habitat

Sporophores grow on dead wood or on living dicot tree trunk in groups.

Habit

Pileus upto 2.0 cm broad, convex to planoconvex, surface dry, squarrose, scaly, yellowish brown when young, dark yellow at maturity margin regular with velar remnants. Lamellae adnate to decurrent upto 2 mm wide, yellowish orange. Stipe upto 3.0 to 4.0 cm, cylindrical, solid, surface covered with recurved scales, reddish yellow, annulus present, persistent. Stipe tissue consisting of subparallel hyphae 2.8-8.4 µm in diameter. All hyphae with clamp connections.

Distribution

Reported from India (Sathe and Sasangan, 1977), Tamil Nadu (Natarajan and Raman, 1983 b), Jammu and Kashmir (Watling and Gregory, 1980).

361. *Piptoporus betulinus* (Bull. ex Fr.) Karst.

Synonym

Polyporus betulinus Fr., *Boletus betulinus* Bull., *Piptoporus suberosus* L. ex. Murs.

Common Name

Birch bracket, Birch polypore, Razor-stop fungus.

Habitat

Grows on dead trees or dead portion of live trees.

Habit

Sporophores sessile or short stipitate, fleshy or corky when fresh, hard when dry (3-15×3-25×1-5 cm). Outer surface white, covered by whitish, brown or tan pellicle which disappear later, glabrous, without zonation, margin inrolled. Hymenial surface white, pore tubes 0.2-0.8 cm long. Pores circular or subangular, short, minute, more or less thick walled. Spores (4-5 µm long) cylindrical or allantoid, white and smooth.

Distribution

Reported from Assam (Bhattacharya and Baruah, 1953).

362. *Pleurotus citrinopileatus* Singer

Synonym

Pleurotus cornucopiae var. *Citrinopileatus* (Singer) Ohira.

Common Name

Golden oyster Mushroom

Habitat

Grows in clusters hosting a high number of individual mushrooms.

Habit

Fruit body highly fragile, easily breaking if mishandled especially along their cap margin, quickly loose their bright yellow luster subseguet to harvest. It is brilliant yellow coloured. Few mushrooms are as spectacular as this one. Its brilliant yellow color astonishes all who first see it. This species forms clusters hosting a high number of individual mushrooms, whose stems often diverge from a single base. Its extreme fragility post harvest limits its distribution to far away markets. Spicy and bitter at first, this mushroom imparts a strong nutty flavor upon thorough cooking. Primordia are yellow at first, especially from strains kept close to their natural origins. Mycelium is dense on grain. Colonization of bulk substrates at first wispy, only becoming dense well after colonization. This mushroom casts a much finer mycelial mat and produces pale pinkish buff colored spores.

Distribution

Reported from Jorhat (Assam) by Gogoi *et al.* (2000), from Kohima, Nagaland (Verma *et al.*, 1995), and from Tamil Nadu (Shivaprakasham and Kandaswamy, 1983).

363. *Pleurotus columbinus* Quel Apud. Bres.

Habitat

Sporophores grow on dead portion of living *Ficus religiosa*.

Habit

Pileus fleshy, 7.5 cm or more in diameter, flabelliform, bluish grey to ochreaceus in colour, smooth, margin entire, flesh brittle. Gills fleshy, decurrent forked, white, margin entire. Stipe fleshy, short and tapering towards the base or sessile, cylindrical, stuffed. Hymenophoral trama irregular. Basidia claviform, thin walled. Basidiospores hyaline, cylindrical non-amyloid, smooth, thin walled. Hyphal system monomitic, septate and clamped.

Distribution

Reported from Thoubal district, Manipur. The same was reported for the first time from Pune (Maharastra) by Singh *et al*. (2001).

364. *Pleurotus cornucopiae* (Paulet ex Pers.) Rolland

Synonym

Pleurotus sapidus (Schulz apud kalchb.) sacc.

Common Name

Branched Oyster Fungi, Tamogitake.

Habitat

Sporophores grow in large clusters on dead tree trunks or on dead portions of living tree trunks on fencing stumps of *Jatropha gladurilfera* and avecanut palm and sometimes in open places.

Habit

Pileus 5.1-12.7 cm in diameter convex at first depressed when old, tapering downward, colour whitish or yellowish or grey or lilac or brown, firm, smooth, margin often splitting. Gills subdistant, deccurrent, white, brown, cracked. Stipe usually lateral, short, sometimes elongated, 2.5-5.1 cm, long, white or whitish, smooth, firm solid in tufts and several jointed at the base. Flesh white. Hymenial tissues well differentiated.

Distribution

This species was for the first time reported in India from Madhya Pradesh (Graham, 1915). It was collected from Churachandpur district in Manipur (Singh *et al.*, 2001) and from Pallimukku, Trivendrum, Mankompus, Allepey district of Kerala (Bhavani Devi, 1995).

365. *Pleurotus cystidiosus* O. K. Miller

Habitat

Grows on living, not autochthonous trees or solitary on dead wood of an unidentified tree.

Habit

Sporophores solitary to imbricate. Pileus 13-17×8.5-9.5 cm, pleurotoid, surface brown to greyish brown with numerous punctiform squamules formed by surface cracking, more numerous toward the margin; margin entire, slightly to much festooned, hardly involute. Stipe brown to greyish, almost lateral 3.6-5×2.2-3.5 cm, tapered to the base. Context white, 15-28 mm broad, fleshy when fresh, compact, corky when dry. Lamellae white when fresh, yellow when dry, 4-10 mm broad, thinner toward the stipe, decurrent and forming a pseudoreticulum.

Distribution

Reported from Punjab (Garcha, 1984), Tamil Nadu (Natrajan and Raman, 1984).

366. *Pleurotus djamor* (Fr.) Boedijn

Habitat

Grows on dead and decaying angiospermic wood in clusters. Quite common on logs stored in saw mills or plywood factories.

Habit

Pileus white, 2.5-4.0×2.3-4.0 cm, spathulate to flabelliform, brown when dry; glabrous, smooth to touch; margin entire, frequently lobate and imbricate when old. Context 0.5-6 mm thick, always less than 1 mm when dry. Lamellae white, 2-4 mm broad, 1.5-1.75 mm in dry state, decurrent, crowded, smooth, margin entire. Stipe 5-6 mm long×3-4 mm diameter, lateral, smooth, rarely absent. Spore print white.

Distribution

Reported from Meghalaya, Manipur, Tripura, Mizoram and Nagaland (Verma *et al.*, 1995).

367. *Pleurotus dryinus* (Pers. ex Fr.) Kummer

Habitat

Sporophores grow solitary, usually in large numbers on the trunks of deciduous trees.

Habit

Sporophores eccentric or sublateral, whitish, fleshy, firm and hard when dry. Pileus usually 5.0-10.0 cm in diameter, convex to flattened with a central depression, white at first and later becoming buff, surface floccose when young, scales near the centre are triangular, smaller scales usually toward the margin, margin involute. Gills not close, decurrent, white when young and yellowish when old. Stipe lateral or eccentric, 2.0-12.0 cm long, 1.0-2.0 cm thick, firm, tough but fibrous, veil distinct in young sporophores only. Spores white, cylindrical, 8-15 µm in length.

Distribution

Reported from Jammu and Kashmir (Butler and Bisby, 1960; Berkeley, 1856), Kerala (Bhavani Devi, 1995).

368. *Pleurotus eous* (Berk) Sacc.

Habitat

Grows on dead and decaying wood in clusters or in tiers.

Habit

Piles 2.3-6.3 cm broad, imbricate, spathulate to flabelliform thin, sessile with narrow point of attachment to the substratum. Fruit body pink coloured when young, lamellae decurrent, pinkish white to cream, thin, narrow and crowded. Spores white in mass, cylindric, inamyloid, thin walled, 6.8×2.5-3.6 µm.

Distribution

Reported from Meghalaya, Nagaland (Verma *et al.*, 1995), Manipur (Singh *et al.*, 2001), Sikkim (*vide* Singh and Rajarathnam, 1977).

369. *Pleurotus eryngii* (DC ex Fr.) Qnel.

Habitat

Sporophores grow wild on dead roots of umbelliferous plants including *Eryngium* and *Ferula* or occasionally on sandy soil.

Habit

Pileus 4-15 cm in diameter, convex, then expanded and depressed, grey to rusty tawny, cuticle glabrous, margin turned downward. Gills subdistant, decurrent, broad, white to ochraceous, soft. Stipe usually 4 cm long, central, white, stuffed, base fusiform. Spores (7-9×3.5 μm) ovate or oblong, white, granulose.

Distribution

Reported from Jammu and Kashmir (Wassen, 1969; Agha, 1974; Jandaik, 1976; Watling and Gregory, 1980).

370. *Pleurotus euosmus* (Berk.) Sacc.

Synonym

Agaricus euosmus Berk.

Habitat

Sporophores grow in group on wood.

Habit

Pileus upto 4 cm broad, flabelliforfm, pastel red, surface smooth, margin thin, entire, plane. Gills decurrent, unequal, pink, fleshy, crowded. Stipe reduced upto 2 mm × 2 mm, lateral, concolorous with pileus, smooth. Spores (7-10×4-6 μm) ellipsoid, hyaline with pink tinge, thin walled, smooth, apiculate inamayloid.

Distribution

Reported from Karnataka (Sathe and Kulkarni, 1980).

371. *Pleurotus fassulatus* (Cooke) Sacc.

Synonym

Pleurotus cretaceus Man.

Habitat

Sporophores usually grow clustered in groups on dead wood or decaying logs.

Habit

Pileus 3.0-14.0 cm in diameter, plano-convex when young, later flabelliform. Colour whitish or creamish. Surface initially glabrous and rimose when mature. Margin involute when young and revolute at maturity. Stipe more or less eccentric,

4.0-7.0 cm long, white. Gills crowded, decurrent, anastomosing toward the stipe, spores ellipsoid, cylindrical and non-amyloid, 8-12×4.5-6.0 µm.

Distribution

Reported from Madhya Pradesh (Vasudeva, 1960), Jammu and Kashmir (Kaul, 1979; Watling and Gregory, 1980), Uttaranchal (Puri *et al.*, 1981).

372. *Pleurotus flabellatus* (Berk and Br.) Sacc.

Habitat

Grows on ground on dead wood or caespitose on decayed wood.

Habit

Pileus brown in colour, intense towards the margin, surface smooth, margin irregular and incurved. Gills distinctly formed, decurrent, white when fresh and yellow when dry, thin, edge smooth. Stipe eccentric, usually 1-3 cm long, 0.1-0.2 cm thick, white rigid and solid, Spores hyaline, cylindrical, smooth, non-anyloid, point of attachment prominent, 6.5-7.5 × 2.7-3.3 µm. Spore mass white.

Distribution

Reported from West Bengal (Watling and Gregory, 1980; Thakur, 1980). South Andman Islands (Butler and Bisby, 1931). Foothills of Himalayas (Jandaik and Kapoor, 1975), Kerala (Bhavani Devi, 1995), Manipur by Singh *et al.* (2001), Maharashtra (Sathe and Deshpande, 1979), Calcutta, Mysore (Nita Bahl, 1988), Tamil Nadu (Natarajan and Raman, 1981), Meghalaya (Shajahan *et al.*, 1988) and from Madhya Pradesh (Vasudeva, 1960), Jammu and Kashmir (Kaul, 1979; Watling and Gregory, 1980), Uttaranchal (Puri *et al.*, 1981).

373. *Pleurotus ostreatus* (Jacq. ex Fr.) Kummer.

Synonym

Pleurotus solignus (Pers. ex Fr.) Kummer.

Common Name

Oyster mushroom

Habitat

Grows in large numbers on decaying stumps of a variety of deciduous trees, old logs and coconut husk. Two or more grows from a common base, attached to the cap laterally or near the centre. Grows very rarely on living trees.

Habit

Pileus 8.0-20.0 cm or more broad, spathulate to kidney shaped, white grey or sometimes yellowish often drying, surface smooth, margin incurved. Gills not crowded, decurrent, anastomosing at the base, broad. Stipe eccentric or lateral 1.0-3.0 cm long, 0.5-2.0 cm thick, firm sometimes hairy at the base. Flesh white, soft, spongy, 0.5-1.5 cm thick near the stipe. Spores white, oblong, 7.0-10.0 µm long.

Distribution

Reported from Punjab (Garcha, 1984), Kerala (Bhavani Devi, 1995), Jammu and Kashmir, Sikkim, West Bengal, U.P. (Bose and Bose, 1940; Ghose *et al.*, 1974; Kaul and Kachroo, 1974), Gujarat (Moses, 1948), Jammu and Kashmir (Nita Bahl, 1988), India (Anonymous, 1950), Lucknow, U.P. (Ghosh *et al.*, 1974) and from Kashmir (Murrill, 1924).

374. *Pleurotus platypus* (Cooke and Massee) Sacc.

Synonym

Agaricus platypus Cooke and Massee.

Habitat

Grows on *Euphorbia*.

Habit

Pileus 7.5-11.5 cm diameter, convex when young and sometimes depressed slightly at the center with yellowish tinge on drying. Surface greyish to dark cream coloured. Lamellae crowded, decurrent, white in colour. Stipe eccentric to central, 205x1.0 cm. Spores white in colour, 7.5-8.8×3.8-5.0 µm in size, cylindrical in shape. Context white, fleshy and consists of thin walled interwoven hypae.

Distribution

Reported from Manipur (Singh *et al.*, 2001), Jammu and Kashmir (Watling and Gregory, 1980), Tamil Nadu (Natarajan, 1978) and from Western Ghats of India (Anandh and Prakasham, 2000).

375. *Pleurotus pulmonarius* (Fr.) Quel.

Synonym

Agaricus pulmunarius Fr.

Habitat

Sporophores grow on deciduous trees.

Habit

Sporophores medium size, sometimes smaller, usually stipitate. Pileus 27-30×15-17 mm flabelliform, whitish to greyish. Surface smooth, glabrous, shiny, gelatinous. Lamellae decurrent, crowded, 2-3 mm broad, cream to yellow when dry, margin entire. Stipe eccentric to lateral, smooth, 2-10 mm long × 2-6 mm diam. Flesh 2-5 mm thick and less than 1 mm when dry, corky. Spores (8-10×2-3 µm) cylindrical, thickwalled, white.

Distribution

Reported from West Bengal (Chakravarty and Purkayastha, 1955;1976).

376. *Pleurotus sajor-kaju* (Fr.) Sing.

Synonym

Lentinus sajor- kaju Fr.

Habitat

Sporophores grow solitary or in groups occurring on decaying wood.

Habit

Pilues 5-14 cm in diameter, oyster shaped to deeply infundibuliform, often lobed and folded having a coralloid appearance at maturity. Colour white to grey, colour intense towards the margin. Surface smooth. Margin irregular and incurved. Gills decurrent, white when fresh and yellow when dry, thin, edge smooth. Stipe eccentric, 1-3 cm long, 0.1-0.2 cm thick, white, rigid and solid. Spores hyaline, cylindrical, smooth, non-amyloid, 6.5-7×2.7-3.3 µm, white in mass.

Distribution

Reorted from Foothills of Himalayas, Kerala, West Bengal, Tamil Nadu (Jandaik and Kapoor, 1975; Sivaprakash and Kundaswami, 1980; Sathe and Daniel, 1980; Butler and Bisby, 1931), Punjab (Garcha, 1980; 1984), Kerala (Bhavani Devi, 1995); South Andman Island, W.B., Foothills of the Himalayas (Nita Bahl, 1988).

377. *Pleurotus salignus* (Pers:Fr.) P. Kumm

Synonym

Pleurotus ostreatus f. *salignus* (Pers.) Pilat.

Habitat

Grows on trunks and stumps.

Habit

Pileus horizontal, at length depressed and strigose behind, margin entire, incurved, pale yellowish-brown or dusky, 5–10 cm. across; gills horizontal, not distinctly decurrent, 3–6 mm. broad, distinct behind, branched midway between base and front, crowded, pale, dingy, margin often broken; spores dingy, elliptic-oblong, slightly curved, 8–10 × 3.5µ; stem always very short, firm, downy or strigose.

Distribution

Reported from Sikkim (Berkeley, 1856).

378. *Pleurotus salmoneo-stramineus* L. Vass

Synonym

Pleurotus djamor (Fr.) Boedijin var. *roses* Corner.

Habitat

Sporophores grow in bunches on living coconut tree.

Habit

Mushroom is deep rose initially which faded to white with maturity. Sporophores ranged from 5-8 cm. Gills deep rose in the beginning and later faded to white. Spores white in colour. Stalk very small with strong mushroom flavour. It is commercially exploited because of short spawn period, higher yield and attractive colour (dark blue, white, cream, brown, grey, rose etc.).

Distribution

Reported from Tamil Nadu (Krishnamurthy *et al.*, 1997).

379. *Pleurotus sapidus* (Schulz.) Quel.

Habitat

Sporophores grow on old and decaying prostrate trunks of a variety of deciduous trees. Grows occasionally on dead or dying trees still standing.

Habit

Sporophores 5-8 cm broad, convex or depressed, often irregular, glabrous, white, yellowish or even brownish, Gills decurrent, whitish, often branched and interconnected. Stipe generally short 2.5-5 cm long, tufted, two or more Grows from a common base. Spores oblong, pale-violet, 9-11.5 µm long.

Distribution

Reported from Maharashtra (Patil *et al.*, 1995) and from M.P. (Graham,1915)

380. *Pleurotus spathulatus* (Pess Fr.) Peck

Synonym

Hohenbuehelia petaloids (Bull) Schulzer.

Habitat

Grows solitary on the decayed wood pieces burried underground. It does not usually grow directly from dead wood or cultivated soil.

Habit

The characteristic petaloid habit often makes it look like a shoehorn with gills or a rolleds up funnel. Pileus 2–7 cm across, shoehorn shaped or rolled into a funnel shape; rubbery and moist, fairly smooth but sometimes with fine white fuzz in places; beige to grayish brown or yellow–brown. Gills running down to stipe, close or crowded, whitish, becoming dull yellowish. Stipe present but hard to define precisely, since it is continuous with pileus; brownish above, white and fuzzy below. Flesh whitish, rubbery. Spore print white.

Distribution

Reported from Trichur, Kerala (Bhavani Devi, 1995).

381. *Pleurotus squarrosulus* (Mont.) Sing.

Synonym

Lentinus subnudus Berk

Habitat

Sporophores grow in clusters on the logs.

Habit

Pileus 2-8 cm wide, subinfundibuliform with a deep centric depression, white or cream, flexible when fresh and becoming stiff on drying, scales minute, margin glabrous. Gills crowded, decurrent, distinctiy formed, unequal and separable. Stipe 4-5 cm long, central, whitish then brown, without annulus and volva. Spores (4.2-6.8×3-3.4 μm) oblong-elliptical, hyaline, thin walled and smooth.

Distribution

Reported from West Bengal (Bose and Bose, 1940; Chandra, 1974), Maharashtra (Chopra and Chopra, 1955), Tamil Nadu (Pegler, 1975).

382. *Pluteus atricapillus* (Bat. ex Fr.) Sathe and Deshpande

Synonym

Agaricus atricapillus.

Habitat

Sporophores grow solitary in garden under *Plumaria alba* L.

Habit

Pileus upto 3 cm in diameter, convex then expanded, subumbonate, brownish black in centre, brown towards margin, surface viscid, glabrous but cracked with age, margin striated, inflexed. Gills free, unequal, white then pale pink, moderately crowded. Stipe upto 6.5 cm long, central, cylindric, concolorous with pileus, exannulate, solid, glabrous. Spores (6.3-8×6-6.8 μm) subglobose, pinkish white, non amyloid, smooth, apiculate.

Distribution

Reported from Maharashtra (Sathe and Deshpande, 1980).

383. *Pluteus cervinus* (Shaeff. ex Fr.) Kummer

Habitat

Sporophores lignicolous solitary or scattered on the ground, rotten wood, decaying stumps or logs or on saw dust.

Habit

Sporophores usually centrally stipitate, veli remnants lacking. Pileus 5.0 to 14.0 cm in diameter, campanulate at first, later becoming convex and finally expanded, often with umbo, white, yellowish brown or sooty, fleshy, smooth, sometimes silky and shining with radiating fibrils, surface occasionally scaly, margin upraised in old fruit bodies at times. Gills not crowded, free, at first white, later becoming flesh coloured, broad, round near the stipe. Stipe uniform in thickness or tapering slightly at the top, 6.0 to 15.0 cm long and 0.6 to 2.5 cm thick (at the top), smooth or with scales, solid, colour almost similar to pileus or whitish near the top.

Distribution

Reported from Arunchal Pradesh (Mishra 1999), Gujarat (Moses, 1948), Jammu and Kashmir (Watling and Gregary, 1980), Maharashtra (Sathe and Deshpande, 1979) and from Lucknow, U.P. (Ghosh *et al.*, 1974)

384. *Pluteus conizatus* (Berk and Br.) Sacc.

Habitat

Grows solitary or in qroups on soil in pine forest.

Habit

Sporophores stipitate, umbrella-like, without ring and volva, cap bell-shaped when young, turning convex to flat at imaturity with radiating cracks, yellow coloured with fine radiating fibrils.

Distribution

Nongstoin and upper Shillong in west and east Khasi Hills, and Nartiang in Jaintia, Hills of Meghalaya (Verma *et al.*, 1995).

385. *Pluteus flavofuligineus* Atkinson

Habitat

Grows solitary on dead organic matter and humicolous soil under different trees like *Melia azadirachta*, *Morus alba*, *Mangifera indica*, *Tectona grandis* etc.

Habit

Pileus upto 5.5 cm broad, convex then expanding, greyish brown, covered with concolourous appressed fibrillose scales, margin regular. Gills free, subdistant to crowded, unequal, not in series. Stipe upto 6.5 cm long, central, cylindrical, solid, white to pinkish, without annulus. Spores (6-7.5×5-6 μm) subglobose to globose, thick walled, smooth, cyanophilic, inamyloid.

Distribution

Reported from Punjab (Kaur and Atri, 2002).

386. *Pluteus subcervinus* (Berk and Br.) Sacc.

Synonym

Agaricus subcervinus B et Br.

Common Name

Coloured Pluteus

Habitat

Grows solitary, caespitose or scattered on decaying wood, saw dust piles around saw mill.

Habit

Pileus 5–9.5 cm. broad, plane, convex to plano–umbonate, white with or without pinkish tints, slight yellow brown at the centre, lucid, glabrous, fleshy, margin inflexed to extended striate context white 3–6 mm thik near the stipe. Lamellae free, flesh colour, or tinged yellow to pink, unequal, edges entire. Stipe central, cylindrical 5–10 cm long 5–14 mm thick at the apex, equal or some what dialated, hollow, white. Clamp connections present.

Distribution

Reported from Trichur, Kerala (Bhavani Devi, 1995) and from South India (Natarajan and Raman, 1983b).

387. *Podabrella microcarpa* (Berk. and Br.) Singer

Synonym

Entoloma microcarpum Berk and Br.; Termitomyces microcarpus (Berk. and Br.) Heim.

Habitat

Grows solitary, occurring in large numbers on termite nests or on soil.

Habit

Sporophores usually small and centrally stipitate. Pileus small with central acute umbo, usually 1.0-2.0 cm in diameter, flattened, pink at the margin and olive brown at the umbo, surface smooth. Transverse section of pileus filamentous. Gills crowded, free, white. Stipe slender 3.0-5.0 cm long, slightly enlarged at the base, white above ground, solid with or without pseudorhiza. Basidia club shaped, 4-spored. Basidiospores hyaline, ellipsoidal, thin walled, smooth, 6.0-7.0×3.0×4.0 µm.

Distribution

Reported from West Bengal (Bose and Bose, 1940; Banerjee, 1947; Chopra and Chopra, 1955), Tamil Nadu (Ram Krishnan and Subramanian, 1952), W.B. (Chakravarty and Khatua, 1979), West India (Natarajan, 1975; Patil *et al.*, 1979), Karnataka (Sathe and Kulkarni, 1980), Kerala (Leelabathy *et al.*, 1983), West Bengal and Tamil Nadu (Nita Behl, 1988), M.P. (Rai, 1997).

388. *Podaxis pistillaris* (L. ex Fr.) Morse

Habitat

Sporophores grow abundantly on sandy soil or in mud, usually under the ground and come out when ripe.

Habit

Sporophores upto 25 cm high, stipitate, head pyriform or rounded. Stipe slender, firm and woody, arising from a volvate base. Experidium dirty white, thin but hard. Endoperidium firm and dehisces by splitting. Gleba copious with reddish brown spore mass. Capillitium aseptate, unbranched and coiled. Spores (7-20×5-15 µm) ellipsoid to subglobose, reddish brown, thick walled, with a prominent germ pore.

Distribution

Reported from Tamil Nadu (Subramanian, 1973), Maharashtra (Nair and Patil, 1978; Bilgrami *et al.*, 1979), Delhi, Haryana, Punjab and Rajasthan (Kannaiyan and

Ramsamy, 1980 and Dey, 1999), Andhra Pradesh, Orissa, Uttar Pradesh (Bilgrami *et al.*, 1979), Chandigarh (Thind and Thind, 1982).

389. *Podoscypha nitidula* var. *warneckeana* (Berk) Pat

Habitat

Grows singly or in groups (in concentric ring) on soil in the vicinity/association of *Saccharum murga*. When grow in closely crowded groups may become confluent and form compound fructifications.

Habit

Sporophores fan to funnel shaped, sessile to substipitate and brownish to white, 0.8-2.2 cm high, 0.25-1.5 (3.0) cm wide, thin, transparent. Pileus when fresh 'pale-pinkish-buff', often ornamented with slightly dark zones. Stipe 0.4-1.0 (2.0) cm high,0.3-2.0 (3.0) mm wide, erect or flattened, especially inflaballate sporophores and varying colours from brown to fawn. Spores thin-walled, hyaline, often monoguttulate, subglobose to very broadly elliptical or ovate, 4.0-5.75 (6.2) × 3.2-4.0 (5.0) mm.

Distribution

Reported from North Eastern Hills (Verma *et al.*, 1995).

390. *Polyporus aquosus* P. Henn

Common Name

Dryadi saddle and Pheasant's back mushroom.

Habitat

Grows on tree trunks.

Habit

Fruit body stipitate. Pileus horizontal, subreniform to flabellate. Icabellin to pallid-coffea colour, margin entire to incised, sublobate, 12–15 cm long, 1.5–2 cm across. Stipe lateral, concolorous with pileus, glabrous, 4 cm long, 1.5–2 cm across. Context pallid; hymenium interrupted with pores. Pores smooth, rounded to angular then lacerate, flaccid, 0.5–1 mm diameter. Spores ablong to ellipsoid, obtuse, hyaline, subflavidus 7–10×4–5 mm.

Distribution

Reported from India (Lloyd, 1912).

391. *Polyporus arcularis* (Batsch) Fries

Habitat

Quite Common on sticks and logs in the woods of sal and other logs, dead and living trees of oaks.

Habit

Pileus coriaceous, tough, convex, subumbilicate, zonate, brown scaly at first, then glabrous, yellowish, pores oblong, thin, entire whitish, fairly large; spores 3×8 µm. It is confusing with *P. brumbalis* from which it can be differentiated in having large rhomboidal pores.

Distribution

Reported from all over India, including Assam and Kashmir (Thind, 1961; Bakshi and Puri, 1978) and from Mussoorie, Uttarakhand (Thind and Chatrath,1959).

392. *Polyporus badius* (Pers.) Schwein

Habitat

Grows on stumps and fallen wood of deciduous trees.

Habit

The pileus round and funnel shaped, very smooth, glossy tawny brown to chestnut. Pores small, whitish, decurrent. Stem narrow, tapered at base. Covered with a blackish crust. Spores white, elliptic 7-8×4 µm.

Distribution

Reported from Meghalaya (Verma *et al.*, 1995).

393. *Polyporus brumalis* (Pers.) Per. Fr.

Synonym

Boletus brumalis Pers.

Common Name

Winter Polypore.

Habitat

Sporophores grow singly or in groups on half buried log of *Shorea robusta*, *Alnus* sp. *Betula* sp. or charred wood.

Habit

Sporophores centrally stipitate or slightly eccentric. Stipe upto 2 cm long, white or dark brown, hairy. Fruit body upto 6 cm in diameter, circular in outline, rigid when dry. Upper surface yellowish brown to brown, minutely scaly, margin thin and incurcved. Hymenial surface white when fresh, light brown when dry, Pores circular or slightly angular. Spores (4.6-6.6×1.5-2.2 µm) oblong, hyaline, slightly apiculate.

Distribution

Reported from Orissa (Vasudeva, 1960) and from West Bengal (Bakshi, 1971; Banerjee,1947).

394. *Polyporus frondosus* Fr.

Habitat

Grows on trunks, sometimes on injured stumps of deciduous trees.

Habit

Pileus greyish or light grey, 0.2-0.7 cm in diameter, fleshy and tough. Stipe short, thick, branched, bearing many overlapping pilli giving a globose appearance. Hymenial surface whitish to yellowish, pore tubes 0.2-0.3 cm long, decurrent, pores angular. Spores (5-7×3.5-5 μm) ovoid or ellipsoid, hyaline, usually smooth.

Distribution

Reported from India by Garcha (1980).

395. *Polyporus montanus* (Quel.) Cost and Duy.

Synonym

Bondarzewia montana (Quel.) Sing.

Habitat

Grows densely, gregarious or aggregated on wood.

Habit

Sporophores 6-25 cm in diameters, stipitate or astipitate, fleshy when fresh, hard on drying, simple or merismoid. Stipe short, stout, extending into tuberous root like underground structure. Upper surface irregular, dark brown, velvety, pubescent. Hymenial surface usually white, dark on drying, pores somewhat angular, thick walled. Spores globose, hyaline and ornamented, 1-2 per mm. Cystidia absent.

Distribution

Reported from India by Bakshi (1971), Bakshi and Puri (1978).

396. *Polyporus picipes* Fr.

Habitat

Grows on dead wood of birch, walnut and Indian oak.

Habit

Pileus upto 20 cm in diameter (usually much smaller); round; convex when young, becoming depressed with age, sometimes funnel-shaped; margin often lobed; surface reddish-brown, becoming blackish with age. Flesh whitish, leathery, rigid on drying. Pore surface white at first, later becoming pale brown; tubes somewhat decurrent on stipe; tube mouths very small, circular to angular. Stipe upto 6 cm long; relatively thin; central or eccentric; black at least on lower half. Spores white, cylindric to ellipsoid, smooth; 6-8×3-4 μm.The rich chestnut-red pileus and the minute (almost invisible) to the naked eye pores readily distinguish it from *P. elegans*.

Distribution

Reported from Jammu and Kashmir (Murrill, 1924), West Bengal (Banerjee, 1947), Himachal Pradesh (Bakshi, 1971; Bakshi and Puri, 1978) and from Mussoorie, Uttarakhand (Thind *et al.*, 1957).

397. *Polyporus rugulosus* Lev.

Habitat

Reported to grow in forest on deed wood.

Habit

Serial section of the sporophore shows minute and regularly spaced pore cap perforations which are of similar dimensions to nuclear membrane pore.

Distribution

Reported from South India, Madras (Lloyd, 1898–1919; Sundararaman and Marndarajan, 1925).

398. *Polyporus sanguineus* Klotzsch

Synonym

Pycnoporus sanguineus (L) Murrill, *Boletus sanguineus* L., *Polystictus sanguineus*, Fr., *Xylometron sanguineus* G Meyer.

Common Name

Orange shelf fungi

Habitat

Grow on ground or on dead trees of *Shorea robusta, Cocos nucifera* and *Phoenix sylvestris*.

Habit

Pileus thin, coriaceous, sessile or spuriously stipitate, imbricate, laterally connate at times, 3–5×4–8×0.4–0.6 cm, surface zonate, finely tomentose to glabrous, bright red, often variegated with yellowish–red zones, fading to pure white in old specimen exposed to sun; margin acute, finely tomentose, yellowish red; context floccose, elastic, yellowish–red, 1–3 mm, thick; tubes annual, very short, bright reddish miniatous, scarcely a mm long, mouths circular to angular, regular, minute 3–5 to a mm, edges thin, firm entire, concolorous with the interior. Spores smooth, hyaline, oblong.

Distribution

Reported from Bombay, Maharashtra (Mundkur, 1938; Uppal *et al.*, 1935) and form Calcutta, W.B. (Banerjee, 1947).

399. *Polyporus squamosus* (Huds.) Per Fr.

Common Name

Scaly Polypore; Dryad's Sadle.

Habitat

Grows solitary on trunks of living trees of Walnut or Elm or on dead stumps of Poplar.

Habit

Sporophores stipitate or substipitate, lateral stalk, black at the base, usually fleshy but corky on drying, 2-5×1-5.5 cm, semicircular, reniform, fan-shaped or infundibuliform. Upper surface pinkish buff, flat or depressed at the centre with broad and conspicuous scales, margin wrinkled on drying. Hymenial surface white when fresh, pinkish buff when dry, pores whitish or cream, angular, pore tubes upto 1 mm long. Spores (9-12×3.9-4.5 µm) oblong elongate, hyaline, apiculate, inamyloid.

Distribution

Reported from Himachal Pradesh (Bakshi, 1956), Darjeeling (West Bengal) and Eastern Himalayas (Bakshi, 1971), North Western Himalaya (Hennings, 1900; Berkeley, 1856) and from Manali, H.P. (Bakshi, 1956).

400. *Polyporus udus* Jungh.

Habitat

Grows on trunks/branches of frondose trees or rotting logs in clusters.

Habit

Sporophores annual greyish-brown, spatulate, leathery with a distinct lateral or central stipe. Pileus conchate to fan shaped.

Distribution

Reported from Manipur (Verma *et al.*, 1995).

401. *Polyporus umbellatus* Pers. ex Fr.

Synonym

Boletus umbellatus Pers.; *Grifola ramosissima* Scop. ex Fr. Murr.; *Polypilus umbellatus* (Pers. ex Fr.) Bandarrew and Sings.

Habitat

Sporophores grow at the base of trees.

Habit

Sporophores stipitate, 7.0-20.0 cm, in diameter, compound, formed by a main stipe with numerous small branches, each bearing a single centrally attached pileus arising from a scleritioid base; pileus whitish to smoky brown, small, round, fleshy, surface fibrillose or smooth. Stipe white, elongated, several stipes of the adjacent pilei united at the base to form a large clump. Context white, thin, hymenial surface white or yellow after drying; pore tubes usually less than 2 mm, pore angular, entire or irregular, thin walled, 2-4 per mm. Spores (9-10×3-4 µm) oblong, hyaline, smooth.

Distribution

Reported from Jammu and Kashmir (Butler and Bisby, 1960; Murrill, 1924).

402. *Poria monticola* Murr.

Habitat

Grows on coniferous timber.

Habit

Sporophores usually evanescent, leathery when fresh, brittle after drying, fruiting body resupinate. Hymenial surface white but turns woody brown later, distinctly pored, pores usually angular but round near the margin, larger when coalesced. Cystidia present, Bosidiospores hyaline, ellipsoid, apiculate, 5.0-6.0 × 3.0 µm.

Distribution

Reported from U.P. (Butler and Bisby, 1960), Western and Eastern Himalayas (Bakshi, 1971) and from Dehradun, Uttarakhand (Bagchee and Bakshi, 1951; Puri, 1955).

403. *Psathyrella candolleana* (Fr.) Maire.

Habitat

Grows on soil among grass, apparently amongst woody debris, gregarious.

Habit

Pileus 10-50 mm broad, convex becoming plano-convex or plane, often with a small umbo, clay pink to clay buff, center being more brownish, becoming pale cream or ivory with age, smooth rarely cracked or split with age, not viscid, flesh-fragile, moist, margin floccose, white, entire, non-striate, incurved beconing split or even upturned in some older specimsns, flesh white, very narrow. Lamellae free, rarely slightly adnexed, close to crowded, easily separable from flesh, 4-6 mm brood, 4 lenghts with entire margin. Stipe 25-70×10-20 mm, central cylindrical, uniform, smooth, striate at very apex, spore print umber with purplish tint.

Distribution

Reported from Sanat Nagar, Srinagar at 1700m (Abraham *et al.*, 1984) and from Tamil Nadu, by Natarajan and Raman (1983a).

404. *Psathyrella coprinoceps* (Berk. and Curt.) Dennis

Habitat

Sporophores grow scattered on buried twigs.

Habit

Pileus 1-2.5 cm diameters, subglobose expanding convex to applanate, surface white in the bud, then mouse grey, hygrophanous, smooth and glabrous. Lamellae adnate, pale brown darkening to fuscous brown at maturity, 2-3 mm wide, crowded with lamellulae of three lengths. Stipe 1.5-2.5 cm×1.2 mm, central, cylindric, slender, hollow, surface white, shiny, smooth and glabrous. Context thin upto 1 mm thick at disc composed of thin walled, hyaline hyphae, inflated to 15 µm diameter with clamp connections. Lamella edge sterile, cheylocystidea crowded. Pleurocystidia absent. Hymenophoral trama regular, hyaline, of parallel hyphae.

Distribution

Reported from Kerala by Vrinda *et al.* (2001).

405. *Psathyrella hydrophila* (Bull ex Merat) Maire

Synonym

Hypholoma appendiculatum Bull.; *Psathyrella condolleana* (Fr.) Maire.

Habitat

Sporophores grow usually scattered or clustered on old tree trumps or logs or sometimes on earth or in groups on litter.

Habit

Sporophores centrally stipitate, very fragile; fruiting body not becoming a black fluid. Pileus 5.0-7.0 cm in diameter, at first ovate, then convex to flattened, usually appears white or brown but pale when old; surface smooth or sometimes covered with numerous, white delicate floccose scales, flirty and thin, margin often elevated, sometimes cracking irregularly or splitting into lobes when old. Gills crowded, distinctly formed, adnate when young but free at maturity, at first white, then brownish with a purplish tinge. Stipe central, 0.4-0.6 cm. thick, white, smooth or with whitish granules at the apex, hollow, veil very delicate, white, distinct only in young and fresh fruit bodies, attached with margin of the cap as fragments but disappears later. Flesh white and thin.

Distribution

Reported from Gujarat (Moses, 1948), U.P. (Hennings, 1901; Butler and Bisby, 1960), Gujarat and U.P. (Nita Bahl, 1988), Kerala (Natarajan and Raman, 1983a).

406. *Psathryrella velutina* (Pers ex Fr.) Sing.

Synonym

Agaricus atrchus Berk; *Agaricus hemisoodes* Berk; *Lacrymaria* sp.

Common Name

Weeping widow.

Habitat

Sporophores caespitose on soil among grass on the way side.

Habit

Pileus 3.0-8.0 cm, broad obtusely conical to broadly companulate, surface initially brownish yellow, often cracking, margin appendiculate with silky brownish yellow or slightly pelar veil remnants, non-striate. Gills adnate, first creamy white then dark brown, crowded with lamellulate of different lengths, edge whitish and often beaded with water droplets. Stipe 7-15 cm × 4-14 mm, central, cylindric, soon hollow; surface dull white with a buff tinge or pale "yellowish brown", longitudinally fibrillose with brownish yellow scales, concentrate towards basal portion, smooth above. Veil fibrillose, leaving remnants on cap margin and an obscure, superior hairy zone on stalk.

Distribution

Reported from Western Ghats of Kerala (Vrinda *et al.*, 2003) from northern part of India by Berkley (1850), Trivedi (1972) and Watling and Gregory (1980).

407. *Ptychoverpa bohemica* (Krombholz) Boundier

Synonym

Verpa bohemica (Kromb.) Sohroet.; *Morchella bohemica* Krombh.; *Morchella bispora* Sor.

Habitat

Sporophores grow scattered on the ground.

Habit

Pileus conical or cylindrical, 1.0-2.0 cm broad, 2.0-3.0 cm long, colour tan or brown, under surface white, margin usually slightly wavy and flaring. Hymenial surface gyrosely furrowed. Entirely fertile, lacking sterile ribs. Stipe cylindrical, 4.0-8.0-1.5 cm, white, surface smooth, hollow or stuffed with loose mycelium. Asci borne on a stalked receptacle. Ascospores lacking internal oil drop and usually two in each ascus.

Distribution

Reported from Jammu and Kashmir by Butler and Bisby (1960) and Cooke (1870).

408. *Pycnoporus sanguineus* (L.) Murrill

Synonym

Boletus reberlam, Boletus sanguineus L., *Coriolus sanguineus* (L) G. Cunn, *Fabisporus sanginneus* (L) Zunitr, *Macroporus sanguineus* (L) Pat, *Polyporus sanguineus* (L) Fr. Zmitr. *Polystichus sanguineus* (L) G Mey., *Trametes cinnabarina* var. *sanguinea* (L) Pilat, *Trametes sanguinea* (L) Imazeki, *Trametes sanguinea* (L) Lloyd.

Common Name

Red fungus.

Habitat

Grows on hardwood logs of various deciduous or ever green trees.

Habit

Pileus thin, coraceous, sessile or spuriously stipitate, dimidiate, conchate or reniform, imbricate, laterally conchate at times, 3–5 × 4–8 × 0.4–0.6 cm., surface zonate, finely tomentose fading to pure white in old specimens exposed to the sun, margin acute, finely tomentose, yellowish–red. Context floccose, elastic, yellowish–red, 1–3 mm thick, tubes annual, very short, bright, reddish miniatous, scarcely a mm long. mouths circular to angular, regular, minute, 3–5 to a mm edges thin, firm, entire, concolorous with the interior, spores smooth, hyaline, oblong.

Distribution

Reported from Vaza (Arunachal Pradesh) by Bisht and Harsh (2001).

409. *Ramaria apiculata* (Fr.) Donk.

Habitat

Sporophores grow on dead wood of *Cedrus deodara* or on decaying palm or at their base.

Habit

Sporophores upto 22 cm long, usually astipitate, brown, paler at the top, pinkish on drying, fleshy, tough, glabrous, highly branched, branches unequal, dichotomous or polychotomous, tips of ultimate branchlets acute and sterile. Flesh pale at the base, concolorous at the top, spores (8-10.5×3-4.5 µm) ellipsoidal or cylindrical, light brown, papillate, verrucose, multi-guttulate.

Distribution

Reported from Himalayan Region (Thind, 1973), Himachal Pradesh (Sharma and Jandaik, 1978).

410. *Ramaria aurea* (Fr.) Quel.

Synonym

Clavaria aurea Fr.

Common Name

Golden Coral.

Habitat

Sporophores grow on soil under oak forest.

Habit

Sporophores upto 9.5 cm high, stipe may be present or absent, orange with yellow apices when fresh, reddish brown on drying, branched, flesh brittle, smooth. Stipe tapering downwards, white, branched, cylindrical, polychotomous below, dichotomous above, tips obtuse or blunt. Flesh firm, white. Spores (8.8-10.4×3.2-4 µm) ellipsoid-elongate, light brown, papillate, aguttulate, minutely verrucose.

Distribution

Reported from Uttranchal (Thind, 1961) and Himachal Pradesh (Sharma and Jandaik, 1978).

411. *Ramaria botrytoides* (Pk.) Corner

Synonym

Ramaria botrytis (Fr.) Ricken; *Clavaria botrytis* (Pers.) Fr.

Common Name

Red tipped *Clavaria*.

Habitat

Grows on the ground in woods or in open places.

Habit

Sporophores usually erect thick stem, branching into a coral like head, stipitate, 5.0-12.5 cm high and 520 cm across, massive, at first white, then ochraceous with red tips. Stipe stout, short, thick, fleshy with numerous elongate branches, branches usually crowded like cauliflower, tips reddish, surface rugose wrinkled, flesh white,moderately brittle. Basidia 2 or 4 spored.

Distribution

Reported from Khasi Hills, Assam by Thind (1961), from Himalaya in India by Sharda *et al*. (1997) and from East Shillong in East Khasi Hills of Meghalaya (Verma *et al.*, 1995).

412. *Ramaria brevispora* var. *albida* (Corner) Thind and Dev.

Habitat

Grows singly on floor of coniferous and mixed forests, on pine needles, twigs etc.

Habit

Fructifications profusely branched, erect with tough pale brown flesh, bodies with distinct acute, light yellow tips fasciculate.

Distribution

Collected from upper Shillong in East Khasi Hills of Meghalaya by Verma *at al.* (1995) and from Mussoorie, Uttarakhand (Corner *et al.*, 1958).

413. *Ramaria calobrunnea* Corner Thind and Anand.

Habitat

Sporophores grow on soil in Oak forest.

Habit

Sporophores densely branched, light yellow in colour with tough flesh and stout stipe.

Distribution

Reported from Meghalaya by Verma *et al.* (1995) and from Mussoorie, Uttarakhand (Corner and Anand,1956).

414. *Ramaria flava* Quel.

Common Name

Yellow Coral Fungus. Local name–Changle.

Habitat

Grows singly on the ground occasionally on very rotten wood in open deciduous woods, especially beach in open coniferous forest on mountainous areas.

Habit

Fruiting body with many short branches from a large base, 10-20 cm high, 6-15 cm thick. Branches cylindrical, erect, crowded, divided at the tips in to tooth-like points. Base white, becoming reddish-brown when bruised or with age, branches bright sulphur yellow, becoming brown with age. Flesh white, soft and slastic.

Distribution

Reported from Himalayas, India (Thind, 1961) and from Garhwal Himalayas (Sharda *et al.*, 1997).

415. *Ramaria flavobrunnescens* (Atk.) Corner

Synonym

Clavaria flavobrunnescens (Atk.).

Habitat

Grows singly or in groups on coniferous forest floors.

Habit

Sporophores upto 20 cm high, profousely branched, branching upto six times, smooth, glabrous,light to deep golden yellow when young, becoming tan with age, with branching thick, base short, thick, white. Flesh white, does not obtained when bruised. Spores yellow to light ochraceous; subcylindrical; minutely roughened; 9-12x3.5 µm

Distribution

Reported from Meghalaya and Manipur (Verma *et al.*, 1995), from Himalaya, India by Thind (1961) and Sharda *et al.* (1997) and from Mussoorie, U.P. (Thind and Sukhdev,1957).

416. *Ramaria flavobrunnescens* var. *aurea* Corner

Habitat

Grows on soil under oak forest, *Rhododendron* or under mixed forests of angiosperm as well as *Pinus*. It may also grow on humicolus soil in pure *Quercus* forest.

Habit

Sporophores erect, small to medium 11×7 cm, base white orange yellow in centre, unchanging on bruising, trunk 2.5 cm long, 2 cm broad, solid, smooth branching polychotomous, profuse, compact, axils narrow, tips minute, clustered flesh light yellow, fleshy fibrous. Hyphal system monomitic, context hyphae inflated, clamped, acynophilous. Basidia clavate, 4 spored.

Distribution

Reported from Uttarakhand by Thind and Dev (1956) and Thind (1961), from Nainital (Khurana, 1977), Shimla, H.P. (Sharma *et al.*, 1978) and West Kemeng in Arunachal Pradesh in Eastern Himalayas (Sharda, 1983), from Garhwal Himalayas (Sharda *et al.*, 1997) and from Mussoorie, Uttarakhand (Thind and Sukhdev, 1957).

417. *Ramaria flavobrunnescens* var. *formosoides* Corner

Habitat

Grows on soil under mixed forest as well as under coniferous forest in association with *Abies* and *Picea* or under predominantly angiospermous forest sometimes under angiospermous forest dominated by oak alongwith *Rhododendron arborium* and *Myriad nagi*.

Habit

Sporophores small to medium measuring upto 14×6 cm high, white, pale yellow to light yellow, unchanging on bruising, branching profuse polychotomous, close type, tips minute, acute to subacute, fleshy fibrous. Hyphal system monomitic. Hyphae inflated upto 14 mm, clamped. Basidia 4 spored.

Distribution

Khurana (1977) reported its occurrence in India from Punj Pulla, Delhousie (H.P.) and Gulmarg, (J.K.). Subsequently Sharda (1983) collected it from Thimphu, Bunakah in Bhutan.

418. *Ramaria flavobrunnescens* var. *typica* (Atk.) Corner

Habitat

Sporophores grow solitary gregarious and in close association on ground under oak forest or under angiospermous forest or under *Quercus* forest.

Habit

Sporophores small to medium measuring upto 9.5×6.5 cm, white at base, light yellow in the centre, darker at tips, unchanged on bruising, indistinct trunk (stipe), branching polychotomous below, dichotomous above, profuse, irregular, tips acute. Flesh whitish, yellow, solid, texture chalky, fragile, hyphal system monomitic, context hyphae inflated, clamped, ornamented, ampullaeform swellings thick walled acynophilous. Basidia clavate, guttulate, 4-spored, cynophilous.

Distribution

Reported from India by Thind and Dev (1956) from Mussoorie Hills (Uttaranchal), Himachal Pradesh by Khurana (1977), Arunachal Pradesh by Sharda (1983) and from Garhwal Himalayas by Sharda *et al*. (1997).

419. *Ramaria formosa* (Fr.) Quel

Synonym

Clavaria formosa Fr.

Common Name

Pale yellow clavaria.

Habitat

Sporophores grow on ground. It is a wide spread terrestrial species.

Habit

Fruiting body upto 20 cm high, with massive base, branching upto six times from base, with lower branches thick, tips rounded or pointed, base white to brownish white, branches pinkish or salmon colour to yellowish tan with age, tips yellow when young. Flesh bruises brownish. Spores dull yellow, ellipsoid, ornamented with large warts, 9-12 × 4.5-6 μm.

Distribution

Reported from Mussoorie Hills by Thind (1961).

420. *Ramaria holorubella* (Alk) Corner

Synonym

Clavaria holorubella Atk, *Clavaria rufescens*–sensu Coker, *Claveria botrytis* sensu Kauffm.

Habitat

Grows singly or in groups on soil in coniferous forest.

Habit

Sporophore yellowish white, looking like the head of ripe couliflower at first sight, 6–16 cm by 8–20 cm. with tips that are very small, short, bifid, rosy or amethyotine, confluent in bigger and bigger branches and finally in a single trunk, thick, short, fleshy, rounded at the base, whitish then ochraceous. Flesh white, firm, brittle. Spores pale ochre.

Distribution

Reported from Upper Shillong in Meghalaya (Verma *et al.*, 1995) and from Ranikhet and Dalhousie (H.P.) by Thind and Rattan (1967).

421. *Ramaria obtussissima* (Pack.) Corner

Synonym

Clavaria sanguinea Coker

Habitat

Grows on ground or logs and stumps under oak forest and *Rhododendron* forest, *Pinus* forest. This species also grows near *Quercus leucotrichophora* in a predominantly angiospermous forest.

Habit

Sporophores large, massive, solitary, measuring upto 18×9 cm, with white base, yellowish white in centre, tips darker, branching irregular polychotomous, axils cornute or lunate, tips acute or subacute, in pairs. Flesh white, flesy and fragile. Monomitic, context hyphae upto 16.4 μm wide inflated, clamped, ornamented, ampullaeform swelling acynophilus. Basidia 4-spored.

Distribution

In India reported for the first time from Chakrata toll, Mussoorie (Uttarakhand) by Thind and Dev (1956), Thind and Anand (1956) and subsequently reported from Kilbury, Nainital (Uttarakhand) by Khurana (1977). Later it was reported from Shillong (Meghalaya) by Sharda (1983) and from Garhwal Himalayas by Sharda *et al*. (1997).

422. *Ramaria sandaracina* Marr. And Stundtz

Habitat

Grows singly in *Pinus* forest and under angiospermous forest.

Habit

Sporophores small to medium measuring upto 11×5 cm, base whitish yellow, clear, orange in centre, trunk indistinct, branching profuse, polychotomous irregular, axis U-shaped, or even lunate, acute tips. Hyphal system monomitic context hyphae inflated, upto 19 μm wide, clamped, cyanophilous. Basidia clavate, cyanophilous.

Distribution

In India reported for the first time from Shillong, Eastern Himalayas by Sharda (1983). Subsequently reported from Garhwal Himalayas (Sharda *et al.*, 1997).

423. *Ramaria songuinea* (Cocker) Corner

Synonym

Clavaria sanguinea Coker

Habitat

Grows scattered on the ground under deodar forest or sometimes under mixed forest.

Habit

Sporophores upto 14 cm high, astipitate, yellow but pale red upwards, brown when dry, fleshy, glabrous, highly branched, showing a cauliflower like appearance, branch tips blunt or sulcate. Flesh generally white, reddish brown when bruised. Spores (8.8-11.2×3.6-4 μm) ellipsoidal, light brown, smooth, aguttulate, papillate.

Distribution

Reported from Uttaranchal (Thind, 1961), Himachal Pradesh (Sharma and Jandaik, 1978), Meghalaya (Verma *et al.*, 1995) and from Mussoorie Hills (Thind and Sukhdev, 1957).

424. *Ramaria stricta* (Fries.) Quelt

Synonym

Clavaria stricta Fr.

Common Name

Upright Ramaria or Closed Clavaria.

Habitat

Sporophores grow gregarious on rotting wood, dead wood or even on stump.

Habit

Fruiting body upto 14 cm high, branched repeatedly with branches slender, vertical, compact and light tan to vinaceous-brown. Stipe short or absent. Spore cinnamon buff, ellipsoid, minutely ornamented, 7-10×4-5 µm.

Distribution

Reported from different localities of Kerala by Bhavani Devi (1995), from Shimla, H.P. by Sharma and Jandaik (1978) and from Mussoorie Hills (Thind and Ahmad, 1956).

425. *Ramaria subaurantiaca* Corner

Habitat

Grows on soil amid mosses in very beautiful clusters and probably in mycorrhizal association with *Eucalyptus*.

Habit

Carpophore upto 120 mm high, orange coloured to pinkish orange which does not alter with age, extensively branched, branches forming coralloid clump or tuft, 5–8 (12) × 6–8 (–10) cm, smooth, solid, ultimate tips usually ending in one or two short, tapered, blunt process, dry. Flesh white, thick at base, and brittle. Stipe non or very reduced and if present whitish. Spores 6–10×4–6 mm, ellipsoid to broadly ellipsoid, finely rough, light brown, spore print ochre–brown.

Distribution

Reported from Mussoorie, U.P. (Thind and Sukhdev, 1957).

426. *Ramaria subbotrytis* (Coker) Corner

Habitat

Grows isolated under oak forest *Quercus* and *Cedrus* as well as under predominantly broad leaved forest.

Habit

Sporophores erect, small to medium measuring 9-12x7 cm, extreme base white, pinkish in centre, tips clear pink, not changing on bruising, trunk solid, branched, polychotomous, lax, tips blunt, bifurcated. Hyphal system monomitic, inflated hyphae upto 16.4 µm clamped. Basidia clavate.

Distribution

In India reported for the first time from Chakrata toll, Mussoorie (Uttaranchal) by Thind and Dev (1956) subsequently reported from Lover'·walk, Dalhousie (Thind and Rattan (1967), Dalhousie (Khurana (1977) and from Garhwal Himalayas by Sharda *et al*. (1997).

427. *Rhodocybe paurii* Baroni, Moncalvo, Bhatt and Stepknson

Habitat

Grows on stumps in patches in forest dominated by *Cedrus deodara* and *Cupressus torulosa*.

Habit

Pileus 1-4 cm broad, conchate or reniform, sometimes spathulate, layered in shelving clusters with several pilei per cluster, opaque, finely tomentose, dark grey with a faint purplish hue, context thin. Lamellae adnate or short decurrent, close, brownish when young and become paler with age. Stipe very short or absent, eccentric or lateral, 0.3-0.5 cm long spores (5.1-6.3 × 4.2-5.4 μm) subglobose or globose, fleshy brown, walls weakly to moderately undulate pustulate, thin.

Distribution

Reported from Uttaranchal (Moncalvo *et al.*, 2004)

428. *Rhodophyllus abortivus* (Berk. and Curt.) Singer

Synonym

Entoloma abortivum

Habitat

Sporophores grow in clumps on the soil.

Habit

Pileus upto 12 cm in diameter, hemispherical, then convex, then infundibuliform or flat, surface dry, smooth, silky, pale when young thin greyish white, margin involute. Stipe 5-8 cm long, greyish purple, solid, fibrous. Spores (7.5-10×6.5-7 μm) irregular.

Distribution

Reported from Kerala (Kaul and Kapur 1983), from Kashmir (Watling and Abraham 1992) and from Vellayani, Kerala (Bhavani Devi, 1995).

429. *Rigidiporus ulmarius* (Saw ex Fr.) Imaz.

Synonym

Fomes geotropus Cooke., *Polyporus* (*Fomes*) *geotropus* Cooke.

Habitat

Sporophores grow solitary, sessile to effuso, reflexed or imbricate on dead wood of coniferous trees and sometimes on hard woods.

Habit

Sporophores soft and flesy when fresh, hard and woody when dry 6.0-15.0 cm in diameter, applanata to ungulate forming multistratous tube layer. Upper surface usually whitish but ochraceous to pale tan or pale bay pubescent later forming grey to greyish brown nodulose crust, margin thick and incurved. Context whitish or buff when fresh, ochraceous on drying, firm, corky to woody when dry. Pore tubes usually

0.5-1.5 cm long, orange yellow, not stratified. Pores very small circular to angular, entire, 5-8 per mm, pore wall thin. Basidia 8.0-9.0 µm in diameter.

Distribution

In India reported from Mundali, Chakrata, Mussoorie, Uttaranchal (Vasudeva, 1962; Nita Bahl, 1988); temprate region of Eastern and Western Himalayas (Bakshi, 1971).

430. *Russula adusta* (Pers) Fr.

Habitat

Grows on humicolous soil in mixed forest of *Quircus incana* and *Cedrus deodara*.

Habit

Pileus upto 13.5 cm broad, young pilli covex with depressed centre and involute margin, finally expanding to a plane depressed disc with raised margin, surface greyish brown, leathery, cracking along the margin, margin irregular, non-striate. Gills subdecurrent, close, unequal, not in series. Stipe upto 8×3 cm, central, slightly broad at the base, pruinose, whitish, solid. Spores (6-9×6-7.5 µm) broadly ellipsoid, subglobose or obovoid, ornamented, apiculate, amyloid.

Distribution

Reported from Uttaranchal (Atri and Saini, 1990).

431. *Russula aeruginea* (Lind. Fr.)

Common Name

Grass green Russula.

Habitat

Sporophores grow on solitary or in groups in mixed woods and coniferous forests.

Habit

Pileus 3-7 cm. in diameter, convex when young, becoming expanded and centrally depressed with age; margin often slightly striate; surface viscid when wet, velvety when dry dark or grass green or greenish grey. Flesh thick; white to greenish. Gills adnate to nearly free; white to cream color. Stipe 3-5 cm in length; moderately thick, smooth, white, firm. Spore pale yellow, subglobose, lightly ornamental with small warts connected by lines. 7-9 × 5.5-7 µm.

Distribution

Reported from Kerala (Bhavani Devi, 1995), Jammu and Kashmir (Gardezi and Ayub, 2003).

432. *Russula alachuana* Murr.

Habitat

Sporophores grow scattered on humicolous soil under *Quercus incana* and *Rhododendron arboreum*.

Habit

Pileus upto 2 cm broad,convex to infundibuliform at maturity, dull liliac, leathery, striate, margin dry punctate with yellowish tinge when young and dark purple at maturity. Stipe up to 4.5×1.5 cm, white, tough, stout, slightly tapering below, fleshy, solid. Spores (7.6-9×7-8 µm), short ellipsoid to globose, hyaline, ornamented, amyloid.

Distribution

Reported from Himachal Pradesh (Saini and Atri, 1982).

433. *Russula albonigra* (Krombh) Fr.

Habitat

Grows solitary on humicolous soil in mixed forest dominated by *Quercus*, *Abies* and *Picea*.

Habit

Pileus upto 10 cm broad, plano-convex, soon expanded with depressed centre, surface subviscid, then dry, smooth, dark brown, margin inrolled, irregular, splitting at maturity. Gills adnate to subdecurrent, crowded, unequal, not in series. Stipe upto 6×2.4 cm, cylindrical, central, smooth, solid, white. Spores (7-9×6-8 µm) obovate-ellipsoidal, warty, ornamented, apiculate, amyloid.

Distribution

Reported from Uttaranchal (Atri and Saini, 1990).

434. *Russula atropurpurea* (Krombh) Britz.

Habitat

Grows on humicolous soil under *Quercus incana*.

Habit

Sporophore having pileus with greyish margin and violaceous, yellowish or brownish creamy areas, fairly peeling cuticle, white unchanging mild flesh, pale yellow subdistant mild flesh, pale yellows subdistant equal Rausellae, pale yellow spore cleposil, white slightly greying veined stipe and gelatinrzed. Pileus cuticle with few pilueystidia.

Distribution

Reported from Uttar Pradesh (Chakarata) by Saini and Atri (1989).

435. *Russula alutacea* (Pers. ex Fr.) Fr.

Habitat

Sporophores usually grow on ground in mixed forests or humicolous soil in *Cedrus deodara* forest.

Habit

Sporophores brittle and juiceless (without milk like fluid). Pileus 5.0-10.0 cm or more in diameter, ovate to campanulate, later expanded usually red in colour but pale at the centre when old fleshy, margin thin and striate, tuberculate, sticky when wet. Gills not crowded distinctly formed free equal white when young but ochraceous when mature broad intervenose. Stipe uniformly thick measuring 1.5-2.5 cm thick, white or partly reddish, solid without ring and volva flesh white very brittle.

Distribution

Reported from Jammu and Kashmir (Butler and Bisby, 1960; Berkley, 1816), Himachal Pradesh (Atri *et al.*, 1991).

436. *Russula amethystina* Quel.

Habitat

Grows on humicolous soil in mixed gymnospermous forest.

Habit

Pileus upto 8.6 cm broad, applanate with a depressed centre, surface slightly viscid, violet brown. Gills adnate, equal, forked near the stipe, yellowish white stipe upto 6.7 cm long, almost equal in diameter with slightly bulbous base, solid, smooth. Spores (6.4-8×5.6-6.5 µm) subglobose, warty, apiculate, amyloid.

Distribution

Reported from Himachal Pradesh (Atri *et al.*, 1993).

437. *Russula aurantacea* (Schaetf) Romagn.

Habitat

Grows solitary to scattered on humicolous soil under fir trees.

Habit

Pileus upto 9.5 cm broad, flattened, depressed, dull scarlet, bright rust or apricot, often reddish at the margin and orangy in the centre. Stipe 5-6×1.2-1.7 cm, white or rose tinted, firm. Spores 8-8.7×6.5-7.5 µm, spine upto 1.2 µm high.

Distribution

Reported from Himachal Pradesh (Kaur and Atri, 2002).

438. *Russula aurata* Fr.

Habitat

Grows scattered on humicolous soil in the coniferous forest.

Habit

Pileus upto 7.3 cm broad, convex when young, expanded, depressed at maturity, surface dry, orange to reddish yellow with golden patches, margin irregular, splitting. Gills adnexed, equal, broad, yellowish white. Stipe upto 7.5 cm long, 1.4 cm broad near the pileus and 2.5 cm broad near the base, central, stout, white, smooth. Spores (9-12.5×8-10 µm), subglobose to ellipsoid, ornamented, apiculate, amyloid.

Distribution

Reported from Jammu and Kashmir as well as Uttarakhand (Saini *et al.*, 1989).

439. *Russula baghensa* Gardezi

Habitat

Grows solitary to scattered or gregarious in mixed woods and under conifers.

Habit

Pileus upto 12 cm in diameter, campanulate then expanded, centre broadly depressed, surface viscid and shiny when wet, white to reddish white, margin obtuse but not true rounded. Gills adnate, white to cream coloured. Stipe upto 10 cm long, equal, slightly inflated towards the base, tough, cylindrical, hollow. Spores (5-11×4-7 μm) clavate to ovoid, hyaline, ornamented, amyloid.

Distribution

Reported from Jammu and Kashmir (Gardezi and Ayub, 2003).

440. *Russula brevipes* Peck.

Habitat

Grows on the ground or Pine litter.

Habit

Pileus 8-20 cm in diameter, margin inrolled, centre depressed, minutely felted, dry, white to buffy, staining yellowish brown, with age. Flesh thick firm white. Gills decurrent, thin close, white often with blue-green tinge near stipe, droplets of liquid sometimes on gills when young, becoming stained with age. Stipe upto 8 cm long, 3.5 cm thick, equal or tapering toward base, solid, dull white. Spores white to light cream, ellipsoid to nearly globoge, ornamented with coarse or reticulated 8-10.5×6.5-8.5 μm.

Distribution

Reported from Jammu and Kashmir (Watling and Gregory, 1980; Abraham *et al.*, 1980; 1981), Himachal Pradesh (Atri *et al.*, 1991) and from Garhwal, Himalaya (Bhatt *et al.*, 1995).

441. *Russula chamaeleontina* Fr.

Habitat

Grows solitary to scattered or gregarious in woods, associated mainly with conifers.

Habit

Pileus upto 10 cm broad, ovoid or reniform, not depressed, surface shiny, viscid when moist, thin, purplish to liliac coloured, margin striate. Gills adnate to adnexed, close, white to yellow, free. Stipe upto 8 cm long, equal, slightly thicker at either ends, white to cream, dry, glabrous, longitudinally rugulose. Spores (4.5-8×4-6 μm) obovate to ellipsoidal, ornamented, amyloid.

Distribution

Reported from Jammu and Kashmir (Gardezi and Ayub, 2003).

442. *Russula comobrina* Dupain.

Habitat

Grows solitary among the mosses under the gymnospermous forest composed of Abies and Picea.

Habit

Sporophore upto 6 cm in height, pileus upto 3.5 cm broad, expanded with depressed center, margin pectinate, surface moist, atomate, brownish red, cuticle peeling, flesh white turns reddish on exposure. Lamellae adnate, unequal, lamellulae few, close, broal, turns, pinkish where bruised, edyes smooth, spore print yellowish white. Stipe upto 5.7 cm long 7.8 mm broad, brownish red, whitish near the base, flesh reddish on exposure, surface smooth and solid.

Distribution

Reported from Jamnotri, Hanuman chatti (Atri and Saini, 1990 b).

443. *Russula congoana* Pat

Habitat

Grows on soil in mixed angiospermic forest.

Habit

Pileus caramine red, viscid. Stipe luteus. Smell pungent. Spore print creamy white.

Distribution

Reported from North West Himalaya (Rawala and Sarwal, 1983).

444. *Russula cyanoxantha* (Schacff. Schew) Fr.

Synonym

Russula furcata Fr.

Common Name

The Charcoal burner.

Habitat

Grows singly or scattered on calcarious soil or on humicolors soil in the broad leaved as well as mixed forests dominated by conifers.

Habit

Pileus upto 15 cm in diameter, globose, then convex, finally expanded with a depressed centre, often with various shades of green, slate-grey or purple, lighter towards the margin, viscid when wet, fleshy and firm. Stipe upto 10 cm long, central, white, firm, solid then spongy and hollow. Spores (6.5-9×6-7.5 μm) ellipsoidal, white, apiculate, warted. It does not yield latex.

Distribution

Reported from Sikkim (Berkeley, 1856), Meghalaya (Verma *et al.*, 1995), Himachal Pradesh (Saini and Atri, 1984; Atri and Saini, 1986, 1990; Atri *et al.*, 1991d), Uttarakhand (Atri and Saini, 1990c; Saini *et al.*, 1993).

445. *Russula delica* Fr.

Common Name

Milk white Russula.

Habitat

Grows solitary or in groups in symbiotic relationship with roots of Sal, *Picea, Pinus, Populus, Rhododendron* and *Quercus* in mixed forests or gregarious on humicolous soil in Gymnospermous forest.

Habit

Pileus upto 18 cm in diameter, convex, soon expanded, usually funnel shaped, dull white, sometimes with rusty patches, surface dry, hairy when young, margin inrolled. Stipe upto 6 cm long, cylindrical, sometimes tapered towards the base, surface glabrous or slightly hairy. Spores (8-12×7.9 μm) globose to ellipsoid, covered with conical warts (ornamented), amyloid.

Distribution

Reported from Uttaranchal (Pomerlean, 1951), West Bengal (Jana and Purkayastha, 1983; Das *et al.*, 2002), Himachal Pradesh (Atri *et al.*, 1991d; Atri and Saini, 1996), Jammu and Kishmir (Saini *et al.*, 1988; Saini and Atri, 1984).

446. *Russula densifolia* (Secr.) Gillet.

Habitat

Grows solitary, scattered on humicolous soil under *Picea smithiana* or in the forest of *Betula*; sometimes in mixed conifer forests.

Habit

Pileus upto 15 cm in diameter, convex, then depressed or cyathiform, whitish with brown centre, surface viscid, margin non-striate and inrolled when young. Gills crowded, whitish, subdecurrent and narrow. Stipe upto 6 cm long, concolorous, turning reddish when bruised, hard. Spores obovate and covered with small warts.

Distribution

Reported from Jammu and Kashmir (Watling and Gregory, 1980; Abraham *et al.*, 1981) and West Bengal (Purkayastha and Jana, 1983; Das *et al.*, 2002), NW Himalaya and Batota (Atri and Saini, 1986, 1990b) from Uttar Pradesh, Nainital and Jammu and Kashmir (Atri and Saini, 1990a).

447. *Russula emetica* (Shaeff. ex Fr.) Pers ex Fr.

Common Name

Emetic Russula.

Habitat

Grows solitary, scattered on the ground in the forests or in open places or on rotten wood.

Habit

Sporophores brittle, when broken and milk like fluid lacking. Pileus 4.0-8.0 cm in diameter, convex to companulate when young, expanded and depressed when old. Pink to red when young, turning to pale red with age, surface smooth and shining, slightly sticky when young, margin marked with streaks, finally somewhat sulcate, cuticle easily peeled off. Gills subdistant adnaxed or free, equal, white, broad, inter venose. Stipe cylindrical or narrowed near the top, 4.0 -7.0 x 1.0 -2.0 cm, white spongy but firm when young, fragile with age, solid, without ring and volva. Flesh usually white, red beneath the cuticle.

Distribution

Reported from M.P. (Atri *et al.*, 1997), Assam and West Bengal (Butter and Bispy, 1960; Nita Behl, 1988), Madhya Pradesh (Rai, 1997), North Western Himalayas (Atri *et al.*, 1997) and from W.B. (Berkeley, 1856).

448. *Russula foetens* (Pers.) Fr.

Common Name

Stinking Russula or Foetid Russula.

Habitat

Grows on humicolous soil under *Quercus incana*. It is also common in oak woodlands and coniferous forests.

Habit

Sporophores upto 12 cm high, pileus upto 10 cm broad, convex when young, applanate with a depression in the centre at maturity, margin sulcate, striate, pellucid, surface viscid, sticky when moist, glabrous brown in the centre, light orange along the margin, flesh colour unchanged when cut, latex absent. Lamellae subdecurrent, crowded, rarely forked, broad, lamellulae few, yellowish white, no change on bruising, edges wavy. Stipe upto 10×2.2 cm, slightly tapering downwards, veined, pruinose, fleshy, solid but hollow at maturity, yellowish white, no change when cut. Basidia clavate, 38.0-47.0×7.6-10.6 μm, 2 and 4 spored.

Distribution

First reported in India from Punjab (Saini and Atri, 1981) subsequently collected from Himachal Pradesh (Atri *et al.*, 1997).

449. *Russula fragilis* (Pers. Fr.) Fr.

Habitat

Grows on humicolous soil in the broad leaved forest under the trees of *Rhododendron arborium* and *Quercus incana.*

Habit

Sporophores purplish red expanded depressed pileus with peeling cuticle, white unchanging acrid flesh, faint odour, yellowish white subdistant lamellae, yellowish white spore deposit, white watery fragile stipe, changing pale on bruising, pileus cuticle divisible in to epicutis and subcutis, and the presence of cystidioid elements on the pileus and stipe surface.

Distribution

Reported from Himachal Pradesh (Shimla) by Saini and Atri (1989).

450. *Russula heterophylla* Fr.

Habitat

Grows solitary on humicolous soil in mixed forest of *Quercus, Abies* and *Rhododendron.*

Habit

Pileus upto 10 cm broad, first convex then plane with a depressed centre, surface dry, smooth, greyish green with darker centre, margin irregular, revolute, splitting. Gills adnate. Stipe upto 9.5×2.3 cm, slightly broad below, central, cylindrical, white, smooth, solid. Sporocarp spotted brown thought out. Spores (7-9 × 6-7 μm) subglobose to ellipsoidal, pyriform or comma shaped, warty, ornamented, amyloid, apiculate.

Distribution

Reported from Uttaranchal (Atri and Saini, 1990). And from H.P. (Bhatt and Lakhanpal, 1988)

451. *Russula hullera* Gardezi

Habitat

Grows solitary to scattered or gregarious in mixed woods and under conifers.

Habit

Pileus upto 18 cm in diameter, deeply infundibuliform expanding cyathiform. White to dull white with a pale brown or pale cream tinge, surface slimy, glabrous or pruinate on the margin, margin inflexed, not striate. Gills white then cream, close, subdecurrent, irregular. Stipe upto 8 cm long, white, solid, slightly tapering upwards, subequal. Spores (11-12.5×5-8 μm) echinate, hyaline, ornamented, mostly cylindric but conic-obtuse also.

Distribution

Reported from Jammu and Kashmir (Gardezi and Ayub, 2003).

452. *Russula kashmira* Gardezi

Habitat

Scattered or in small groups in hardwood and conifers.

Habit

Pileus upto 26 cm broad, convex soon expanded often umbonate with incurved margin when young, soon plane to plano-depressed, margin striate, surface viscid and shiny when wet, granular, white to pale yellow. Gills adnate, close to medium, white to cream colour. Stipe upto 7 cm long, equal, surface dry, brittle, glabrous, longitudinally rugulose. Spores (8-11×2-5.5 µm) elliptical, fusiform hyaline, ornamentation reticulate, amyloid.

Distribution

Reported from Jammu and Kashmir (Gardezi and Ayub, 2003).

453. *Russula lepida* Fr.

Habitat

Grows solitary in mixed or coniferous forest on the ground, associated with *Quercus semecarpifolia, Cedrus deodara* and *Abies pindrow.*

Habit

Sporophores usually smaller, centrally stipitate, brittle when broken, latex lacking. Pileus 6.0-8.0 cm in diameter, convex at first, later becoming plane, bright red, becoming pale with age, sometimes whitish near the centre, texture silky, surface not shining, cracking when mature, margin not marked with furrows. Gills crowded, rounded near the stem, usually white, edge red near the pileus surface, forked. Stipe uniform in thickness 8.0 cm. long, 1.0-2.0 cm. thick, white or red, firm solid.

Distribution

Reported from N. W. Himalaya by Atri *et al.* (1997), Darjeeling, Kodaikanal, Tamil Nadu (Butler and Bisby, 1960, Nita Behl, 1988), West Bengal, Darjeeling (Berkeley, 1816), U.P. Chakrata, Deoban (Saini and Atri, 1984), H.P. Dalhousie on way to Ranikhet (Atri *et al.*, 1991d), Shimla (Kumar, 1987), Shailo Mulona Road, Narkanda (Saini *et al.*, 1993), Tamil Nadu, Kodai Kanal (Rama Krishan *et al.*, 1952), Kashmir (Gardezi and Ayub, 2003).

454. *Russula lilacea* Quel.

Habitat

Grows on humicolous soil.

Habit

Pileus upto 6 cm broad, campanulate, surface greyish red with dark brown tone in the centre, margin pellucid. Gills adnate, mostly equal, narrow, smooth, light yellow. Stipe upto 6-6.8 cm long, slightly obclavate, yellowish white, smooth. Spores (8-9.7×5.6-7.5 µm) subglobose, warty, warts pointed, apiculate, amyloid.

Distribution

Reported from Himachal Pradesh (Atri *et al.*, 1993).

455. *Russula lutea* (Huds ex Fr.) Fr.

Habitat

Grow solitary scattered, associated with *Cedrus deodara, Picea smithiana, Pinus wallichiana, Quercus incana* and *Rhododendron arboreum.*

Habit

Pileus 4-7 cm in diameter, convex, applanate at maturity, shallowly depressed at the centre, glabrous, deep yellow to golden yellow, slightly striated plane margin at maturity. Gills adnexed or free, thin, close, equal in length, yellow to orange. Stipe 3-5 cm long, central, cylindrical, equal, smooth, yellowish white. Basidiospores (8-10×7.5-9 μm) broadly ellipsoid to subglobose, yellowish, amyloid and ornamented.

Distribution

Reported from Himachal Pradesh (Bhatt and Lakhanpal, 1988), NW Himalaya (Atri and Saini, 1986).

456. *Russula mustelina* Fr.

Habitat

Grows solitary on humicolous soil in the forest of *Pinus roxburghii.*

Habit

Pileus upto 8.5 cm broad, first convex with inrolled margin, finally applanate with depressed centre, surface moist, slightly viscid, reddish brown, margin regular, involute, feebly splitting. Gill adnate, unequal, not in series, highly forked. Stipe upto 8×4 cm, slightly broad at the base, central, stout, tough, forms rhizomorphs, white. Spores (7-10×6.5-9.6 μm) broadly obovate, warty, apiculate, amyloid.

Distribution

Reported from Himachal Pradesh (Atri and Saini, 1990).

457. *Russula nauseosa* Fr. in Epicr.

Habitat

Grows abundantly in mixed forest of *Pinus wallichiana* and *Picea smithiana,* on humicolous soil in *Cedrus deodara* forest.

Habit

Pileus upto 7 cm in diameter, convex, then depressed, variously coloured with darker or lighter centre, surface shining, delicate, thin and often fragile. Gills crowded, saffron coloured, adnexed, fragile and thin. Stipe upto 7.5 cm long, central, white, smooth, dry, fragile and slightly enlarged at the base. Basidiospores obovate to ellipsoid, covered with isolated spines.

Distribution

Reported from Jammu and Kashmir (Watling and Gregory, 1980; Abraham *et al.*, 1980), Himachal Pradesh (Atri *et al.*, 1991d)

458. *Russula nigrainitallis*

Habitat

Grows solitary to scattered in small groups on soil in ectomycorrhizal relationship with sal tree (*Shorea robusta*).

Habit

Fructification upto 8 cm in height. Pileus upto 7 cm in diameter, plano-depressed to depressed deeply in the centre, black to dark black, surface dry, smooth, tough, margin entire and soft. Gills adnate to decurrent, crowded. Stipe upto 8 cm long, clavate, cylindrical, sometimes tapering downwards, central, solid, glabrous, dry, greyish. Spores (6-8×4-6 µm) spherical to globose, whitish, granulate.

Distribution

Reported from Madhya Pradesh (Rahi *et al.*, 2003).

459. *Russula nigricans* (Bull ex Fr.)

Habitat

Grows solitary on humicolous soil in gymnospermous forest dominated by Abies pindrow spach and *Picea smitheana.*

Habit

Sporophore upto 8 cm in height. Pileus upto 12.5 cm broad, expanded with depression, margin regular, rarely splitting, surface dry, whitish brown, finally coal black, epicutis peeling feebly. Flesh white, turns to reddish and finally black. Lamellae adnate, equal, lamellae not in series, thick, distant, moderately broad, creamy white, then sooty brown from edges inward, turns greyish brown and finally black when bruised, stipe stout, first white, then blackish brown and finally black. Spore deposil white.

Distribution

Reported from Jammu and Kashmir, Gulmarg (2700 on) by Saini *et al.* (1988).

460. *Russula nitida* (Pers.) Fr.

Habitat

Grows scattered among mosses under *Abies pindrow*, on humicolous in *Cedrus deodara* forest.

Habit

Species is characterized by infundibuliform red coloured pileus, white, mild, unchanging flesh, adnexed subdistant yellowish white highly interreined gills, pale orange spore deposit and the presence of dermate cystidia on the pileus and stipe surface.

Distribution

Reported for the first time from India by Atri and Saini (1986), from Himachal Pradesh (Dalhousie) by Atri *et al.*, (1991d) and from Shimla by Saini and Atri (1989a)

461. *Russula ochroleca* (Pers.) Fr.

Habitat

Grows solitary on humicolous soil in the broad leaved forest dominated by *Quercus* and *Rhododendron* trees under *Qincana.*

Habit

Sporophore upto 9 cm in height. Pileus uptp 10 cm. broad, infundibuliform, margin regular, slightly sulcate at maturity, surface dry, rough, leathery, pale orange yellowish, unchanging. Lamellae subdecurrent almost equal, few, some of the lamellae branched, close, broad, cream coloured, unchanging when bruised, edges slightly fimbriate. Stipe upto 3.5 cm. long and 2.5 cm broad near the pileus, slightly tapering downwards, white greyish where bruised, fleshy, first solid then hollow.

Distribution

Reported from Himachal Pradesh, Shimala and Uttar Pradesh, Chakrata by Saini *et al*. (1989).

462. *Russula ochreleucoides* Kauff.

Habitat

Grows scattered on humicolours soil under *Abies pindrow.*

Habit

Fructification upto 8.5 cm in height. Pileus upto 8 cm broad, first convex, finally expanded with a depressed centre, margin pectinate, striate, slightly splitting at maturity, surface moist, slimy, yellow with orange yellow tone in centre; flesh white, unchanging, firms. Lamellae adnate, equal, few forked, closed, broad, yellowish white, unchanging when bruised, edges smooth. Spore deposit yellowish white. Stipe upto 7 × 3.2 cm. slightly broad below, stought, tought, surface veined, brilliant yellow, unchanging when bruised, first solid, then hollow. Surface and flesh turn pinkish in $FeSO_4$.

Distribution

Reported from Shimla, Narkanda, Himachal Pradesh (Atri and Saini, 1990).

463. *Russula olivacea* (Schaeff ex Secr.) Fr.

Habitat

Grows frequently singly or in groups on calcareous soil under conifers or oak trees plantation.

Habit

This handsome mushroom is recognized by its concentrically zoned cap and bright yellow gills. Sporophores attaining a diameter of 35 cm. Pileus brownish-red, olive-brown or even yellow, 6-16 cm in diameter, subglobose, soon becoming flattened and depressed in center. Margin inrolled with concentric cracks when cuticle peels up to one third of the way to the center. Gills finally deep egg-yellow, adnate, broad distant. Stipe stout thicker at base, 5-10 cm long and 1.5-4 cm thick.

Distribution

Reported from NW Himalaya (Atri *et al.*, 1997).

464. *Russula paludosa* Britz.

Habitat

Grows among grasses in mixed forest.

Habit

Pileus upto 8 cm broad, applanate, surface greyish red, margin regular. Gills adnate, equal, close, forked near the stipe. Stipe upto 8.3 cm long obclavate, white. Spores (8-10.5×6.4-8.7 µm) subglobose, warty, warts small, apiculate, amyloid.

Distribution

Reported from Himachal Pradesh (Atri *et al.*, 1992).

465. *Russula parazurea* Schaeff.

Habitat

Grows on humicolous soil in the mixed forest of *Cedrus deodara*, *Quercus incana* and some angiospermic plant.

Habit

Fructification 6-8 cm in height. Pileus more than 6 cm broad, cuticle with highly septate to moniliform projecting elements forming a turf. Stipe 5×2 cm white to yellowish white with pinkish tinge, slightly attenuated towards base, unchanging, surface with projecting cystidial elements. Spores 7.5-9×5.2-7.5 µm.

Distribution

Reported from Himachal Pradesh (Atri *et al.*, 1991; 1997).

466. *Russula paclinoides* Peck

Habitat

Grows on humicolous soil among grasses.

Habit

Sporophore upto 10.3 cm in height. Pileus upto 7.4 cm broad, infundibuliforming margin tuberculale, pellucid surface yellowish brown towards the margin and dark brown in the center, cuticle fully peeling, flesh yellowish grey underneath lamellae adnate, forked near the stipes, moderately broad, smooth edged, yellowish grey. Spore print yellowish white. Stipe upto 7.3 cm long, clavate, surface orange grey, smooth.

Distribution

Reported from Himachal Pradesh, Dalhousie (Atri *et al.*, 1993).

467. *Russula quelitii* (Fr.) Champ.

Habitat

Grows solitary on humicolous soil in mixed forest predomirahed by *Quercus semecarpifolia*, *Abies pindrow*, and *Picea smithiana*.

Habit

Sporophore with plano convex dark ruby pileus, feebly peeling to half peeling cuticte, white unchanging acrid flesh, fruity odour, adnate, subdistant to close, cream coloured lamellae, yellowish white spore deposit, whitish ruby tinted stipe, cuticle divisible in to epicutis and subcutis and the presence of dermatocystidia on the pileus and stipe surface.

Distribution

Reported from Uttar Pradesh (Chakarta) by Saini and Atri (1989).

468. *Russula rosea* Quel.

Habitat

Grows scattered on loamy soil under *Quercus incana*.

Habit

Pileus upto 11 cm broad, first convex with incurved margin then expanded with depressed centre and raised margin, surface moist, adomate, sticky, pale red, margin splitting at maturity, pectinate. Gills adnexed to adnate, close, broad, yellowish white. Stipe upto 10 cm long and 2 cm broad near the pileus and 3 cm broad in the middle and tapering towards the base, white, smooth, solid. Spores (6-7.5×5-7 µm) short ellipsoid, ornamented, apiculate, amyloid.

Distribution

Reported from Jammu and Kashmir, Himachal Pradesh and Uttarakhand (Saini *et al.*, 1989).

469. *Russula sanguinea* (Bull. ex St. Amans) Fr.

Synonym

Russula rosacea (Pers. ex Secr.) Fr.

Habitat

Grows in pine forests.

Habit

Pileus upto 10 cm in diameter, convex then flat and depressed, purplish red or blood red with white patches, surface with radial ridges, margin almost smooth. Gills crowded, cream coloured, narrow, forked or anastomosed. Stipe upto 10 cm long, cylindrical, white, pink or red with yellow or brown spots at maturity, wrinkled or striated. Basidiospores creamy white, warted or spiny.

Distribution

Reported from Sikkim and Uttaranchal (Butler and Bisby, 1931).

470. *Russula vesca* Fr.

Common Name

Bare edged *Russula*

Habitat

Grows in deciduous forests.

Habit

Cap 5-10 cm, convex then expanded, fleshy, firm, colour very variable, usually shades of pale reddish brown, buff, pinkish brown to flesh, olive or almost white, cuticle will also peel to half way. Gills narrow, crowded forking near stipe, white to pale cream. Stipe firm, white, base often pointed, stained rust brown. Spores white, ovate, 6-8 × 5-6 µm.

Distribution

Reported from Kashmir (Gardezi and Ayub, 2003).

471. *Russula virescens* (Schaeff) Fr.

Common Name

Green Cracking *Russula*

Habitat

Grows solitary on humicolous soil in the forest dominated by *Quereus incana*.

Habit

Fructifications upto 10 cm in height. Pielus upto 12 cm broad, expanded with depressed centre, surface dry, greenish white, leathery, margin irregular and splitting. Gills adnexed, equal, close, broad, creamy, edge smooth. Stipe whitish, slightly browning tough, solid with persistent pith. Spores (6-8×5-7 µm) subglobose, ornamented, amyloid.

Distribution

Reported from Uttaranchal (Saini *et al.*, 1988; Saini and Atri, 1984; Atri and Saini, 1986; Saini *et al.*, 1993) and Himachal Pradesh (Atri *et al.*, 1991).

472. *Russula xerampelina* (Schaeff. secr.)Fr.

Habitat

Grows solitary on hunicolous soil in the forest of *Abies pindrowspach*.

Habit

Sporophore upto 7 cm height. Pileus upto 9 cm broad, convex with slight depression, margin irregular, feebly splitting, surface mealy, dull green with almost dark brown to black center, mosic along the margin due to dark and light greenish patches, epictis peeling. Flesh white, mild, pinkish on exposure, dark green in $FeSO_4$. Lamellae adnate, lamellae few, distant broad, pale yellow, edges tinged yellowish.

Distribution

Reported from Jammu and Kashmir by Saini *et al.* (1988).

473. *Sarcoscypha coccinea* (Scop. Ex Fr.) Lamb

Common Name

Scarlet Cup

Habitat

Reported to grow on unidentified rotten stump buried in soil.

Habit

Fruit bodies spherical later shallowly saucer or cup shaped with rolled in rims, 2–5 cm in diameter. The inner surface of the cup is deep red fading to orange when dry and smooth, while the outer surface is whitish covered with dense matted layer of hairs. The stipe when present is stout upto 4 cm long by 0.3–0.7 cm thick and whitish with tomentum. Colour variants of the fungus exist that have reduced or absent pigmentation, these forms may be orange, yellow even white. Spores 26–40 by 10–12 mm, elliptical, smooth, hyaline, having small lipid droplets concentrated at either end. The asci are long and cylindrical and taper into short stipe like base, 300–375 mm by 14–16 mm. Ascospores located in basal parts of the ascus. Paraphyses (hymanium) about 3 mm wide and contain red pigment.

Distribution

Reported from Darjeeling (W.B.) by Kar Chakrabarti (1977).

474. *Sarcodon imbricatum* (Linn. Ex Fries) Karsten

Synonym

Hydnum imbricatum Linn.

Common Name

Imbricated hydnum.

Habitat

Grows on coniferous wood.

Habit

Sporophores gregarious. Pileus 5-15 cm in diameter, stipitate, greyish at maturity, covered with numerous large, brown, overlapping scales, margin involute. Stipe short, central, white or grey, smooth, fibrillose. Spines decurrent, thin, crowded, grey, upto 1.2 cm long. Spores broadly elliptical, yellowish and minutely warted.

Distribution

Reported from Assam (Bhattacharya and Baruah, 1953).

475. *Schizophyllun communae* Fr.

Common Name

Split-gill fungus.

Habitat

Grows in groups on hardwoods, bomboo, tree trunks/branches, stacked timber, etc.

Habit

Pileus upto 4 cm in diameter, semicircular or circular, greyish white, surface tomentose, margin incurved. Stipe often absent, if present, in rudimentary and concolorous. Gills distinct, white or greyish white, radiating from the attachment of fruit body, branched. Spores (5.5-7×2.5-3.5 μm) oblong with obtuse ends, hyaline, smooth, non-amyloid.

Distribution

Reported from Jammu and Kashmir (Watling and Gregory, 1980), Arunachal Pradesh, Meghalaya, Manipur, Mizoram, Nagaland, Tripura and Sikkim (Verma *et al.*, 1995), Maharashtra (Sathe and Deshpande, 1979), from Allahabad, U.P. (Singh *et al.*, 2001) and from Dehradun (Vijayan and Rehill, 1990).

476. *Scleroderma aurantium* Persoon

Synonym

Scleroderma citrinum.

Common Name

Common Earth ball.

Habitat

Grows solitary, rarely gregarious on soils of coniferous as well as frondose wood, quite common on road cuts along the forests. Usually it grows beneath the surface of the ground in a net like manner and is exposed only by erosion, the burrowing of animals or rooting of pigs. Sometimes it comes up above the surface at maturity.

Habit

Fruit bodies 4.0-8.0 cm in diameter, nearly spherical, hard, sessile, dirty peridium, wall thick, olive yellow to brownish with scaly cracks. Gleba greyish to purplish black. Glebal cavity traversed by white tramal plates. Spores are embedded within a matrix, capillitium absent, rudimentary, basidiospores in mass are black with a purple tinge, reticulated and globose, blackish brown, 12-16 μm. Spores escape through cracks in the peridium. Old Sporophores damaged by beetles.

Distribution

Verma *et al*. (1995) collected this species from Thangling hills in south, Kumbi in Central and Senapati in North districts of Manipur, Jowai and Nartians in Jaintia Hills; Mairang and Nangstoin in West Khasi Hills and Umiam and Tdomlsiang in East Khasi Hills district of Meghalaya. Ahmad (1942) reported it from Arnigadh, Mussoorie (Uttaranchal) and Mishra (1999) from forests near Mobo and Gelling, Arunachal Pradesh.

477. *Scleroderma citrinum* Persoon

Synonym

Scleroderma aurantium Persoon *Scleroderma vulgare* Fries

Common Name

Common earth ball

Habitat

Grows singly or gregarious on the ground.

Habit

Sporophore irregular, rounded 4–8 cm across with a thick scaly, cracking outer skin, pale yellowish to tawny orange. The solid interior spore mass is first white, then purplish black with a strong rather, unplesant odors. Spore rounded, 8–13µm with a fine reticulate network on surface.

Distribution

Reported from Trivandrum district of Kerala (Bhavani Devi, 1995).

478. *Scleroderma flavidum* Ell. and Ev.

Habitat

Grows on clay bank.

Habit

Cap width 40 mm, total height 35 mm, colour mustard yellow, surface dry, irregularly roughened, slightly scaly, roughened interior. Shaped like pointed petals–7 points. Interior lighter yellow. Petals egg–yolk colour. Exterior finely cracked.

Distribution

Reported from Khasi Hills, Assam (Berkeley, 1856)

479. *Scleroderma lycoperdoides* Schw.

Habitat

Grows on soil in the forest.

Habit

Fructification gregarious or caespitose, 1.3-4 cm. broad, surface light brown, dotted all over with dark brown separated scales, globose to subglobose sessily flat below and abruptly rooted by a short embedded stalk of varying length which is soon dissipated into few short strands. In some cases pseudo-stalers present. Peridium single, thick when immature, becoming thinner with maturity. Capillitium absent, glebal chambers very small and delicate and straight hyphae constitute the thin plates of the glebal chamber. Dehisceune irregular through the pores on the upper portion of the fructification.

Distribution

Reported from Solan by Sohi *et al.*, (1964).

480. *Scleroderma verrucosum* (Bull.) Pers.

Habitat

Grows solitary or in groups on floor of coniferous and broad-leaf forests or in mixed woods especially on sandy soils.

Habit

Sporophores globular to depressed globose, 4.0–11.0 cm in diameter with a distinct stem like rooting base which binds together a mass of soil, smooth and rounded. Peridium thick, leathery, fragile when dry, dehiscing by a large irregular stroma, look like earth-star, externally ochraceous or umber sometimes tinged purple. The skin is pale brownish with fine wart like scales. Spore mass finally olive-brown. Spore 10-14 µm with spines and ridges.

Distribution

Reported from Manipur and Meghalaya (Verma *et al.*, 1995), Himachal Pradesh and Uttaranchal (Hennings, 1901; Ahmad, 1942), Mussoorie, U.P. (Ahmad, 1942; Cunnigham, 1942; Lloyd, 1904–1919 and Mathur, 1936).

481. *Sepultaria arenicola* (Loveille) Massee

Habitat

Usually grows on ground or sometimes on sandy soil developing a hollow spheric just below soil surface with upper portion protruding.

Habit

Apothecia at first subterranean, becoming completely superficial, larger, usually sessile, cup shaped, soft outer surface covered with long septate undulating hairs; scarcely differentiated from mycelial hyphae. Asci dehiscent. Ascopores elliptical to fusiform, smooth, usually with one or two oil drops, 23.0-28.0×14.0-16.0 µm.

Distribution

Reported from H.P. (Soni *et al.*, 1965a), India (Kaul, 1971).

482. *Sinotermitomyces cavus* Zong.

Habitat

Grows gregarious to scattered on termiteria under *Bougainvilea glabra*. These mushrooms come up in gregarious clusters on underground termitaria.

Habit

Fruictication upto 23.5 cm in height. Pileus upto 2.3 cm in diameter. Surface creamish white, conical convex to campanulate, not opened, viscid, glabrous, non hygrophanous with papillate obtuse umbo; cuticle fully peeling; flesh white, unchanging, upto 5 mm thick lamellae free to adnexed, unequal crowded, bifurcate moderately broad (upto 4 mm broad), white, never exposed, covered with pellicular veil which extends on to the stipe in the form of zig zag patches covering the gills. Spore deposit white. Stipe central upto 3 cm long and 1.6 cm wide, concolour with pileus, leathery, tapering downwards to form upto 20 cm long pesudorrliza with discoidal base, stipe fleshy, first solid then hollow having zig-zag patches or scales on surface.

Distribution

It was collected from Punjabi Univ. campus, Patiala (Atri and Kaur, 2003).

483. *Sparassis crispa* Wulfen ex Fries.

Synonym

Sparassis radicata.

Common Name

Wood cabbage, Coat of wood; Cauliflower fungus; Brain fungus.

Habitat

Usually grows at the base of conifers.

Habit

Sporophores large (upto 50 cm in diameter), erect, densely branched from a short stout rooting stipe, flattened and curled lobes resemble a cauliflower head, whitish but black with maturity. Stipe very short, stout, white then black. Spores pip-shaped and white.

Distribution

Reported from Himachal Pradesh (Thind, 1973) and from Arunachal Pradesh (Bisht and Harsh, 2001).

484. *Steccherinum pulcherrinum* (Berk. and Curt.) Banker

Synonym

Hydnum pulcherrinum Berk. and Curt.

Habitat

Grows on wood.

Habit

Sporophores sessile, semicircular and lobed, fibrous, spongy, hairy. Surface with concentric zones. Margin thin. Hymenium covered with acute awl shaped spines or teeth.

Distribution

Reported from West Bengal (Butler and Bisby, 1931).

485. *Stereum hirsutum* (Willd. : Fr) Gray

Common Name

False Turkey Tail

Habitat

Grows on dead wood, decaying trunks, stumps, branches and logs of *Shorea robusta* and other angiospermic as well as *Cedrus deodara* trees.

Habit

Fruit body resupinate when young, forming thin, leathery overlapping shelves at maturity, 1–3–5 cm wide and up to 8 cm long when fused with adjacent shelves, upper surface hairy, undulate, lobed, banded, orange brown to yellow–brown. Older tissue gray to grayish–brown, lower fertile, surface smooth, orange–buff to pale–buff, if zoned, less conspicuously than the upper surface; flesh 0.5–1.0 mm thick, pliant when young, tough in age, stalk absent. Spores 5.5–7×3–3.5 mm, cylindrical smooth.

Distribution

Reported from Darjeeling, W.B., Mussoorie, Nainital, Khandala (Maharashtra), N.W. Himalaya and Sonmarg, Kashmir (Berkeley, 1856; Hennings, 1901; Lloyd, 1904–1919), Calcutta, W.B. (Banerjee, 1947), Kolhapur, Maharashtra (Prandekar, 1964), H.P. (Bagchee and Bakshi, 1954) and from India, (Anonymous, 1950).

486. *Strobilomyces floccopus* (Vahl. ex. Fr.) Karst.

Synonym

Strobilomyces strobilaceus (Scop. ex Fr.) Berk.

Common Name

Old man of the woods or cone like boletus.

Habitat

Grows scattered on ground or in woods.

Habit

Pileus upto 15 cm in diameter, hemispherical to convex, covered with thick coat of black or blackish brown mycelium which breaks up in to scales with the expansion of pileus, margin with scales and fragments of veil. Stipe tapering at the top, solid, tomentose or scaly, often grooved at the top, ring present. Spores (10-13×8-10 μm) spherical, purple black with reticulations.

Distribution

Reported from Jammu and Kashmir (Murrill, 1924) and Uttar Pradesh (Pathak and Gupta, 1982).

487. *Stropharia coronilla* (Bull ex Fr.) Quet

Habitat

Grows solitary or scattered on the ground in lawns, pastures and grassy woods.

Habit

Cap yellow–ochre, sometimes bright yellow, 2–6 cm in diameter, convex becoming flattened. Margin at first white and sometimes wavy. Cuticle slimy when wet smooth and minutely powder when dry. Flesh white, thick, firm. Gills pale brownish, the dark brown with edges. Adnate but wavy next to the stipe, broad, rather crowded. Stipe stout narrowed at the base, 5–7 cm long, 5–10 mm thick white, becoming yellow with age, minutely wooly above the narrow, radially grooved ring, fibrillose, finally shiny below. Flesh faintly yellow, solid.

Distribution

Reported from Alleppy, Kerala (Bhavani Devi, 1995).

488. *Stropharia delipata* (Pers. ex Fr.) Karst.

Habitat

Grows on soil rich in cow-dung. Sporophores scattered or in groups of 2-3.

Habit

Pileus 8-12 cm in diameter, white or dull white, compact and thick towards centre, smooth and viscid. Gills broadly attached to stipe. Stipe central, white, smooth, firmly attached to pileus, broader at the base, annulate. Annulus white, girdling the stem. Basidiospores (10-14×6-7.5 µm) ellipsoidal and brown.

Distribution

Reported from Himachal Pradesh (Sharma and Thakur, 1978).

489. *Stropharia rugosoannulata* Farlow

Habitat

Grows on partially decomposed cow-dung during rainy season.

Habit

Pileus upto 19 cm in diameter, obtusely conic then obtusely infundibuliform with a depression in the centre, brown, surface glabrous, not viscid, margin inrolled at first. Gills adnate, crowded, creamish white. Stipe upto 25 cm long, central, cylindrical, base swollen, slightly tapering above and thicker again in the upper region, solid, longitudinally striated, floccose, annulus conspicuous. Spores (6-7×11.2-13 µm) broadly ovoid, purple, smooth inamyloid.

Distribution

Reported from Himachal Pradesh (Upadhyay and Sohi, 1987).

490. *Suillus brevipe* (Peck) Kuntze

Habitat

Grows on ground in pinus plantalion solitary near Pillar Rocks.

Habit

Pileus upto 7 cm broad, convex to plano-convex, surface glabrous to glutinous, light orange, margin incurved, veil absent. Tubes adnate, 4 mm deep, 1 mm wide, vivid yellow. Stipe upto 3-1.8 cm, glabrous, solid, vivid yellow spore print colour brownish orange, spore hyaline in 10 per cent KOH, yellow in Melzer's reagent. All hyphae without clamp connection.

Distribution

Reported from Kodaikanal, Tamil Nadu (India) by Natarajan and Raman (1983).

491. *Suillus granulatus* (L. Fr.) Kuntze

Synonym

Boletus granulatus L.; *Lyocopmus granulatus* (Fr.) Quel.

Habitat

Grows solitary or in groups under Pines in mixed stands of *Pinsu wallichiana*, *Cedrus deodara* and *Quercus* and ectomycorhizal with *P. wallichiana* or along pine forest paths.

Habit

Sporophores typical of boletes with a cushion like cup, a central solid cylindrical stipe without a ring and distinct pores underneath; cap first reddish brown or rusty, turning to orange–yellow and slimy at maturity.

Distribution

Reported from upper Shillong and Mairang in East and West Khasi Hills and Nartiang in Jaintia Hills of Meghalaya (Verma *et al.*, 1995) and from Himachal Pradesh (Lakhanpal, 1996).

492. *Suillus placidus* (Bononden) Singer

Synonym

Boletus Placidus Mohl's.

Habitat

Grows scattered to gregarious under pines in mixed canopies of *Pinus walichiana*, *Cedrus deodara* and *Quercus incana* along with some occasional trees of *Pinus roxburghie* associated with *Pinus walichiana*.

Habit

Pileus 20-100 mm broad, convex becoming almost plane in age, surface glabrous to viscid, whitish first then brownish yellow to light yellowish in age, light brownish towards margin when mature. Context whitish, pale yellowish near the tubes, light reddish brown on bruising. Tubes 4-10 mm deep, adnate to subdecurrent, pores boletinoid, roundish, minute first but later boardening to about 1-2 mm, yellowish to olive yellow, glandulate dotted. Stipe 4-10 mm thick, equal, solid, white to pale white. Veil absent. Flesh whitish becoming yellowish in age. Spore was light brown, hyaline to pale in KOH.

Distribution

Reported from Himachal Pradesh, Shimla by Lakhanpal (1996).

493. *Suillus plorans* (Roll.) Singer

Habitat

Grows among fallen needles in mixed coniferous wood/land.

Habit

Pileus cream coloured, white with sulphur yellow or brown tints, context slowly discolouring pink with ammonia.

Distribution

Reported from North West Himalayas (Watling and Gregory, 1980).

494. *Suillus sibircus* (Singer) Singer

Habitat

Gregarious under conifers in mixed forest of *Pinus wallichiana, Cedrus deodara, Picea smithiana, Quercus* sp. and *Rhododendron arboretum*.

Habit

Pileus 2.5 to 1.0 cm in diameter, convex to planoconvex, pale yellowist when young to olive–yellowish in age, context 4-6 mm deep, firm, surface viscid to glutinous, spotted, overall covered by brownish appressed scales; stipe 3.0–8.5×7.0–15.0 mm, solid, pale yellow, then yellow brown, glandular, dotted overall, occasionally with a ring grandulate darken in mature specimens; tubes adnate, turning decurrent, 5-10 mm deep pale yellowish to yellowish, pores radially arranged, angular, 1-2 mm broad, pale yellow to yellowish brown; basidia 4-spored, clavate, basidiospores light brown, 8-11×3.5-4.0 μm, narrowly elliptic, inamyloid.

Distribution

It was collected from Ziro, Arunachal Pradesh by Mishra (1999) from Pahalgam, Kashmir by Watling and Gregory (1980), Himanchal Pradesh (Lakhanpal, 1996).

495. *Suillus subluteus* (Peck) Snell ex Slipp. and Snell.

Habitat

Grows on ground in *Pinus patula* plantation solitary in pine regeneration area.

Habit

Pileus upto 7 cm broad, convex to plano convex, surface glabrous, glalinous, brownish orange, margin incurved. Tubes adnate 3 mm deep, 1 mm wide, vivid yellow. Stipe upto 3 × 1.5 cm, solid with a thick baggy annulus, surface light yellow. Spore print brownish orange basidia turnal by alive in 10 per cent KOH and Melzer's reagent. Spore cluster turns yellowish in 10 per cent KOH and yellowish in Melzers reagent. Context upto 1 cm thick, white at first, turning yellow, gelalinized. Hyphae without clamp connection.

Distribution

Reported from Kodaikanal, Tamil Nadu by Natarajan and Raman (1983).

496. *Termitomyces albuminosus* (Berk.) Heim

Synonym

Collybia albuminosa (Berk.) Petch.

Habitat

Usually grows in termite nests.

Habit

Pileus 9.0 cm in diameter, subconical to expanded, obtusely umbonate, greyish or brownish, fleshy, surface covered with wrinkles. Gills adnexed, white. Stipe attenuated upward, cartilaginous outside, with root like extensions; annulus evanescent, volva absent.

Distribution

Reported from West Bengal (Bose, 1923), Central Province, Berar (Chopra and Chopra, 1955).

497. *Termitomyces badius* Otieno.

Habitat

Grows in groups on grassy soil under *Pongamia glabra* tree.

Habit

Pileus upto 2.3 cm broad, convex with a prominent pointed, brown umbo, surface brownish cream, dry, glabrous, striate on drying margin entire. Gills free, crowded, unequal. Stipe 0.2-0.3 cm broad and upto 3.6 cm long, cylindrical, solid, tapering at both ends, white and without annulus. Spores (6-8.2×4.5-5.2 μm) ovoid or ellipsoid, smooth, thin walled, hyaline and inamyloid.

Distribution

Reported from Tamil Nadu (Natarajan, 1975), Punjab (Atri *et al.*, 2005).

498. *Termitomyces cartilagineus* (Berk.) Heim

Habitat

Grows in groups, always in association with termite nests.

Habit

Pileus upto 12 cm in diameter, convex, then expanded with an umbo, brown. Gills white or cream, free. Stipe upto 12 cm long. Basidiospores ellipsoid, hyaline, smooth walled and non-amyloid.

Distribution

Reported from Western India (Patil *et al.*, 1979).

499. *Termitomyces clypeatus* Heim

Habitat

Grow solitary or scattered on sandy soil or on shaded ground in association with termite nests.

Habit

Sporophores umbrella like, stipitate. Pileus 5.5-7 cm in diameter, conical then expanded with sharply pointed umbo, greyish orange, surface silky. Stipe 7-9 cm long, central, white to dirty brown, fleshy, fibrous, hollow, with pseudorhiza. Basidiospores (5.5–8.5×4–5.5 µm) obovoid to ellipsoidal, smooth, hyaline and thin walled.

Distribution

Reported from Calcutta by Burkill, Kerala by Leelavathy *et al.* (1983), Trivendrum by Wilson and from shillong by Cerma (in Pegler and Vonhaecke, 1994), from Madhyapradesh, Roy (1997) Tamil Nadu (Natarajan, 1975), Uttar Pradesh (Pathak and Gupta, 1982) and Imphal valley, Manipur and Mizorum by Verma *et al.* (1995), from Trivendrum district of Kerala by Bhavani Devi (1995); from Punjab Plains (Atri, *et al.*, 2005) and from Midanapure District W.B. (Das *et al.*, 2002).

500. *Termitomyces eurhizus* (Berk) Heim

Synonym

Rajapa eurhizus (Berk) Sing.

Habitat

Grows on termite mounds usually solitary, sometimes in groups in humus rich soil.

Habit

Sporophores grey to brownish grey, with scales, centrally stipitate, white but olivaceous near the umbonal region. Stipe very long upto 20 cm long, 1.5-2.5 cm thick, solid, tough, white tapering into a hollow and fibrous pseudorrhiza and with a persistent annulus. Pileus 3.0-9.5 cm in diameter, at first convex, later expanded with prominent umbo, scale present on the surface, firm, margin regular, not incurved. Gills crowded, distinctly formed, free to subadnate, pliable, white, entire. Flesh white. Hymenophoral trama truly and strictly regular; hyphae non-amyloid, without clamp connections. Basidia clavate, 4-spored, 22.9-25.5×6.8-7.6 µm.

Distribution

This species was reported earlier from India by Burkill from Tamil Nadu and from Calcutta by Bahatt and Harsh, by Nagchan, from Manipur and Verma from Meghalaya (in Pegler and Vanhaecke, 1994), Atri *et al.*, (2005) from Punjab Plain; from Rajasthan (Doshi and Sharma, 1990). Chandernagore and Midnapore, West Bengal (Purkayastha and Chandra, 1975), Moirang, Nambol, Imphal valley (Manipur), Shillong and Tadomsiang (Meghalaya) and Tadong (Sikkim) (Verma *et al.*, 1995), Punjab Batra and Batra, 1963), Kerala (Leelavathy *et al.*, 1983), Midanapore Distt., W.B. (Das *et al.*, 2002).

501. *Termitomyces globulus* Helm and Goossens.

Habitat

Grows on the ground.

Habit

Pileus upto 19 cm in diameter, subglobose or expanding to plane, with a poorly defined umbonate perforatium, milky coffee to pale sepia, umbo remaining dark, surface smooth, glabrous, margin entire or splitting. Stipe upto 13 cm long, cylindrical, solid, expanding at the base before tapering as black pseudorrhiza. Spores (7.5-8.5 × 4.5-5.5 μm) ovoid to ellipsoid, hyaline, thin walled.

Distribution

Reported from Kerala (Leelavathy *et al.*, 1983).

502. *Termitomyces heimii* Natarajan

Synonym

Termitomyces albigenosus (Berk.) Heim.

Habitat

Grows in groups, in association with termite nests.

Habit

Pileus upto 8.5 cm in diameter, plano-convex with an umbo, whitish grey at the umbonal region, margin incurved and cracking. Gills crowded, whitish and free. Stipe upto 18.5 cm long, central, cylindrical, cream white, stuffed and white. Basidiospores ellipsoidal, hyaline, smooth, thin walled non-amyloid.

Distribution

Reported from West Bengal (Bose, 1923; Purkayastha and Chandra, 1975), Madhya Pradesh (Chopra and Chopra, 1955; Rai, 1997), Tamil Nadu (Natarajan, 1977, 1979) and Karnataka (Sathe and Kulkarni, 1980; Bhavani Devi, 1995). Leelavathy reported this species from Kerala and Rao from Andhra Pradesh (in Pegler and Vanhaecke, 1994). Atri *et al.* (2005) reported it from plains of Punjab and Rai (1997) from Madhya Pradesh.

503. *Termitomyces indicus var. patialenses* (Natarajan) Atri

Habitat

Grows in groups on grassy soil among mosses under *Bougainvillea glabra* hedge.

Habit

Pileus 1.2-4.8 cm broad, convex then planoconvex, creamish white, smooth, surface moist, margin straight, reflexed with maturity. Gills adnexed to free, white then pinkish, moderately crowded. Stipe 3.6-7.8 cm long and upto 0.5 cm broad, solid, with bulbous base and creamy surface. Spores (4.6-6×3.7-4 μm) ellipsoid, smooth, thin walled, hyaline and inamyloid.

Distribution

Reported from Punjab (Atri *et al.*, 2005).

504. *Termitomyces letestui* (Pat.) Heim.

Synonym

Lepiota letestui (Pat.) Heim

Habitat

Grows in close group in the "Termite Nests" during collection the soil is required to be digged carefully to track the pseudorhizal connection with the termite nest.

Habit

Pileus 100-150 mm in diameter, convex to flattened with a blunt cylindric perforatorium, large, fleshy, surface light and coloured, cuticle smooth, viscid with large thin scales arranged in circular fashion, margin fleshy, thin, fissured, striated, involute. Context white, fibrous, 2-6 mm thick at the central region. Lamella free, crowded, equal in length, whitish but turning to pink at maturity, edge entire and smooth. Stipe 100-130x30-50 mm, above ground level, equal, cylindric, long, central, slender, solid at the upper portion, gradually becoming hollow near the base. Sheathing, membranous annulus strongly attached to the upper half of the stipe. The stipe expands 100-110x20-30 mm below ground level and then attenuated to form a radiciform base by which carpophore is burried on the ground of termites honey comb nurseries. Flesh thick, firm, white with pleasant smell. Basidia long, cylindric, fusiform 20.8-31.2x5.2-7.8μm.

Distribution

Reported from West Bengal (Roy and Samajpati, 1981).

505. *Termitomyces mammiformis* Heim

Habitat

Grows in scattered groups on ground in association with epigeal termite maunds or on sandy soil under bamboo tree.

Habit

Sporophores fleshy, centrally stipitate. Pileus upto 7 cm in diameter with a sharply differentiated umbo, margin bent inward. Gills white, crowded, free. Stipe upto 9 cm long, solid, central, persistent annulus present. Flesh white. Basidia club shaped. Basidiospores ellipsoidal, hyaline, smooth and inamyloid. Spore print pink. Hyphae without clamp connections.

Distribution

Reported from Maharashtra (Patil *et al.*, 1979), Uttar Pradesh (Pathak and Gupta, 1982) and Assam, Jorhat (Gogoi *et al.*, 2000), Kerala (Leelavathy *et al.*, 1983; Kaul and Kapur, 1983).

506. *Termitomyces medius* Heim and Grasse

Habitat

Grows scattered in groups on sandy, grassy or humicolous soil in open or under angiospermic plants such as *Tectona grandis*.

Habit

Pileus 1-3 cm in diameter, conical to convex, greyish white with brownish grey centre, glabrous, surface moist, margin incurved and incised. Gills free, unequal, crowded, white to orange white. Stipe upto 8.5 cm long (with pseudorrhiza) cylindrical, solid, white, without annulus. Spores (6.7-7.6×4.5-5.6 µm) broadly ellipsoid, thin walled, smooth, hyaline, inamyloid.

Distribution

Reported from Karnataka (Natarajan and Purshothama, 1986), Punjab (Atri *et al.*, 2005).

507. *Termitomyces microcarpus* (Berk and Br.) Heim

Synonym

Entoloma macrocarpus (Berk and Br.); *E. Intermixum* (Ber and Br.) Sacc.; *Podabrella microcarpa; Agaricus microcarpus* Berk and Br.; *A. intermixtus* Berk and Br.; *A. seiolus* Kalchhr.; *Mycena sciola* (Kalchhr.) Sacc.; *Collybia microcarpa* (Berk and Br.) V. Hohn.; *M. microcarpa* (Berk and Br.) Pat.; *M. temitum* Beeli.; *Gymnopus microcarpus* (Berk and Br.) Van.; *Hygrophorus obrusseus* Singer Berk.

Habitat

Grows in ceaspitose clusters solitary in large number on termites nests or on soil.

Habit

Sporophores 2.5-4.0 cm high. Pileus upto 1 cm broad, convex to expanding with depressed center, surface dry, yellowish white, margin irregular, splitting at maturity, inrolled, cuticle not peeling, flesh upto 0.1 cm thick, yellowish white, unchanging on exposure. Lamellae subfree to adnexed, subdistant, narrow 0.2 cm. deep, unequal, not arranged in series, cream coloured, unchanging when bruised, gill edges serrate. Stipe upto 3.5 cm long, 0.1 cm broad, concolour with pileus, equal in diameter throughout except slightly bulbous base, surface dry, smooth, unchanging when bruised. Veil and pseudorhiza absent. Basidia 16.9-23.1×4.6-5.3 mm, clavate, 4-spored, pigmented.

Distribution

First reported from Tamil Nadu in India (Natarajan, 1975, 1977a), West Bengal (Bose and Bose, 1940; Singer, 1961; Ramkrishnan and Subramanian, 1952, 1952a; Butler and Bisby, 1931 and Das *et al.*, 2002), from Chandigarh, Punjab (Rawla *et al.*, 1983), and from Punjab plain, Patiala (Arti *et al.*, 1995, 2005) Tamil Nadu (Natrajan, 1975; Natrajan and Raman, 1981) and from Karnataka (Sathe and Kulkarni, 1980).

508. *Termitomyces poonensis* Sathe and Deshpande

Habitat

Grows in group on termite nests.

Habit

Pileus upto 21 cm in diameter, hemispheric to plane with prominent dark brown perforatorium, surface glabrous, cracked, viscid when moist, margin cracked, inflexed when young, veil absent. Gills sinuate, unequal, cream coloured, fleshy, crowded. Stipe upto 12 cm long, central, white with black pseudorrhiza (6 cm bellow the ground), exannulate, solid, fibrose, glabrous. Spores (6.4-10.7×3.7-7.5 µm) elliptical, hyaline, smooth, thin walled, apiculate inamyloid.

Distribution

Reported from Maharashtra (Sathe and Deshpande, 1980).

509. *Termitomyces raburii* Otieno.

Habitat

Grows solitary on grassy or humicolous soil under angiosperms.

Habit

Pileus upto 6.5 cm broad, first convex then expanded with prominent umbo, brownish orange with greyish brown centre, smooth and dry surface, margin irregularly lobed, splitting radially. Gills free, crowded and unequal. Stipe upto 9 cm long, cylindrical, solid, whitish, attenuated into long slender dark brown pseudorrhiza, annulus absent. Spores (6-9×3.7-4.5 µm) ovoid, thin walled, guttulate.

Distribution

Reported from Tamil Nadu (Natarajan, 1977), Punjab (Atri *et al.*, 2005).

510. *Termitomyces radicatus* Natarajan

Habitat

Grows on lawn, ground or on clayey soil.

Habit

Pileus 3.5 cm in diameter, convex then expanded to plain, sharp, pointed and perforated umbo, orange-white to orange grey, smooth. Stipe upto 4.5 cm long, equal, cylindrical, yellowish white, solid, tapering below the soil line. Basidiospores (6-9×4-5 μm) cylindrical, hyaline, smooth and non-amyloid.

Distribution

Reported from Jammu and Kashmir (Natarajan, 1977), Kerala (Bhavani Devi, 1995; Pegler and Vanhaecke, 1994) and from Punjab (Atri *et al.*, 2005).

511. *Termitomyces robustus* (Beeli) Heim

Synonym

Schulzeria robusta Beeli.; *Termitemyces fulginosus* Heim.

Common Name

Uppukoon, Mazatthandun, Nilmmulappan.

Habitat

Sporophores grow solitary or scattered in smaller groups of 2-3 on the red laterite soil above termite mounds.

Habit

Sporophores large and robust, cream to brownish grey in colour. Pileus convex 6.2 to 22.0 cm in diameter at maturity with a slight brown perforatorium, pileus brownish cream when young, becoming greyish on maturity, smooth and viscid. The margin of the pileus inflexed when young, flat reflexed and lacerate when old. Gills free to subadnate, white with a decurrent tooth and thickly crowded. Context easily separable from epicutis. Basidia clavate, upto 21.0 μm long and 5.2-7.2 μm broad at the anterior end bearing four sterigmata.

Distribution

Collected from different places throughout the state of Kerala (Bhavani Devi, *et al.*, 1980; Leelavathy *et al.*, 1983), from Imphal Valley in Manipur (Verma *et al.*, 1995) and from Poona and Kerala (Patil *et al.*, 1979., Devi *et al.*, 1980).

512. *Termitomyces striatus* (Beeli) Heim

Habitat

Grows on termite nests and on soil under the shades of coconut plantation or grassy and open fields.

Habit

Pileus upto 7 cm in diameter, convex with prominent umbo, brown, central region darker, margin incurved. Gills crowded, free and orange coloured. Stipe upto 7 cm long, central, concolorous, attenuated below the ground level. Basidiospores (16-21×4-5 μm) hyaline, with a dot like attachment mark (hilum).

Distribution

Reported from Uttaranchal (Butler and Bisby, 1931; Bakshi and Puri, 1978), Chandigarh, Punjab (Rawla *et al.*, 1983), Kerala (Leelavathy *et al.*, 1983; Bhavani Devi, 1995), West Bengal (Roy and Samajpati, 1980, 1981), West Bengal, Birbhum (Gupta, 1984).

513. *Termitomyces striatus* var. *annulatus* Heim

Habitat

Grows scattered on humicolous soil.

Habit

Pileus 5-6.8 cm broad, subglobose to convex to expanded with maturity, pale with pale orange centre, surface moist, glabrous, margin striate, splitting at maturity. Gills free, crowded, unequal, whitish. Stipe upto 8 cm long and 0.4 to 2.6 cm in diameter, cylindrical, solid, whitish, fibrillose below annulus. Spores (6-8.2×4.5-5.2 µm) ellipsoid, thin walled, hyaline and inamyloid.

Distribution

Reported from Punjab (Atri *et al.*, 2005).

514. *Termitomyces tyleranus* var. *macrocarpa* (Otieno) Atri

Habitat

Grows scattered on humicolous soil under *Clerodendron inerme* and *Grevillea robusta*.

Habit

Pileus 4-10 cm broad, convex to campanulate, creamish, surface dry, glabrous, margin cracking radially. Stipe upto 16 cm in length (including pseudorrhiza) concolorous, solid, expanding slightly before entering into the ground. Spores (6.7-9×4.5-5.2 µm) ovoid to ellipsoid, thin walled, guttulate, inamyloid.

Distribution

Reported from Punjab (Atri *et al.*, 2005).

515. *Termitomyces umkowaani* (Cooke and Mass.) Reid

Synonym

Agaricus umkowaani (Cooke and Mass); *Schulzeria umkowaani* (Cooke and Mass) Sacc.

Habitat

Grows solitary to scattered on ground near or over termite nests.

Habit

Pileus 8-15 cm diameter, fleshy, hemispherical expanding convex, and then shallowly convex, finally with an upturned margin with a small pointed to broadly conical perforatorium, surface yellowish grey or greyish yellow or brownish orange, darkening to soot brown or greyish brown at the centre, smooth and glabrous or radially rugose, viscid when moist, margin straight, uneven finally refluxed, splitting with age. Lamellae free, white to orange-white, upto 8 mm broad, venticose, crowded with lamellae of different lengths. Stipe 9.5-13×1-2 cm above ground level, cylindric to compressed, solid expanding to a swollen base (upto 2.5 cm diameter) before tapering downwards to a long, firm, tough pseudorrhiza, striate above and black below soil level. Annulus absent. Context white, upto 12 mm thick at the disc consisting of septate 3-22.5 µm wide hyphae, lacking clamp connections. Spore print flesh pink. Basidia 19.5-27×6.5-9 cm, clavate, bearing 4 sterigmata.

Distribution

Reported from different areas of Western ghats, Kerala (Vrinda *et al.*, 2002).

516. *Trametes cubensis* (Mont.) Sacc.

Habitat

Grows on hard wood stump of *Cocos nucifera, Shorea robusta* and *Eucalyptus maculata* var. *citriodora.*

Habit

Cap 150–250 mm wide, weakly lumpy and zoned, glabrous, clammy to touch, varying shades of plain old brown, margin broadly rounded (upto 10 mm thick). Flesh white, firm but very watery, exudes water when squeezed, faint concentric rings visible. Pore pure–white, 2–3 per mm, holes small relative to walls, smooth, 0.5 mm deep.

Distribution

Reported from Bombay, Dehradun, Bengal and Assam (Bose, 1934), Calcutta, W.B. (Banerjee, 1947), from Central India (Bagchee, 1953) and from India (Anonymous, 1950).

517. *Trametes versicolor* (L : Fr.) Pilat

Synonym

Polyporus versicolor, Coriolus versicolor.

Common Name

Trky Tail.

Habitat

Grows on hardwood fencing posts in tiled layers.

Habit

Fruit body has offset or indistinct cup with pore on decurrent hymenium and lack stipe. The top surface of the cap shows typical concentric zones of different colours. Flesh 1-3 mm thick, leathery texture. Cap with rust-brown or darker brown, sometimes blackish zones. Older specimens can have zones green-algae of growing them, thus appearing green. Cap flat, up to 8 × 5 × 0.5-1 cm, often triangular or round, with zones of fine hairs. Pore surface whitish to light brown, pores round and with age twisted and labyrinthine. 2-5 pores per mm. Spore print white or yellow.

Distribution

Reported from Bhalukpong (Arunachal Pradesh) by Bisht and Harsh (2001).

518. *Trametes hirsuta* (Wulf ex Fr.) Pilat

Synonym

Boletus hirsutus Wulfen, *Nigromarginatus* Schwein., *Velutinus* Planer, *Coriolus hirsutus* (Wulfen) Pat., *Nigromarginatus* (Schwein.) Murrill, *Vellereus* (Berk.) Pat., *velutinus* P. Karst., *Daedalea polyzona* sensu auct., *Fomes gourliei* (Berk.) Cooke, *Hansenia hirsuta* (Wulfen) P. Karst., *vellerea* (Berk.) P. Karst., *Microporus galbanatus* (Berk.) Kuntze, *hirsutus* (Wulfen) Kuntze, *nigromarginatus* (Schwein.) Kuntze, *vellereus* (Berk.) Kuntze, *Polyporus cinerescens* Lév., *cinereus* Lév., *fagicola* Velen., *galbanatus* Berk., *gourliei* Berk., *hirsutus* (Wulfen) Fr., *vellereus* Berk., *Polystictoides hirsutus* (Wulfen) Lázaro Ibiza, *Polystictus cinerescens* (Lév.) Sacc., *galbanatus* (Berk.) Cooke, *hirsutus* (Wulfen) Fr., *nigromarginatus* (Schwein.) P.W. Graff, *vellereus* (Berk.) Fr., *Scindalma gourliei* (Berk.) Kuntze.

Habitat

Grows on hardwood stumps or dead wood of deciduous trees.

Habit

Cap (Pileus) shows good deal of variation in its appearance and pore surfaces making identification of this mushroom uncertain. Cap 1.5–10 cm wide, 1.5–6 cm long and upto 2 cm thick per surface densely hairy, grayish to yellowish or brownish, often with brownish margin, zonate or not usually concentrically grooved pore surface white to tan or grayish.

Distribution

Reported from Namsai (Arunachal Pradesh) by Bisht and Harsh (2001).

519. *Tremella fuciformis* Berk.

Common Name

White Jelly fungus; Silver ear fungus; Shirokikurage.

Habitat

Grows on logs of *Shorea robusta* or *Bambusa anundinacea*.

Habit

Sporophores tough, gelatinous, translucent to opaque, white when young but dirty with age, cerebriform or foliose or tuberculate. Thin, yellowish and horny when dried. Basidiospores subglobose to obovate, white or yellowish, narrow at one end, granulate.

Distribution

Reported from West Bengal (Banerjee and Ghosh, 1942; Banerjee, 1947).

520. *Tremella reticulata* (Berk.) Farl.

Habitat

Grows on decaying moist logs of *Mangifera indica* and *Eugenia jambolana*.

Habit

Pileus upto 8-12 cm in diameter, sessile, tough gelatinous when young and soft gelatinous at maturity, tuberculate, caespitose, irregularly branched with leaf like coalescing, erect, hollow, rounded, lobes, glossy white. Spores (6.4-10.4 × 4.8-6.4 µm) subglobose to broadly ovate, hyaline, smooth, depressed on one side.

Distribution

Reported from Maharashtra (Kundalkar *et al.*, 1983).

521. *Tricholoma caligatum* (Viviani) Ricken

Synonym

Amrillaria caligata (Vivani) Gillet.

Common Name

Matsutake.

Habitat

Grows in group, gregarious under coniferous trees in forests.

Habit

Pileus 8.0-18.0 cm in diameter, convex, later expanded and finally become plain with age, yellowish–ochraceous or reddish brown, surface silky, squamulose. Stipe long, tapering downwards, white. Gills crowded sinuate, narrow white; eroding, becoming free. Basidia clavate, four spored. Basidiospores white, elliptical and smooth 6-7.5 × 4.5-5.5 µm inamyloid, cystidia absent. Flesh thick, firm and white.

Distribution

Reported from Ziro, Arunachal Pradesh (Mishra, 1999); Maharashtra (Sathe and Rahalkar, 1975).

522. *Tricholoma crassum* (Berk.) Sacc.

Habitat

Grows on open grassland and also at the base of the tree namely *Cocos nucifera*, *Ficus benghalensis* and *Polyalthia longifolia* Benth and Hook.

Habit

Pileus 5.0-11.5 cm in diameter, fleshy, convex to hemispheric, sometimes expanded to flat, without any umbo, occasionally obtusely umbonate, surface white when young, changed to creamish-white at maturity, cuticle easily peeled with separable epicutis, margin regular, entire, incurved, smooth, fleshy, context fleshy, 10-40 mm thick at the centre, whitish, soft. Lamellae sinuate, adnexed to adnate, regular, easily separable. Stipe 10-28 × 1-8–12 cm, central, usually inflated at the base and attenuated towards the pileus, sometimes may be inflated in the middle, white when fresh but becoming creamish white, solid, without volva and annulus, flesh soft, fleshy, white, somewhat fibrous. Basidia cylindric to clavate, tetrasterigmatic. Spare print white.

Distribution

Reported from different localities of West Bangal (Roy and Samajpati, 1979).

523. *Tricholoma equestre* (L) P. Kum

Synonym

Agaricus auratus Paulet, *equestre* L. *flavovirens* Pers. *Tricholoma auratum* (Paulet) Gillet, *Tricholoma flavovirens* (Pers.) S. Lundell

Common Name

Man on Horseback, Yellow Knight, Yellow Tricholoma, Canari (French) Chevalier (French), Grünling Pilz (German)

Habitat

Scattered to densely gregarious in grassy or shrubby areas in soil and in coniferous woods.

Habit

Stipe is yellow, usually 4–10 cms long with an uneven diameter. Gills are also yellow coloured and the spores are white. The cap is usually yellow to yellow green sometimes with a tounch of brown reddish colour. The diameter is usually from 5–12 cm, in length, The skin layer covering the cap is sticky and can be peeled of. Cap is flat. Gills on hymenium is adnexed. Stipe is bare, upto 8 cm tall and 2 cm thick, equal or slightly thicker at base, buffed to tinged yellow, dry, smooth.

Distribution

Reported from Phaltan (Maharashtra) by Patil *et al.*, (1995).

524. *Tricholoma gambosum* (Fr.) Kumar

Synonym

Lycphyllum georgil (L.ex Fr.) Kunh and Romagn.

Common Name

St. George Mushroom.

Habitat

Grows in pastures, forest meadows and near hedges, often forming fairy ring.

Habit

It is a mycorhizal fungus (have symbiotic relationship with free roots). Cap 5-15 cm across, subglobose, then expanding, often irregularly wavy and sometimes cracking, dry and fleshy and remains domed shaped throughout its development, smelling strongly of mest. The cap has irrdled edges when young but flatten at maturity. The stipe is ringless and tends to have a bulbous base, 20-40×10-25 mm white soft (1) flesh white, soft, Gills narrow, very crowded whitish. Spore print white. Mushroom is smooth with no scale.

Distribution

Reported from Arunachal Pradesh (Mishra, 1999) and from Orissa (Sinha and Padhi, 1978).

525. *Tricholoma georgii* (Fr.) Quelt

Synonym

Tricholoma gambosum (Fr.) Kummer

Common Name

St George's Mushroom.

Habitat

It is humicolous, gregarious. Grows on lawns and forest floors, in groups.

Habit

Pileus flat, with acute or acuminate umbo, 41-60 mm in diameter, margin striate and sulcate. Cuticle white, smooth, without any scale, context homolomerous, cream coloured, fleshy smooth, dry about 6 mm thick, tapering towards periphery, gills white, free, crowded, thin pen-knife shaped, both ends acute, margin smooth, of different dimensions 6-25×2-5 mm. Gill edge fertile, interlamellar space fertile, subhymenium hyphal, trama thick walled regular. Stipe centric, fleshy, cylindrical, solid later becoming hollow 30-60×7-10 mm. Stipe cuticle smooth, without any scale. Basidium clavate, four sterigmate, 12–6×4–5 mm.

Distribution

Collected from Ziro, Arunachal Pradesh (Mishra, 1999), Also reported from Bhubaneshwar and Kapilash forests, Orissa (Sinha and Padhi, 1978).

526. *Tricholoma giganteum* Massee.

Habitat

Grows in groups on decaying wood in the vicinity of *Ficus religiosa* tree.

Habit

Sporophores very large in size, fleshy mouse-grey colured, several fruit bodies joint at the base to form a fleshy disc, cap convex to planoconvex with drooping margins. Stipe central, straight, thick and stout.

Distribution

Reported from Manipur (Verma *et al.*, 1995) and from Western Ghats of India (Anandh and Prakashham, 2000).

527. *Tricholoma imbricatum* (Fr. and Fr.) Kummer

Habitat

Grows isolated or in groups on soils. It is gregarious and is fairly common in coniferous forests.

Habit

Sporophores centrally stipitate, brownish-yellow to deep-yellow turning to reddish-brown on drying. Cap dry, conical, breaking into imbricating scales, scales, smells faintly mealy. Pileus 5-10 cm, in diameter, convex when young with margin inrolled, becoming companulate, often with a distinct umbo with age, surface dry and minutely fibrillose, reddish brown. Flesh thick, firm white becomes reddish when cut. Gills adnexed, white but soon staining reddish-brown. Stipe 4-10 cm long and 2 cm thick, reddish brown. Spore white, ellipsoid, smooth, 5.5-7×4.5-5 μm.

Distribution

Reported from upper Shillong, Mowkdok, Mairang and Nartiang (Meghalaya) by Verma *et al*. (1995).

528. *Tricholoma lascivum* (Fr.)

Common Name

Rogess Mushroom

Habitat

Grows in group on the ground in deciduous forest, in coniferous woods and in mountainous areas.

Habit

Cap 4-7 cm across, convex then flattened and finally slightly depressed at the center, whitish to pallid tan, silky, smooth with margin almost white, stipe 75-110×10-15 mm, white decolourising pale brownish.Flesh white smell pleasant, sweet scented. Gills crowded, whitish then cream. Spore print white.

Distribution

Reported from Jorhat, Assam (Gogoi *et al*., 2000)

529. *Tricholoma leucocephalum* (Fr. Sensu) Lange

Synonym

Lyophyllum leucocephalum (Fr.) Singer.

Habitat

Humicolous, gregarious, grows on sand forest floor.

Habit

Pileus globose, later expand and flatten with a depressed centre, 26-60 mm in diameter, margin smooth and striate, cuticle white but later turns cream, smooth without scale, crack irregularly at maturity, epicutics hyphal, context fleshy, smooth, white, dry, homoiomerous, 4 mm thick tapering towards periphery, not distinct from epicutics. Gills thin, crowded, free, white, pen knife shaped, acute at both ends, margin smooth, of different dimensions, 16-26×3-6 mm. Gill edge fertile, interlamellar space fertile, subhymenium pseudoparenchymatous. Stipe centric, cylindrical, 16-30×7-10 mm. Stipe cuticle white or cream coloured, smooth, without any scale. Basidia cylindrical, four sterigmate, 30–35×5-6 μm.

Distribution

Reported from Tamil Nadu (Natarajan and Majula, 1978).

530. *Tricholoma lobayense* Heim

Habitat

Grows in clusters on soil.

Habit

Pileus upto 22 cm in diameter, convex then flattened, white, scales towards the centre, margin incurved. Gills white, decurrent. Stipe upto 28 cm long, unequal, solid, attenuated at the apex, enlarged at the base. Spores (4-6×3.5-6 μm) ellilpsoid, hyaline, smooth, thin walled.

Distribution

Reported from West Bengal (Chakravarty and Sarkar, 1982) and from Western Ghat of India (Anandh and Prakasham, 2000).

531. *Tricholoma sulphureum* (Bull. ex. Fr.) Kummer

Synonym

Tricholoma bufonium (Fr.) Gillet.

Common Name

Sulphur tricholoma.

Habitat

Grows on the ground.

Habit

Sphorophores sulphur yellow. Pileus 4 to 8 cm in diameter, convex then flattened, sulphur yellow, surface smooth. Stipe central, long fibrillose, hollow at maturity. Spores white, elliptical, nonamyloid.

Distribution

Reported from West Bengal (Mondal and Purkayastha, 1983).

532. *Tricholoma terreum* (Schaeff. ex Fr.) Kummer

Habitat

Grows solitary or in groups of 7-8 in coniferous or mixed forest soils.

Habit

Sporophores large measuring 3-7 cm in diameter centrally stipitate, fleshy and grey coloured. Pileus 3.5 cm in diameter, light grey, broad, convex to flattened or companulate, margin incurved, floccose squamules, Gills distant, adnexed, emarginate, greyish white when young becoming pale grey with age, broad thick. Stipe upto 5 cm long, slender subequal, white to pale grey, fibrillose; veil fungaceous or absent, Basidiospores 6.5–8 × 3–5.5mm, elipsoidal, white, non amyloid.

Distribution

Reported from Jammu and Kashmir (Watling and Gregory, 1980 and Abraham and Kaul, 1985); Upper Shillong, Mairang, Nongstoin and Nartiang in Meghalaya (Verma *et al.*, 1995).

533. *Tricholomopsis rutilans* (Schaeff ex Fr.) Singer

Habitat

Grows on humicolous soil and on Angiospermous stump.

Habit

Pileus 3–12 cm broad, convex to plane, ground colour yellow overlaid with dark red or purplish red fibrils, dry margins smooth, entire. Gills adnate to notched, close, yellow to pale yellow. Stipe 2.5×10 cm long, 1–2.5 cm broad, typically round above, flattened below, smooth, dry, yellow with reddish to purplish red fibrils. Veil absent. Spore print white.

Distribution

Reported from Teesta, Darjeeling, West Bangal (Sarwal, 1984).

534. *Trogia infundibuliformis* Berk and Br.

Habitat

Grows on decaying wood.

Habit

Fruiting bodies appear like translucent petals of pink colour. It decorates a decaying tree trunk growing in clusters over it. The underside of the cap is arranged into fold–like ridges instead of gills and old cap often split into a number of petal like lobes.

Distribution

Reported from Kidu Forest, Karnataka (Natarajan and Purushotama, 1986) and from Kerala (Kumar and Manimohan, 2009).

535. *Tuber indicum* Cke. and Mass.

Habitat

Grows on the ground, subterranean.

Habit

Sporophores 2-3 cm in diameter, globose, surface rough, blackish, gleba fleshy. Asci 2-4 spored, subglobose, indehiscent. Ascospores (15-18×10-12 μm) ellipsoidal or truncate, brownish.

Distribution

Reported from Uttaranchal (Butler and Bisby, 1931; Bakshi and Puri, 1978) and from Mussorie, U.P. (Zhang and Minter, 1988).

536. *Tuber magnatum* Vitt.

Common Name

White or piedmont Truffle.

Habitat

Grows burried in soil beneath oaks and other broadleaved trees, mycorrhizal association with the roots of local medicinal plant missing in the forest area.

Habit

Fruiting body irregularly top shaped and resembling a knobbly tuber, 3-15 cm in diameter, yellowish to clay colour with darker scurf, flesh soapy, rlsy-white to grey with whitish marble like veins, smells of gaslie or chese.

Distribution

Reported from Kollar forests area, Trivendrum, Kerala by Bhavani Devi (1995).

537. *Tulostoma brumale* Pers.

Common Name

Stilt Puffball

Habitat

Grows on sandy soil among moss or short grass.

Habit

Fruit body is like lolipop and is easily overlooked among snail shell and other pale debris. The rounded "puffball" sits on a twig like stipe and both are white or greyish, the spore released through hole (peristome) surrounded by a reddish brown ring. Puffball diameter 0.5 to 1 cm; total height 2–5 cm. The stipe is 0.2 to 0.3 cm in diameter and 1.5–4 cm tall, surface smooth and grey or more often fibrous and mottled grey brown.

Distribution

Reported from Gurdaspur (Punjab) by Ahmad (1939) and Cunnigham (1942).

538. *Tyromyces palustris* (Berk and Curt.) Murr.

Synonym

Polyporus palustris Berk and Curt.

Habitat

Sporophores grow on living trees.

Habit

Fruiting body dimidiate or effuses-reflexed, sessile, 2.5–6.0 cm in diameter, attached to substratum by a broad base, tough or corky when fresh, hard and rigid after drying. Upper surface at first white, then pale-yellowish in fresh condition, reddish brown when dry, hairy to smooth when fresh, pelliculose and rough after drying, margin more or less thick and rounded. Context watery to waxy, white, 0.5-2.0 cm thick. Hymenial surface whitish or light tan; pore tube 0.5-1.0 cm long, white to light yellow, pores circular or angular, entire or slightly toothed, more or less thin walled, 4-5 per mm. Cystidia absent.

Distribution

Reported from Jorhat, Assam (Vasudeva, 1962).

539. *Vascellum pratense* (Pars.) Kreissel

Common Name

Leibok-marum (in Manipuri).

Habitat

Grows solitary or in groups on grassland.

Habit

Sporophores 1-4 cm in diameter, globular, sessile or sub sessile, white or creamish white, brownish minute scales at the centre, basal portion attached to the substratum by rhizomorphs, endoperidium pale, dehiscing by small torn stroma at the top. Basidiospores (3-4 µm in diameter) globose, pale, echinulate to smooth, guttulate.

Distribution

Reported from Meghalaya and Manipur (Verma *et al.*, 1995).

540. *Verpa bohemica* (Krombh) Schroat

Common Name

Early Morel.

Habitat

Grows on ground and subteranean, often found in moist grassy situations along streams or at the margin of mountain meadows.

Habit

Sporophores like an inverted cone, 1-2 cm wide at the base with longitudinal anastomosing foldes, pale to dark yellow-brown. Pileus with longitgudinal ribs. Stipe white or cream coloured, becoming pale tan, 6-8 cm long, hollow or loosely filled. Spores very large measuring 60-80μm in length.

Distribution

Reported from Jammu and Kashmir (Butter and Bisby, 1931).

541. *Volvariella bombycina* (Schaeff: Fr.) Singer

Synonym

Volvaria bombycina (Schaeff) Quel.

Habitat

Grows on ground and subterranean, solitary or in groups on decaying logs, on living trees, frequently lignicolous or solitary on dead mango tree.

Habit

Sporophores usually large, centrally stipitate. Pileus 8.0-20.0 cm, in diameter, globose when young, then campanulate, finally expanded, subumbonate when old, white, surface usually covered with fibrilose scales giving a silky appearance or red, rarely smooth at the top. Gills crowded, distinctly formed, free, white at first, turning to flesh colour with age. Stipe 5.6-15.0 cm long, 1.0-1.5 cm thick, tapering slightly at the top, base enlarged, usually white but tan toward the base, surface smooth solid, volva large, bag like, thick, loose, whitish. Flesh white, thin. Basidia clavate, 8.0-42.0×6.0-12.0 µm.

Distribution

Reported from Jammu and Kashmir (Butler and Bisby, 1931), Lucknow (U.P.) by Gosh *et al*. (1967) and Pathak *et al*. (1976), and from Kerala by Pradeep *et al*. (1998), Himalayas (Lakhanpal *et al*., 1986; Kumar, 1987).

542. *Volvariella castanea* (Massee) Rath.

Synonym

Volvaria castanea.

Habitat

Grows on soil.

Habit

This species is characterised by the presence of white, viscid, non-striate, 5-8 cm broad cap. Volva fleshy and white. Spores 8-10×8 µm. Cystidia present. Rath (1962) described its two forms, one white volvate form and another black volvale form.

Distribution

Collected for the first time from Calcutta in India by Burkill (in Pegler and Vanhaecke, 1994) and from Lucknow (Rath, 1962).

543. *Volvariella delicatula* Massee

Habitat

Grows on ground.

Habit

Pileus membranous, dry, conical to companulate, expanded, subumbonate, silky, 1-1.5 cm broad. Stipe filiform, fibrillose, 1-1.5 cm long. Volva free, fringed, white. Lamellae free, crowded, spores subglobose, pink-rose in colour, 4.0×3.0 μm.

Distribution

Reported from Calcutta by Burkill (in Pegler and Vanhaecke, 1994) and Massee (1912).

544. *Volvariella diplasia* (Berk. and Br.) Sing.

Synonym

Volvaria displasia Berk and Curt.

Common Name

Straw Mushroom or Banana Mushroom.

Habitat

Grows on rotten paddy straw.

Habit

Sporophores without any pigmentation. Pileus 8.0 cm in diameter, hemispherical, sub fleshy, surface silky-cottony. Gills distant, white, ventricose. Stipe 8.0 cm long, narrowed upward, 2.5 cm broad at the base, 0.8 cm at the top, solid, volva bilobed, brown descending, margin curved, smooth. Basidiospores pink.

Distribution

Collected from West Bengal (Bose, 1920), Baroda (Moses, 1948), West Bengal (Chopra and Chopra, 1955; Butler and Bisby 1960, 1931; Rath, 1962), India (Gupta *et al.*, 1970; Prasad, 1971), U.P. and West Bengal (Ghosh *et al.*, 1967; Singer, 1961), Coimbatore, Tamil Nandu (Rangaswami, 1956; Rangaswami *et al.*, 1970; Vasudeva, 1957 and Thomas *et al.*, 1943) and from Maharastra (Sathe and Deshpande, 1979).

545. *Volvariella earlei* (Murr.) Shaffer.

Habitat

Grows scattered among grasses under the angiospermic trees.

Habit

Pileus 2-6 cm broad, flattened, yellowish white to reddish grey, surface dry, margin irregular, splitting at maturity. Gills free, close, unequal, not arranged in series. Stipe upto 11.5 cm long, cylindrical with slightly bulbous base, concolorous, solid, glabrous, volva vaginate and white. Spores (12.3-16×7.7-10 μm) ellipsoid to ovoid, smooth, double walled with thick outer wall, apiculate, guttulate, inamyloid.

Distribution

Reported from Punjab (Atri *et al.*, 1996).

546. *Volvariella esculenta* (Masse) Sing.

Habitat

Grows solitary on paddy straw, heaps of coconut shell waste, decaying logs.

Habit

Sporophores centrally stipitate, fleshy. Pileus upto 14.0 cm in diameter, globose but turning to ovoid cylindrical and finally becoming umbonate, colour brown near the umbonate region and gradually light coloured toward the margin, surface velutinous, covered with violet black punctatious, sometimes with silver grey streaks, margin involute at first, irregular and apendiculate when old. Gill crowded, distinctly formed, free, reddish pink. Stipe 8.0-12.0 cm long, white with fine streaks, volva cylindrical, greyish brown, membranous, irregularly torn, persistent. Flesh of pileus white, stipe creamish, sometimes fibrous. Basidia clavate, 4-spored.

Distribution

Reported from Cannanoore, Nearamangalorm, Kerala (Bhavani Devi, 1995), Tamil Nadu (Kandaswami, 1974 vide Purkayastha and Chandra, 1985); Coimbtore (Krishnamohan, 1975).

547. *Volvariella hypopithys* (Fr.) Shaffer

Habitat

Grows on sandy soil under *Eugenia jambnolana*.

Habit

Pileus upto 4.5 cm in diameter, yellowish white, convex, margin irregular, pectinate, striate, surface mostly with squarrose scales. Stipe upto 4.2×0.5 cm, broader at the base, fleshy, solid, concolorous, volvate with free colour. Spores (6-9×4.6-6 µm) ellipsoid to ovoid, with an eccentric pore, smooth, pinkish, inamyloid.

Distribution

Reported from Punjab (Saini *et al.*, 1983).

548. *Volvariella indica* Pathak, Ghosh and Singh

Habitat

Grows on dead wood.

Habit

Pileus 12-18.5 cm in diameter, campanulate, becoming obtuse-expanded, fuliginous becoming darker, fibrillose, margin non striate. Stipe 10-13.5 cm long and 1-2 cm in diameter, whitish becoming slightly greyish, smooth, solid, bulbous at the base, volva large, 5-8 cm in length, tomentosa, 4-lobed, white becoming quite thick in some specimens. Lamellae, free, slightly distant from stipe, ventricose, whitish becoming pink at maturity, flesh white. Spore print pink. Basidia 34.2-41.88×9.7-11.4 µm, clavate, tetrasporic. Cystidia numerous. Gill trama inverse. Subhymenium pseudoparenchymatous. Pileus cuticle made of parallel, septate hyphae. Clamp connection absent.

Distribution

Reported from Lucknow (Pathak *et al.*, 1976).

549. *Volvariella liliputiana* (P. Henn.) Rath.

Synonym

Volvariella liliputiana P. Henn.

Habitat

Grows on soil.

Habit

Fruiting body of small size with a viscid, white, non- striate cap carrying spores (12×6-8 μm). Cheilocystidia present. Singer (1955), however, considers this species as a Synonym of *V. lepicotospora.*

Distribution

Collected From U. P. (Hennings, 1901; Rath, 1962).

550. *Volvariella media* (Schum. ex Fr.) Singer.

Synonym

Volvaria media Schum. ex Fr.

Habitat

Grows solitary on grassy ground.

Habit

Pileus 3 cm or more in diameter, white, centre dark brown, fibrillose, edge splitted. Gills crowded, reddish, adnexed, thick with variable length and easily detachable. Stipe 5 cm or more long, cylindrical, fibrillose, base subbulbous, volva white and well developed at the base. Basidiospores (8-10×8 μm) oval, yellowish brown, thick, smooth walled and non-amyloid.

Distribution

Reported from Uttar Pradesh (Hennings, 1901; Rath, 1962; Pathak *et al.*, 1978).

551. *Volvariella pusilla* (Pers. ex Fr.) Singer

Synonym

Volvaria pusilla (Pers. ex Fr.) Schroet.

Habitat

Grows solitary to scattered on forest floor among litter or in open areas among grass.

Habit

This is one of the small sized species of *Volvariella* with a white, slightly convex, cup having a diameter of 0.5-3 cm and striate margin. Spores small (4.5-7 × 3.5-5.5 μm). Cystidia present.

Distribution

Reported from Kerala (Pradeep *et al.*, 1998); Himalayas (Lakhanpal *et al.*, 1986; Kumar, 1987) and from Lucknow, U.P. (Pathak *et al.*, 1976; Ghosh *et al.*, 1967).

552. *Volvariella speciosa* (Fr. ex Fr.) Singer.

Synonym

Volvaria speciosa Br.; *Volvariella gloeocephala* (Fries) Gillet.

Common Name

Handsome volvaria, Rose gilled grissette.

Habitat

Grows solitary on sandy soil among the fallen leaves of *Eugenia jambolana* and *Mangifera indica*. It has been also reported to grow on richly manured ground or on dung or on forest floor.

Habit

Sporophores centrally stipitate, robust, with viscid white or greyish cap. Pileus 8.0-14.0 cm in diameter, globose then campanulate and later becomes umbonate, smooth, fleshy, very sticky. Gills free, flesh coloured to red. Stipe cylindrical, slightly attenuated upward, 10.0-12.0 cm long, hairy when young, smooth at maturity. Volva large, flabby, grey in colour, very close to the stipe, edge free, torn irregularly, flesh white soft. Basidia clavate, 4-spored.

Distribution

Reported from Mannar, Mavelikkara, Kerala (Bhavani Devi, 1995), from Lucknow, U.P. (Pathak *et al.*, 1976; Ghosh *et al.*, 1967), Patiala, Punjab (Saini and Atri, 1993) and Kerala, Idukki (Pradeep *et al.*, 1998).

553. *Volvariella terastius* (Berk and Br.) Singer

Synonym

Volvaria terastia Berk and Br.

Habitat

Grows on ground amongst rotten paddy straw heaps.

Habit

Pileus 10.0 cm in diameter, at first hemispherical, later campanulate to expanded, purple, flesh, surface provided with streaks, softly cracked, margin, split. Gills free, ventricose, pale flesh. Stipe white, attenuated at the top, self, volva very broad, broken irregularly, descending.

Distribution

Reported from West Bengal (Butler and Bisby, 1931; Singer, 1961), Bengal (Bose, 1918). West Bengal (Bose and Bose, 1940; Banerjee, 1947; Chopra and Chopra, 1955; Rath, 1962); Kerala (Manimohan *et al.*, 1988).

554. *Volvariella volvacea* (Bull ex Fr.) Sing

Synonym

Volvaria volvacea (Bull) Secc.

Common Name

Paddy Straw Mushroom.

Habitat

Grows solitary or scattered groups on rotting paddy straw heaps, oil palm pericarp waste, spent tea waste and paddy straw.

Habit

Sporophores large fleshy, umbrella like with base of the stipe in a bag like volva, but without ring, cap bell shaped when young, becoming convex and flat with a hump at the center. Grey to dark grey in colour. Centrally stipitate. Pileus 5.0-12.3 cm in diameter, companulate at first, later becoming umbonate, greyish sepia, but sepia near umbo and margin, distinct sepia radial streaks upto middle of the pileus, soft, margin sometimes split. Gills crowded, distinctly formed, free, rosy buff when young, flesh coloured with reddish tinge at maturity. Stipe central, cylindrical, attenuated upward, 8.0-14.0 cm long, whitish, ending below with solid bulbous base. Volva well developed, persistent usually buff but sepia above, bilobed. Flesh white soft. Hymenophoral trama inverse, Basidia clavate, tetrasterigmatic, 18.7-28.9×6.8-11.9 μm.

Distribution

Moirang, Bishenpur and Imphal valley of Manipur; Tura in west Garo Hills of Meghalaya, Kolasib in Mizoram and Dimapur in Nagaland (Verma *et al.*, 1995), Hills of U.P. and West Bengal (Singer, 1961; Butler and Bisby, 1931). S.W. Monsoon from Anchal, Kesavadara-puram in Kerala (Bhavani Devi, 1995). From different parts of India (Gupta *et al.*, 1970, Prasad, 1971; Munjal and Chatterjee, 1971; Purkayastha and Chandra, 1975a), U.P. (Hennings, 1901), Himalayas (Lakhanpal *et al.*, 1986; Kumar, 1987) and from Maharastra (Sathe and Deshpande, 1979).

555. *Volvariella woodrowiana* Massee

Habitat

Grows on ground.

Habit

Pileus fleshy, companulate, dry margin striate, glabrous, 8-12 cm. in diameter. Lamellae free, crowded, salmon coloured, edge entire spores spherical to ellipsoid (10.0×8.0 μm). Stipe central, solid, glabrous, 9-15 cm long, 1.0 cm. in diameter, white. Volva large, variously fringed free and white.

Distribution

Collected from Poona by Massee (1912).

556. *Xylaria polymorpha* (Pers.) Grev.

Synonym

Xylospharera polymorpha (Meral) Dumortur.

Common Name

Dead man's finger; Wood club fungus.

Habitat

Grows in groups on dead wood on old logs of *Shorea robusta* in shade.

Habit

Sporophores consisting of irregularly cylindrical or finger like head, grey to black with age, fleshy when young and woody at maturity, surface rough. Flesh white, solid or fibrous. Stipe generally short, black, narrowed at the base, sometimes branched. Ascospores fusiform with a flat side.

Distribution

Reported from West Bengal (Bagchee, 1953) and Madhya Pradesh (Saksena and Vyas, 1962).

There are 12 additional taxa which are edible and are reported from India but their taxonomic feature could not be collected. They are presented below along with their habitat,distribution in India and synonym as well as common name, if any:

557. *Boletus veripes* Peck

It grows in forest and has been reported from Shimla, H.P. (Lakhanpal and Sagar, 1989).

558. *Coriolus consors* (Berk.) Imazeki

Synonyms

Irepex consor and *Xylodon consors.*

It grows under forest tree and has been reported from India by Anonymous (1952) and Butler and Bisby (1931).

559. *Fomes melanoporus* (Mont.) Cooke

Synonyms

Nigrofomes melanoporus (Mont.) Murrill.; *Polyporus melanoporus* Mont.; *Phallinus melanoporus* (Mont.) Cnn.

It has been reported to grow in Dooar Forests on rotten branches of *Shorea assamica, S. robusta* and *Termanalia tomentosa.* It has been reported from Jalpaiguri, W.B. (Bose, 1919–1928) and from India (Anonymus, 1950).

560. *Favolus tessellates* Mont.

It grows on *Barringtonia acutangula.* It has been reported from Saharanpur, U.P. (Butler and Bisby, 1931; Hennings, 1901).

561. *Lactarius princeps* Berk

It grows on wood and has been reported from Khasi Hills, Assam (Berkeley, 1856).

562. *Lentinus subnudus* Berk

Synonyms

L. squarrosulus, L. cretaceus, L. pergausenius, L. tigrinus, E. squarrosulus and *Pleurotus squarrosulus.*

It grows on tree stumps and can be cultivated on wood logs of various hardwood trees. It has been reported from West Bengal by Singer (1961)and by Bose and Bose (1940).

563. *Lycoperdon asperum* (Lev.) Speg

Synonym

Bovistella aspera (Lev.) Lloyd.

It grows on sloping lawn or on ground in coniferous forest. Its occurrence in India has been reported by Cunningham (1942).

564. *Microporus affinis* (Fr.) Blume and Nees.

Common Name

Yellow Coral.

It grows on hardwood logs and its occurrence has been reported from Namsai, Arunachal Pradesh by Bisht and Harsh (2001).

565. *Micropsalliota brunneosperma* (Singer) Pegler

It grows on soil. It has been reported from Calicut, Kerala (Leelavathy and Little Flower, 1986).

566. *Polyporus gramocephalus* Berk

It grows on erythrima logs, on the stem of *Ficus religiosa* and other hardwood logs and stumps of deciduous trees and dead wood.

It has been reported from Chessa, Arunachal Pradesh (Bisht and Harsh, 2001), Sagar, M.P. (Saksena and Vyas, 1962–64), Nainital (Bhargava and Sehgal, 1954; Mitter and Tandon, 1937), Dehradun, U.P. (Thind and Anand,1956), Raipur, M.P. (Saxena, 1960) and Panhala, Maharashtra (Parndekar, 1964).

567. *Polyporus tricholma* Mont.

Synonyms

Polyporus similes Berk, *P. stipitarius* Berk and M.A.Curtis.

It grows on dead hardwood. It is pore bearing fungus. Its occurrence has been reported from Burdwan, W.B. (Anjali and De, 1977).

568. *Termitomyces narohiensis* Otiero

Common Name

Arikoon.

It grows in large number on the ground and lawns. It has been reported from all districts of Kerala (Bhavani Devi, 1995).

Attempt has been made by the authors to consolidate the fragmentary floristic reports of the edible mushroom taxa reported from varied ecological and geographical niches of India under a single cover. It, however, can not be claimed a complete inventory of Indian Edible Mushroom. Escapes can not be denied. Coordinated efforts of regional contributors in the field are desirable at regular intervals to update this inventory.

Chapter 6

Cultivation of Mushroom

Mushrooms have been eaten and appreciated for their flavor, economic as well as ecological values and medicinal properties. Their edibility is known since time immemorable, but technology of its cultivation is comparatively of late origin. It was first attempted around AD 600 in China by Growing *Auricularia auricula* on logs. In India, Newton was first to cultivate an unknown species of mushroom in the year 1886 (in Ranga Swami, 1956). But possibilities of mushroom cultivation on commercial scale in India was realised after lapse of more then half century by Bose (1921). It was followed by successful farming of mushroom at Coimbatore by Thomas (1943). Since then commendable progress has been made in its technological know how and it has been now successfully transferred to the farmers. Few progressive farmers in Himachal Pradesh and Kashmir started growing button mushroom on commercial scale in late sixties. In seventies medium scale commercial mushroom growing was taken up by many more mushroom growers in Kashmir, Himachal Pradesh, hilly regions of U.P. and Tamil Nadu. By early eighties cultivation of button mushroom was also adopted by farmers in Haryana, U.P. and in area around Delhi. In early nineties export oriented units were put up by corporate houses/industrialists in length and breadth of the country with use of advanced machinery/computerised monitors for mushroom growing. Consequently mushroom production in India increased 6-7 fold to a tune of 40,000 tons in a span of 5-6 year in nineties (Dhar, 1997).

Mushroom cultivation is a potential biotechnological process where in the waste plant materials or negative value crop residues may be converted into valuable food. Pope (2000) reported that there are about 200 kinds of waste in which edible mushroom can be produced. Depending on the nature of substrate mushroom can be classified in the following five categories.

1. Those which grow on fresh or almost fresh plant residues
   Lentinus, Flammulina, Auricularia, Pholiota, Tremella, Pleurotus.
2. Those which grow on slightly composted material
   Volvariella, Stropharia, Coprinus.
3. Those which grow on very well composted material
   Agaricus.
4. Those which grow on soil and humus
   Lapiota, Morchella.
5. Mycorrhizal fungi
   Tuber, Morchella, Lactarius, Amanita and *Cantharellus.*

There are mushrooms which are edible and delicious but can not be cultivated with existing knowledge and technology *e.g. Termitomycetes* spp. and *Tricholoma* spp.

There are at least 12000 species of fungi that can considered to be mushrooms, with at least 2000 species showing various degrees of edibility (Chang, 1999). Furthermore, over 200 species have been collected from the wild and used for various traditional medical purposes, mostly in the far East. Above 35 mushroom species have been cultivated commercially and of these, around 20 are cultivated on industrial scale. In India, mainly three types of edible mushroom are cultivated on commercial scale. These are the common:

1. White button mushroom: *Agaricus bisporus.*
2. Oyster mushroom: *Pleurotus* spp.
3. Tropical paddy straw mushroom: *Volvariella* spp. recently cultivation of Milky mushroom–*Calocybe indica* has been started.

Most of the commercial and export oriented units grow white button mushroom for sale in domestic and distant markets both in fresh and canned form because of having maximum acceptability. Oyster mushroom cultivation is confined to small and marginal mushrooms growers and the produce is sold fresh/dehydrated in the local market, This mushroom is gaining popularity because of:

1. Wider range of growing temperature (20-30°C)
2. Easy method of cultivation
3. Low technology infrastructure
4. Grows on wide variety of base materials/agro wastes with good productivity.

In spite of its lesser acceptability, third most commonly grown mushroom in hot/humid areas of peninsular India is the *Volvariella.*

Production Technology in Use

Cultivation of White Button Mushroom (*Agaricus* spp.)

Due to the low temperature requirement, cultivation of white button mushroom is restricted to the cool climate areas and to the winter in the plains of Northern India which lasts less than four months.

The basic requirements for growing this mushroom are compost, spawn and casing soil. Traditionally compost is prepared all over the world either from the horse manure or from the wheat straw, but now other straws like paddy, barley and chicken manure, saw dust, rice bran along with other ingredients are employed successfully. The chief function is to provide cellulose, hemicelluloses and lignin in bulk. These materials also provide proper physical structure to the mixture to ensure the necessary aeration for the built up of microbial population and the subsequent spawn growth in the compost.

There are so many variations in compost formulations. The basic primarily the cost and availability of the ingredients and suitable supplement in the particular growing area. Widely used ingredients for compost preparation are:

Ingredients	*Quantity*
Wheat straw (chopped 3–6 cm long)	1,000
Wheat bran	80
Urea	10
Ammonium sulphate/Calcium ammonium nitrate	10
Gypsum	40–50
Optional Supplements:	40 kg molasses diluted 10 times with water and 100–110 kg poultry manure can be added at the beginning of the composting.

Preparation of Compost

Three methods of making compost are being practiced.

Long Method

In most of the small seasonally growing mushroom farms, compost is prepared by traditional long method not followed by steam pasteurisation. Compost ingredients are mixed and wetted for 24–48 hr and stacked with rectangular heap using wooden boards. The turnings are given periodically and start with first turning on 4th–6th day of stacking when temperature reaches 70–75° C in the middle of the heap. The subsequent turnings are given on 3–4 days interval. Gypsum is added on 3rd turning as the practice in short method. The compost is ready in 6–7 turnings generally. The quantity of water added to the compost is directly related to mushroom production. A practical test to check the right quantity of water is to take the compost between the two palms and when pressed droplets should collect between the fingers but should not trickle out like water from the tap.

Some pesticides may also be mixed with the compost.

Well prepared compost is brown to dark brown in colour and should be free from the smell of ammonia. The pH of the compost should be 5.8.

This method is not good for three reasons:

1. Quantity of compost available finally after processing is reduced increasing the cost of production.
2. Nutrition value of the compost is substantially decreased.
3. Since pasteurization of the compost is not done, fair amount of insecticides, fungicides and nematicides are used to prevent contaminations, which has a harmful side effect.

Container System

This comprises of 8 days phase I on the composting platform and then additional 8–10 days in the container. This system of compost preparation also involves various preventive treatments with their consequential side effects, however, the period of composting is reduced and the yield of finished compost is greater than in the long method. A recent improvement has been to treat the container to a certain amount of steam pasteuristaion, in order to eliminate the use of harmful chemicals.

Peak Heating

This is the standard method practiced with small variations. It involves 10 days of Phase–I and 7 days of Phase–II. This method is undoubtedly superior. The yield of finished compost is the greatest. The control of disease and insects by pasteurisation is maximum resulting in a healthier crop with quicker and larger flushes. Pasteurisation is the process to destroy undesirable moulds and pests. The formulation being used in the synthetic and horse manure compost, are more or less standard now but, following points must be taken in to consideration while preparing the compost through this method.

Phase–I

This phase involves preparation of compost and their stacking.

1. *The Dimension of the Stack*: Various size of stacks are being made up in India but standard size is 6ft x 5 ft to 6ft high for the right effect.
2. *The Moisture Content*: At the time of stacking moisture content of compost should be 75 to 77 per cent. Accordingly pre-wetting should be done before day "O", the day on which the stack is to be put up.
3. *The Intervals between Stacking*: Oxygen is absolutely essential for correct fermentation. When stack is made, usually it is highly porous. So, 3 or even 4 days before the first turn is recommended. Within a few days when the stack will be consolidating leaving less space for oxygen, an interval of two days between subsequent turnings is required.
4. The compost filled in the trays should be as uniform as possible in depth and density of compression which is move acceptable to mushrooms than

any other fungus.

5. There should be enough space between the surfaces of each tray of compost to allow free circulation of air within the room.

Phase–II

Can be described as peak heating and its function is to complete the process started outside in phase–I by subjecting the compost to more closely controlled environment.

Very high humidity in the air is maintained during this period to avoid excess drying out of the compost. Humidity near saturation, and an acceptable temperature in the air, within 6 hr are achieved by releasing live steam in to the room. There are various methods in the actual process of peak-heating, however, the one which have been found to be satisfactory is given below:

Steam is allowed to enter the room and the compost temperature is rapidly allowed to rise to 60°C in the first 24 hrs. This is probably the safest way to kill unwanted moulds. Sufficient air is then introduced to bring the compost temperature down slowly over the course of the next three days, by which time the production of ammonia will have ceased. Some growers prefer to bring the temperature down quickly to 52°C and to hold it there until ammonia is no more detected. Some other growers hold the compost at 54°C for several days. However, deammonification occurs most effectively at 53.5 to 54.5°C.

Sequential temperature in the compost in phase–II function as follows:

1. *High Phase: above 56°C*: The thermophilic bacteria develop intensely in transforming fixed nitrogen to ammoniacal nitrogen. Therefore sufficient ventilation is necessary to avoid the risk of exceeding the level of CO_2 above 10 per cent which would encourage the growth *Chaetonium*, the olive green mould.
2. *Middle Phase: 52°–55°C*: The thermophilic actinomycetes develop mostly in this range. Some are beneficial to the production of good compost but other are harmful. So, this phase should not be prolonged.
3. *Low Phase: 48°–52°C*: This is the ideal range for themophillic fungi such as *Humicola* and *Torula*, Their presence is indicated by a sort of moss or grayish web on the surface of the compost. They transform the ammonical nitrogen into proteins acceptable to the mushroom.

The pre-wet heap machine has been designed to achieve a quick homogenous mix of water and the raw materials used in phase–I of the culture of *Agaricus bisporus*, to improve the efficiency of the composting process by reducing the amount of time required to achieve the same or better quality compost. Some benefits of using the pre-wet compost turners include improving compost quality, homogenous blending of up to 200 t/h better odor control, less water run off, raw material savings, reducing of phase–I time by over 50 per cent, reducing operating and capital cost, and increasing production yields, crop uniformity and profit potential. Recently a new tunnel/bunker filling system has been developed, which consists of filling the bunkers or tunnels using overhead layering techniques, but only from one end of the bunker/tunnel and not through the holes in the tunnel roof, with the use of overhead and out-of-sight conveyers/elevators (Traymater Machinery).

New methods to produce and increase the productivity of a wide variety of exotic mushrooms have been developed. The Mycocell system (Mycocell technologies, UK) is a method based on micro wave sterilization of pre–packed substrate to which the spawn, nutrients and other supplements can be added. In this case, radiation used in the treatment of the substrates changes the cellulose and increases case of breakdown by the mycelium. Several nutrients can be added and mixed thoroughly, there are no risk of substrate contamination, and colonization of the substrate by the mycelium considerably increases. The commercial benefits of this production system include lower cost, low energy requirements, automation, low labour demand, lack of down-time, light and cheap transport, low contamination risk and long self life. The Mycocell system allows the successful culture of exotic mushrooms such as *Lentinus edodes, Pleurotus ostreatus, P. pulomonarius, P. eryngii, P. djamor, P. cystidiosus, Pholiota* sp., *Hypsizygus* sp., *Flanulina velutipes, Agrocybe aegerita, Ganoderma lucidum. Psilocybe* sp., *Grifola frondosa, Hericium* sp. and *Auricularia*.

Spawning

Spawning refers to the planting of mushroom mycelium growing on a suitable substrate, in the compost. Spawn (mycelium) can be obtained from spores, from a mycelial piece of the specific mushroom or from several germplasm providers (American type culture collection, National Centers for Agricultural Utilization Research, etc.) To obtain inoculum, the mycelium is developed on cereal grain *e.g.*, wheat, rye or millet, which is usually called the spawn. Once developed, it may also be purchased from several companies. The purpose of the mycelium coated grain is rapidly colonised in bulk on the specific growth substrate. Already prepared or purchased spawn is spread uniformly on the tray when half filled with compost and again, after the tray is filled completely.

The tray is then covered with newspaper sheets and the latter is kept moist by sprinkling water. The trays are now staked.

Running of Spawn

The water is continued to be sprayed at intervals throughout the cropping period to maintain the humidity of the bed. After 14-21 days the mycelium of the mushroom appears on the upper surface of the bed. This stage is followed by casing. The room temperature during the above operation is maintained at 24–25°C.

To check the contamination or infestation of beds by undesirable fungi, insects or pests, the beds should be sprayed once a week with 0.2 per cent Cyathion. This discourages the development of diseases in mushrooms.

Casing

Casing means the covering of growing mycelia on the compost by a thin layer of soil or a special type of soil mixture known as casing medium.

Several casing materials are used in mushroom industry. These are loamy soil of good structure supplemented with 5 per cent organic matter, peat (a mass of partially carbonized vegetable tissue formed by partial decomposition) neutralized or made alkaline with calcium carbonate (acid peats are usually found to be free of the known pathogens, therefore, peat-lime stone mixtures are not normally pasteurized), weathered spent compost, sisal waste, ground coconut husk, decomposed farmyard manure, clay soil and forest soil.

Contradictory reports are not uncommon regarding the selection of soil. According to Delmas (1978) virgin soil taken from deep layers of soil (protected from soil contamination) is better than the top soil which contains active microflora. This is necessary for gaseous exchange. Besides, it has better water holding capacity. Good casing soil is that which has water holding capacity but is porous, (pH 7.0–7.5) and free from pathogens and pests.

In India the casing soil used consist of:

1. Well rotten cow dung, mixed with soil in 3:1 ratio, or
2. Farm Yard Manure and gravel in 4:1 ratio
3. Farm Yard Manure and loam (4:1)
4. Soil Peat mixture (2:1)
5. Spent compost, sand and slacked lime (4:1:1) and nematicide mixture. The pH of the casing soil is kept at 8–8.5

To increase the yield of button mushroom, denatured supplements were added at the casing. Supplements were cotton seed cake, groundnut cake, cotton seed meal, soybean meal, wheat bran and mustard cake. Cotton seed cake showed increase in yield when supplemented in spawn run compost as shown in Table 1 (Nita Behl, 1987).

Table 1: Mushroom Yield at 2 per cent Rate of Supplementation at Casing

Treatments	*Average Yield (kg/100 kg compost)*
Cotton seed cake	18.80
Groundnut cake	13.10
Cotton seed meal	16.73
Soybean meal	16.50
Wheat bran	16.10
Mustard cake	2.90
Control	15.40
SEM ±	0.47
CD (0.05)	1.42

Raking of spawn run compost before casing also increased the yield of button mushroom.

Casing is necessary because it allows diffusion of CO_2 produced by the mycelia in the compost. High concentration of CO_2 causes inhibition of fruit-body production. It maintains appropriate water relations between compost and surrounding atmosphere. The most important role of casing soil is that it induces rapid change from vegetative to reproductive growth (fruit-body formation) of mushroom.

When the white cottony mycelium covers the entire compost surface, the casing mixture is applied as a layer (6.0 cm thick) on it. Uniform thickness of casing layer is important for better yield. Uneven casing leads to the formation of stroma. Depth of casing should be checked at regular intervals if done by hand. The casing layer must have a smooth surface. Hand pressing of casing layer is always avoided and casing must be done in proper time because productivity depends largely on the above mentioned factors.

The beds are sprayed over the casing soil with water to maintain relative humidity between 70 to 80 per cent.

A few weeks after casing, the fruiting body of the mushroom begins to appear for two months. There is an interval of 8–10 days between flushes.

Cultivation of Oyster (Dhingari) Mushroom (*Pleurotus* sp.)

Dhingari grows well at higher temperature between 20-30°C. Therefore, it is suitable for cultivation in most of the places in India. This mushroom is grown both on wheat and paddy straw with good yield. The straw is chopped, wetted and treated at 80-95°C in hot water for thirty minutes. After draining the water, the substrate is spawned at 2 per cent of wet weight put into a polypropylene bag and incubated for 10-20 days at 28-30°C for spawn run. After spawn run is completed, the bag is cut open and discarded.

The spawn run straw blocks are put on a rack at a distance of 15-20 cm for fruiting. The humidity in the cropping room is maintained by sprinkling water on straw blocks, floor and even walls, 2-3 times a day. Aeration is given for 1-2 hrs daily. Sufficient amount of diffused light is required for normal fruiting. Mushrooms will appear within 7-10 days, which can be harvested after two days. Straw blocks of 10 kg wet weight are the standard sized blocks used and 2.5 kg of dry straw makes into 10 kg blocks after wetting. The yield on an average is about 1.5 kg of fresh mushrooms.

Cultivation of Paddy Straw Mushroom (*Volvariella volvacea*)

Cultivation method of this mushroom vary in different countries. In India it is grown at Coimbatore, Lucknow, Jammu and New Delhi. Traditionally it is grown under partial shade or in the open field. The Common substrate used is paddy straw which is soaked in water for 24 hrs before being folded and arranged into a stack on a soil base and spawned in layers. No casing material is used. The spawn run is achieved (covering the bed with polythene sheet) at 35°C and RH of 95 per cent. On complete spawn run in a week, the sheet is removed and beds aired for fructification on an average 12 per cent conversion on fresh weight basis is obtained in two weeks of crossing.

Cultivation of Milky Mushroom (*Calocybe indica*)

This mushroom grows well at higher temperatures of 30-40°C. It grows both on wheat and paddy straw with good yield. The straw is chopped, wetted/treated at 80-95°C in hot water for thirty minutes. After draining the water, the substrate is spawned at 3 per cent of wet weight basis, put into polypropylene bags and incubated for 25-30 days at 30–40°C for spawn run. After spawn run is completed, the bag is opened from mouth.

The spawn run straw blocks are put on racks. The block/bags are then cased with sterilized casing soil as done in button mushroom cultivation. The humidity in the cropping room is maintained by sprinkling water on bags, floor and even walls, 2-3 times a day. Aeration is given for 1-2 hrs daily. Mushrooms will appear within 15-20 days which can be harvested after attaining good size. Bags of 10kg wet weight are the standard size used and 3kg of dry straw makes into 10kg bags after wetting. The yield on an average is about 2-2.5 kg of fresh mushrooms.

There are other mushrooms also cultivated on pilot scale at experimental stations but these are not being grown commercially anywhere in our country *e.g.*, shiitake (*Lentinus edodes*), Black Ear mushroom (*Auricularia* spp.) and others. Though shiitake mushroom is collected from nature in North Eastern region of India for export but its cultivation has not been exploited so far in India for commerce. Morels (*Morchella* spp.) collected from nature in Northern area of India, are exported after dehydration.

Recently mushroom culture has moved toward diversification. The cultivation of Shiitake mushroom (*Lentinus edodes*) which was traditionally grown on wood logs, are now grown on artificial log. Artificial logs are prepared by enclosing heat treated supplemented substrates (based on sawdust) in plastic bags. This method of cultivating Shiitake mushroom is called "bag log cultivation." The main advantages of this method are short time to complete a crop cycle and the higher yields.

The edible mycorrhizal mushrooms include some of the worlds most expensive foods and have a global market measured in US billions dollars. Despite this, few have been cultivated with any degree of success. The main obstacles to their cultivation are their need to be associated with a host plant to successfully grow and fruit, contamination with other ectomycorrhizal fungi both before and after establishment of plantations, and a general lack of understanding of each mushroom's trophic relationship, and biotic, edaphic and climatic requirements. However beginning in the late 1960s and early 1970s methods were devised to artificially infect plants with *Tuber* spp. in green houses following inoculation with sporal suspensions or pure cultures prepared either from fruiting bodies or mycorrhizal root tips. Although dozens of nurseries around the world produce seedlings inoculated with *Tuber melanosporum* and *Tuber uncinatum* that ensure success, such as the amount, quality and treatment of inocula as well as watering, fertilisation, temperature, light levels, potting medium formulation and pH, remain trade secrets. Another crucial factor is the control of contaminating saprobic and ectomycorrhizal fungi such as *Thelephora* spp., *Sphaerosporella brunnea* and *Pulvinula constellation*. Consequently, poor seedling quality remains a serious problem in *Tuber*-inoculated plants. Despite these problems, more than half of all *Tuber melenosporum* truffles are now harvested from artificial plantations.

Artificial plantation technique has also led to the formation of edible mycorrhizal mushroom fruiting bodies in the field including *Lactarius deliciosus, Lycophyllum shimeji, Rhizopogon rubescens, Suillus granulatus* and various species of *Tuber*. But despite these successes, fewer than a dozen of the many hundreds of edible mycorrhizal mushrooms have ever been cultivated with any degree of success and this includes important commercial species such as *Boletus edulis, Tricholoma matsutake* and *Tuber magnatum*. Consequently, suppliers of most commercially important edible mycorrhizal mushrooms are still restricted to those that can be harvested from the wild during autumn or winter.

Chapter 7
Mycelial Culture of Mushroom

Of the two thousand known edible species, only a few have been cultivated on a commercial scale. Cultivation technology of most of them is still unknown although they are growing in their natural habitats. The growth and development of sporophores of these species are largely dependent upon natural conditions and therefore, yield is unpredictable. Such a problem could be partially solved by growing their mycelia in culture. Mushroom mycelia are now being considered as a potential source of proteins, amino acids, vitamins and minerals (Mc Connell and Esselen, 1947; Humfeld and Sugihara, 1949; Reusser *et al.*, 1958; Falanghe *et al.*, 1964; Litchfield, 1964; Wargel *et al*, 1974; Jennison *et al.*, 1955; 1957; Jandaik, 1976a; Kaul, 1978a). It is also possible to enhance protein content of mycelia by cultural amendments. Worgan (1968) demonstrated that the amino acid composition of protein in sporophores and mycelia was more or less similar. In addition to it, nutritive value of mushroom mycelium has been also proved potent source of many life saving medicines. LEM protein bound polysaccharide showing anti-cancer activities is derived only from the mycelium of *Lentinus edodes.* A new anti tumour glycoprotein has been isolated from the mycelium of *Flammulina velutipes. Pleurotus* sp. are excellent producer of levostin which has cholesterol lowering ability. Its production on commercial scale from fruit bodies is not viable because of variability in fruit-body composition. Its mycelial cultivation could be the way ahead. The Caterpillar fungus of Tochukaso is not a mushroom type fungus. Anticancer polysaccharide has been isolated from its several species. Its fruiting structure can not be cultivated or cultured buy its mycelium now can be easily cultivated in fermentor and is attracting considerable interest as an agent to treat fatigue and improve motor function (Mizuno, 1999). Therefore, large scale production of mushroom mycelia could provide a promising source of proteinaceous food and many life saving drugs.

Plate 1

1. *Agaricus arvensis,* 2. *A. bitorquis*, 3. *A.campestris,* 4. *A.placomyces,* 5. *Agrocybe praecox* and 6. *Aleuria autrantia* (*Courtesy* Dr. Danielle Castronovo © California Academy of Sciences)

Plate 2

1. *Amanita vaginata*, 2. *Armilleria mellea,* 3. *Crepitotus mollis* (credited with Ron Wolf), 4. *Astraeus hygrometricus, 5. Auricularia auricula* (*Courtesy* Dr. Danielle Castronovo © California Academy of Sciences), and 6. *Bolbitius vittelinus* (*Courtesy* Dr. Ron Wolf)

Plate 3

1. *Boletus edulis* (credited with Dr. Robert Thomas and Dorothy B. Orr © California Academy of Sciences), 2. *B. erythropus* (credited with Ron Wolf), 3. *Bovista plumbea* and 6. *Chlorophyllum molybdite* (credited with Alfred Brousseau © St. Mary's College of California), 4. *Cantherelus cinnabarius* and 5. *C. cibarius* (courtesy Taylor F. Lockwood)

Plate 4

1. *Clavaria vernicularia,* 2. *Clavariadelphus pistillaris,* 3. *C. truncatus* and *5. Clavatia gigantia* (credited with Alfred Brousseau © St. Mary's College of California), 4. *Clavatia cyathiformis* (credited with R. Thomas and Dorothy B.Orr © California Academy of Sciences) and 6. *Claviocorana pyxidata* (courtesy Taylor F. Lockwood)

Plate 5

. *Clavulina rugosa* (courtesy Dr.Danielle Castronovo), 2. *Clavulinopsis fusiformis* and . *Clitocybe nibularis* (courtesy Dr. Ron Wolf), 4. *Clitocybe prunulus* (credited with Dr. lfred Brousseau © St. Mary's College, California), 5. *Collybia confluens* (courtesy Dr. Nick urzenko) and 6. *Collybia dryophila* (credited with Robert Thomas and Dorothy B.Orr © alifornia Academy of Sciences)

Plate 6

1. *Coprinus atramenfarius* (credited with Robert Thomas and Dorothy B. Orr. © Californ Academy of Sciences), 2. *Coprinus comatus* and 5. *Cortinarius violaceus* (courtesy Daniel Castronovo © California Academy of Sciences), 3. *Coprinus dissaminatus* (courtesy Joh W.Wall), 4. *Coprinus plicatilis* (courtesy Amadej Trukoczy) and 6. *Craterellus corniucopioide* (courtesy Taylor F. Lockwood)

Plate 7

. *Flamulina velutipes* and 6. *Hericium erinaceus* (courtesy Dr. Nick Kurzenko), 2. *Gomphus* *lavatus* and 5. *Hericium coralloides* (credited with Robert Thomas and Dorothy B.Orr. © California Academy of Sciences), 3. *Helvella crispa* and 4. *H. lacunosa* (Courtesy Ron Volf)

Plate 8

1. *Heterobasidion annosum,* 4. *Hygrocybe miniata* and 5. *H. punicea* (courtesy Ron Wolf)
2. *Hohenbuehelia petaloides* (credited with R.P. Olowin © St. Mary's College, California)
3. *Hydnum imbricatum* (courtesy Dr. Danielle Castronovo © California Academy of Sciences)
6. *Hygrophoropsis aurantiaca* (Credited with Robert Thomas and Dorothy B.Orr © California Academy of Sciences)

Plate 9

1. *Hygrophorus eburneus* (courtesy Ron Wolf), 2. *Hypholoma capnoides* (credited with R.P. Olowin © St. Mary's College, California), 3. *Hypholoma fasciculare* (credited with Albert P. Baker © California Academy of Sciences), 4. *Laccaria amathystina* (credited with Robert Thomas and Dorothy B.Orr © California Academy of Sciences), 6. *Lactarius controversus* (*courtesy* Dr. Danielle Castronovo) 5. *Laccaria laccata* (credited with Jo and Ordeus)

Plate 10

1. *Lactarius deliciosus* (courtesy Dr. Danielle Castronovo), 2. *L. volemus,* 4. *Leccinum scabrum* and 5. *Lentinus lepidius* (credited with Robert Thomas and Darothy B. Orr. © Figs. 1,2,4,5 California Academy of Sciences; 3. *Lactiporus sulphurous* (courtesy Ted Swiecki) and 6. *Leucoagaricus naucinus* (credited with R.P.Olowin © St. Mary's College, Califomia)

Plate 11

1. *Leucopaxillus giganteus* (credited with R.P. Olowin) and 6. *Marasmius oreades* (credited with Bn Alfred Brousseau © St. Mary's College, California), 2. *Lycoperdon perlatum* and 3. *L.pusilum* (credited with Robert Thomas and Dorothy B.Orr © California Academy of Sciences), 4. *L. umbrinum* (Courtesy Ron Wolf), 5. *Macrolepiota procera* (courtesy Dr. Nick Kurzenko)

Plate 12

1. *Melanoleuca melaleuca,* 2. *Morchella angusticeps,* 3. *M. esculenta*, 4. *Mycena haematopus* and 5. *M. pura* (credited with Robert Thomas and Dorothy B. Orr), 4. *Otidia leporina* (credited with Jo and Ordeno) © 1-6–California Academy of Sciences

Plate 13

1, *Oudemansiella mucida* (courtesy Dr. Nick Kurzenko), 2. *Panaeolus foenisecii,* 4. *Peziza badia,* 5. *P. vesiculosa* and 6. *Pholiota aurivella* (credited with Rober Thomas and Dorothy B.Orr © California Academy of Sciences), 3. *Paxillus involutus* (credited with R.P. Olowin © St. Mary's College, California)

Plate 14

1. *Pholiota squarrosa* (credited with Robert Thomas and Dorothy B.Orr © California Academy of Sciences), 2. *Daedalea quercina* and 3. *Pleurotus ostreatus* (courtesy Ron Wolf), 4. *P. pulmonarius* (courtesy Dr. Nick Kurzenko), 5. *P. sapidus* and 6. *Pluteus cervinus* (credited with R.P. Olowin © St. Mary's College, California)

Plate 15

1. *Podoxis pisfillaris* (courtesy Michel Cloud Hughes), 2. *Polyporus frondosus* (courtesy Ron Wolf), 3. *P. squamosus* (courtesy Mark Radz), 4. *Psathyrella hydrophylla* (credited with R.P. Olowin © St. Mary's College, California), 5. *Ramaria apiculata* and 6. *R.obtussima* (courtesy Dr. Danielle Castronovo © California Academy of Sciences)

Plate 16

1. *Sparassis crispa* (credited with R.P. Olowin © St. Mary's College, California), 2. *Gomphus clavatus,* 3. *R. subbotrytis,* 4. *Russula aeruginea* (credited with Robert Thomas and Dorothy B.Orr © California Academy of Sciences), *5. Russula cyanoxantha* and 6. *R. emetica* (courtesy Ron Wolf)

Plate 17

1. *Russula foeteus,* 4. *Sarcodon imbricatus*, 5. *Schizophyllum commune* and 6. *Scleroderma aurantium* (credited with Robert Thomas and Drothy B.Orr © California Academy of Sciences), 3. *Lentinellus cochleatus* (courtsey Taylor F. Lockwood), and 2. *Russula sanguine* (credited with R.P. Olowin © St. Mary's College, California)

Plate 18

1. *Sparassis crispa* (credited with R.P. Olowin © St. Mary's College, California), 2. *Suillus grannulatus* and 5. *Tricholoma terrum* (courtesy Dr. Danielle Castronovo), 3. *Suillus sibircus* and 4. *Tricholoma imbricatum* (credited with Robert Thomas and Dorothy B. Orr), Figures 2–5 © California Academy of Sciences, 6. *Xylaria polymorpha* (courtesy Amdez Trnkoczy)

Plate 19

1. *Lepiota cristata*, 2. *Strobilomyces floccopus*, 3. *Termitomyces* sp., 4. *Tremella fuciformis*, 5. *Verpa bohemica* (courtesy Taylor F. Lockwood) and 6. *Calocera cornea*

Plate 20

1. *Hypholoma fasciculare* (courtesy Ron Wolf), 2. *Ptychoverpa bohemica* (courtesy Larry Evans) 3. *Laccinum scabrum* and 4. *Cortinarius violaceus* (credited with Robert Thomas and Dorothy B. Orr © California Academy of Sciences) 5. *Lentinula edodeds* and 6. *Lenzite betulina* (courtesy Taylor F. Lockwood)

Chapter 8

Harvesting of Mushroom Fruits (Sporophores)

Picking and Packing

It constitute the most expensive part of the overall production process. Harvesting of the mushroom fruits is done mechanically by employing labour. Mushrooms are picked at different maturation stages depending upon the species and upon consumer preference and market value. The temperature and humidity also influence the time of picking. Usually picking is done at the button stage of the sporophore *i.e.* when the cap is still being tight to its stalk. In harvesting, the cap is held with forefingers slightly pressed against the soil and twisted out. Specially designed wooden box is used for collection.

Since the picking and packing constitute the most expensive part of the overall production process of mushroom, several studies have been carried out to develop automated mushroom–harvesting machines ranging from small modular semiautomatic picking aids to fully automated robotic systems designed for line picking. Reed *et al.* (1995) evaluated the automated harvesting of *Agaricus bisporus* by machine at the laboratory level and the resulting pilot harvester was successfully tested on a commercial mushroom farm. The apparatus combines several handling systems and mechatronic technologies. Mushrooms are located and sized using image analysis and a monochromatic vision system. An expert–selection algorithm then decides the order in which they should be picked and what picking action (bend, twist or both) should be used (Tillett and Batchelor, 1991; Reed *et al.*, 1997). One of a pair of suction cup mechanisms attached to the single head of a Cartesian robot is then developed which can delicately detach individual mushrooms and place them gently into a specially designed, compliant finger conveyor. After high speed trimming, a gripper mechanism is finally used to remove mushrooms from the conveyor into packs at the side of the machine (Reed *et al.*, 1995, 2001).

Picking is also important because sooner the beds are completely cleaned, the sooner the next flush appears. Even a few mushrooms left until tomorrow, will delay it. After picking a complete flush, the concentration of the carbon-dioxide distinctly increased. By picking the still fully active and growing fruit body, an intensely respiring part, an exhaust pipe is cut off from the mycelium. The main metabolism is now shifted to the mycelium which is filled with water and nutrient of the fruit bodies that could explain the high carbon-dioxide. It is supposed that this causes the initiation of the next flush.

The harvested mushroom is stored at 4°C for a few days before marketing.

Threat of Heavy Metals in Mushroom Cultivation

Heavy metals occur as the natural constituents of the earth crust. Use of metal containing agricultural sprays, pesticides, herbicides and fertilizers further add to the contamination of soil. The metals like Arsenic, Cadmium, Chromium, Copper, Iron, Mercury, Manganese, Molybdenum, Nickel, Lead, Selenium, Vanadium and Zinc mostly come from industrial sources. Extensive work has been done on the uptake of heavy metals by micro fungi but very little is known about the uptake of heavy metals by macro fungi including mushrooms and its effect on their growth, productivity and nutritional quality (Gadd, 1986; Pighi *et al.*, 1989; Starling and Ross 1991; Wakatsuki *et al*, 1988). Whatever information are available on this topic relate to the cultivated species of mushroom which is far less compared to the wild edible mushroom. Works done on heavy metal uptake by mushroom from various substrates and effect of heavy metals on it have been reviewed by Mitra and Purkayastha (1995).

It has been demonstrated with *Agrocybe aegerita* that Cd applied to its fruit body is translocated to the successive harvests via the substrate and that the translocation of metal from the substrate to mycelium complex is increased in the presence of fruit body. It is greatest in *Pleurotus flabellatus* with recovery of 75 per cent of the applied Cd from the fruit bodies but least in case of *Agaricus bisporus* where only 1.3 per cent of Cd was recovered (Brunnert and Zadrazil, 1986).

Accumulation of metals occurs more in the pileus than in the stipe of the fruit body. (Yasui *et al.*, 1988). *Pleuotus ostreatus* grown on artificial compost containing mercuric nitrate has been reported to accumulate mercury 65-140 times higher than normal in fruit body. This has been also confirmed in case of wild mushrooms. Besides substrates or soil, mushrooms have also been reported to absorb radio active heavy metals from the atmosphere.

Observation with the metals *e.g.*, Pb, Cd, Zn, Mn, Co, Cu, Ni and Fe has reveled two important facts:

1. Different species of mushroom has preferential choice of metal accumulation. Thus *Agaricus bisporus* absorbed most of the Hg but less of Zn, Cd and least of Pb. On the contrary, *Pleurotus ostreatus* showed maximum uptake of Cd, minimum of Hg and Zn and no uptake of Pb.

2. The uptake of heavy metals such as Cd, Hg and Pb depends upon the concentration of heavy metals present in the substrate and also on the ability of mushroom species. The safe limit of Cd is 0.07 μg/kg for *Agaricus silvicola* and 0.05 μg/kg for *Clitocybe neoularia* but the maximum deposition by them was recorded 18.6 and 10.37 μg/kg fresh weight respectively. Similarly, *Agaricus compestris* was seen with maximum deposition of Pb as 2.44 μg/kg against its safe limit 0.5 μg/kg of fresh weight.

When the wild mushrooms *viz., Termitomyces robustus, Tricholoma lobayensis* and *Volvariella esculenta* were studied for their trace mineral content with reference to Cr, Co, Ni and Zn, it varied with species and parts of fruiting body rather than the stages of development. Zinc was most abundent, almost 5 times more than that of other three metals (Alof and Folaranmi, 1991).

The uptake of heavy metals from the substrate by mushroom mycelia and fruit body also depend largely on the nature of the species of the metallic compound because the availability of cations depends to a great extent on the respective anions. Hence, the presence of lead as acetate is sometimes more efficiently absorbed by lead sulphate may be due to its solubility. The availability of heavy metals could be greatly affected by the interaction of other metals present in the medium. It also depends upon the valence state of the individual metal. Yusui *et al.* (1988) reported that the growth or the uptake of essential nutrients *viz.* Zn^{2+}, K^{+}, I^{3+}, Mg^{2+},Mn^{2+} and Ca^{2+} by Oyster mushroom was not affected due to supplementation of Cu, Cd, Pb^{2+} or Ge^{2+} in the medium. But Poitou and Olivier (1990) observed that the mycelia of *Suillus granulatus, Lactarius deliciosus, Tuber melanosporum* and *T. brumale* rapidly accumulated Cu and the uptake of Cu affected the absorption of K and Mg ions which are essential for the growth of mushroom. Uptake of Cs, lower the amount of essential cations in the soil specially K^{+}. The interaction of Cu and Cd at low concentration (0.5 mg ml^{-1}) significantly reduced Cd uptake by *Volvariella volvacea* and *Pleurotus sajor-caju* but enhanced Cu^{2+} uptake. Cd was highly inhibitory while Cu was least to mycelial growth of aforesaid two mushrooms.

The heavy metals are also known to interfere with the growth of mushroom. Thus, high concentration of Hg^{2+} in the compost strongly inhibited the growth of *P. ostreatus* (Bressa *et al.*, 1988). Similarly the mycelial growth of *Suillus granulatus, Lactarius deliciosus, Tuber melonosporum* and *T. brumale* was strongly inhibited when exposed to different cupric salts at high concentration. Inhibition of mycelial growth has been also observed through 6 μg ml^{-1} of Cd in *V. volvacea* and through Cu and Cd both in *P. sajor-caju*.

The higher concentrations of heavy metals reduce the biological efficiency of fruit production. Sublethal concentration of Co, Cd, Cu, Pb and Zn reduced the fruit production in *V. volvacea* and *P. sajor-caju*. Co and Pb reduced the fruit formation in the two mushrooms respectively but it is increased through Cd in *Armillaria lutea. Thelophora terrestris* is more sensitive to Cu than Zn, toxicity not only checked the fungal growth but also caused abnormal morphological changes in the fungus. (Jain *et al.*, 1988).

Comparative study of edible and non-edible mushroom for their radio-cesium content has revealed considerable variations. Average As^{3+} content has been recorded 0.97 μg/kg (dry wt.) in edible and 1.16 μg/kg in non-edible mushrooms while the average quantity of Hg is 1.07 μg/kg in edible and 0.67 μg/kg in non-edible.

The growth factors like pH and $C/_N$ ratio have been reported to influence the uptake of metals by mushrooms. Mycelial uptake of Cd by *Pleurotus sajor-caju* and *Volveriella volvacea* increased with the gradual drift in pH of the medium from acidic to the neutral range. $C/_N$ ratio of the medium also markedly influence Cd uptake which increased with the increase in $C/_N$ ratio (Mitra, 1994).

The nutritive value of mushrooms usually depends upon the high protein content. Marked reduction in mycelial protein content of *A. bisporus* and *P. ostreatus* due to the uptake of heavy metals like Hg^{2+}, Cd^{2+}, Pb^{2+} and Zn^{2+} has been observed. Lead and Mercury have been reported to cause reduction in mycelial protein (36 per cent) and in fruit body of *P. sajar-caju*. The heavy metal induced reduction in protein content of mycelia and fruit of *V. volvacea* has also been reported after treatment with heavy metals. Aforesaid protein reduction may be assured due to chelation or binding of protein with heavy metals (Trevors *et al.*, 1986).

The works done so far on the heavy metal toxicity of mushroom provides only preliminary information. Very little is known regarding the mechanism of metal uptake by mushrooms, mode of action, preferential uptake, site of accumulation, molecular basis of toxicity and toxicity. Probability of metal contamination of cultivated species is insignificant provided the substrates and environment are kept free from excess toxic metals. But uptake of higher amount of toxic metals by wild edible mushrooms growing in polluted environment and subsequent consumption of these metal contaminated mushrooms by mammals may cause health hazards if adequate precautionary measures are not adopted in time.

Chapter 9

Pests of Mushroom and their Control

In spite of varied agro climatic conditions and huge agro and industrial wastes available in India, the pace of growth of mushroom industries is slow as compared to other mushroom growing nations. Among the several other factors, occurrence of various pests and diseases significantly affect the production and also the interest of the growers of this enterprise. This is more so because a large number of small and satellite growers cultivates mushroom either on long method (unpasteurised) compost or on pasteurized compost under natural climatic conditions and many growers abandon mushroom growing after 2-3 crops because of several pests and diseases (Sharma,1994-1995; Sharma and Vijay, 1996; Sandhu, 1995). Different pests which have been recorded in India have been summarized in Table 2 along with symptoms and economic importance.

Table 2: Pests of Mushrooms in India

Sl.No.	*Pests*	*Symptoms*	*Loss (per cent)*
		A. Insects	
1.	**Sciarids** *Bradysia tritici* *Lycoriella auripilla*	Restricted spawn run, mycelial attachment of pinheads damaged, brown and leathery pinheads, hollow and spongy pinheads, tunneling of stipe.	50
2.	**Phorids** *Megaselia agarici* *Megaselia* spp.	Yellowish brown caps, tunneling of stipe and leathery brown pinheads. Formation of clean wet zone around the vent.	46

Contd...

Table 2–*Contd...*

Sl.No.	*Pests*	*Symptoms*	*Loss (per cent)*
3.	**Cecids** *Heteropezina cathiseus*	Vertical grooves in the stipe, discolored Tiny pustules of black fluid on stipe. Reduction in mushroom size.	19
4.	**Springtails** *Lepidocyrtus* sp. *Xenylla* sp. *Seira iricolor*	Disappearance of mycelium, slight pitting and browning of caps destruction of gills linings. Formation of webby structure, scrapping of stalk, withered primordia.	–
5.	**Beetles** *Staphylinus* sp. *Alphitobious laeoigatus* *Cyllodes whiteii*	Small irregular holes in hymenium and stipe.	–
6.	**Moth**	Tunnels and pits in mushrooms.	–
		B. Mites	
	Tyrophagus putrescentiae *Tyroglyphus berlesei* *Rhizoglypus echinopus* *Histiostoma heinemanni* *Hypoaspis miles*	Red brown or rust brown stipe. Feeds on mycelium and mushroom tissue. Loose attachment of mushroom to the substrate. Shrunk cap and brown rusted spots.	90
		C. Fungi	
1.	**Parasitic**		
	Verticillium fungicola	Light brown superficial spots on cap, splitting of stipe, tilted cap.	38
	Mycogone perniciosa	Distorted mass of mushroom tissue, swollen stipes, whitish mouldy, growth on fruit bodies, oozing of brown to amber colored liquid.	–
	Cladobotryum dendroides	Mycelium is grayish to white turning reddish on maturity, envelopes the fruit body.	66.6
2.	**Competitor moulds**		
	Papulospora byssina	Whitish patches on casing, compost turning to rust color.	59.87
	Scopulariopsis fimicola	Dense white patches of mycelium on casing soil.	–
	Coprinus spp.	Rapid appearance of long slender, thin caps, disintegrate into black slimy mass of spores.	94.4
	Chaetorium spp.	Grayish white mycelium on compost, turning olive green.	53.63
	Dactylium dendroides	Cobweb disease	
	Verticillium fungicola	Dry bubble disease of fruit.	–
	Myceliophthora lutea		
	Chrysosporium merdarium		

Contd...

Table 2–*Contd...*

Sl.No.	Pests	Symptoms	Loss (per cent)
	Sepedonium chrysospermum	Brown yellow patches with a white fluffy edge generally on the bed, corky layer between casing and compost.	100
	Sporendonema purpursecens	Whitish gray mouldy fluff with little white bells on straw. Color changes to cherry red on spore formation.	–
	Dichliomyces microsporus	White fluffy mycelium, turning creamy yellow thick solid wrinkled mass resembling brains like structure.	–
	Tricoderma aggressivum *T. aggressivum* *T. europaeus* *T. Harziannum* *T. koningii* *T. lignorum* *T. viride* *Penicillium* spp. *Aspergillus* spp.	Small blue green coushions on spawned and cased trays, caps turn brown, lop sided.	81
	Doratomyces stemonitis	Mycelium dark and forms black or gray brittles.	–
	Peziza ostracoderma (*Chromelosporium fulvum*)	Round white mycelium on casing layer, saucer shaped yellowish brown fruit bodies.	–
	Cephalothecium roseum	White mouldy growth turns pink in due course in compost or casing soil.	–
		D. Bacteria	
	Pseudomonas fluorescens *P. agariici* *P. alcaligens*	Brown to black lesions on cap. Drippy gills, blotches of varying sizes and colors on pilei, slimy appearance of fruit body yellowing, water soaked area, yellow brown discoloaration of young sporophores.	100 in *Pleurotus*
	P. stulzeri *Pleurotus*	Brown spots in substrate.	27–61 in
		E. Viruses	
		Mycelium does not permeate or disappears after normal spread, dense cluster of mushrooms. Delayed appearance of first flush, formation of fruiting priomordia below casing, slow development of pinheads, watery stipes, streaking in the stipes, hard gills, abnormal veils, musty smell.	

When one speaks of mushroom pest, nematode deserves special mention because it is highly sensitive to the attack of fungal feeding nematodes. Presence of mushroom nematodes in the cropping system poses a serious threat and a most difficult problem in cultivation to solve, as these are only ectoparasite in whole nematode kingdom which have adapted themselves so well with the environment of cropping pattern of mushroom that they are capable of causing frequent and total crop failure.

The nematodes associated with mushroom can be classified into five groups on the basis of their nature of parasitism. Of these, the main culprits are the myceliophagous forms which finish all of mycelium by direct feeding and not allow the one set of sporophore formation in extreme cases.

Two orders namely Tylenchida and Aphlenchida represent this category. The nematodes of the orders attacking mushroom have been summarized in the Table 3.

Table 3: Myceliophagous Nematodes Associated with Mushroom

Sl.No.	Nematode Species	Pathogenicity	References
A.	**Tylenchida**		
	*Ditylenchus myceliophagous**	+++	Newstead, 1906; Goodey, 1960; Arrold and Blake, 1968; Sharma *et al.*, 1985; Khanna, 1993.
B.	**Aphelenchida**		
	*Aphelenchoides agarici**	+++	Seth and Sharma, 1986; Khanna and Sharma, 1988a
	*A. asterocaudatus**	–	Bahl and Prasad, 1985
	A. bicaudatus	+	Mc Leod, 1967
	A. coffeae	+	Mc Leod, 1968
	*A. composticola**	+++	Franklin, 1957; Cayrol, 1967; Khanna and Sharma, 1988b; Khanna, 1991
	A. cyrius	+	Paesler, 1957
	A. dectylocercus	++	Hooper, 1958
	A. helophilus	–	Paesler, 1957
	A. limberi	++	Cayrol, 1962; Choleva, 1969
	*A. minor**	–	Seth and Sharma, 1986
	*A. myceliophagus**	++	Seth and Sharma, 1986; Khanna and Sharma, 1988a
	*A. neocomposticola**	++	Seth and Sharma, 1986; Khanna and Sharma, 1988a
	A. parietinus	–	Paesler, 1957
	*A. sacchari**	+++	Paesler, 1957; Janowicz, 1978; Sharma *et al,* 1981
	A. saprophilus	++	Hooper, 1958; Mc Leod, 1968.
	A. spinosus	–	Paesler, 1957
	A. subtenuis	–	Hooper, 1962
	*A. swarupi**	++	Seth and Sharma, 1986
	Aphlenchus avenac	+	Hooper, 1962a; Mc Leod, 1968; Rao *et al.,* 1992 a
	Paraphlenchus myceliophthorus	++	Goodey, 1958a; Goodey, 1960
	P. pseudoparietinus	–	Farkas and Korenczy, 1974
	Seinura winchesi	++	Paesler, 1957; Farkas and Balazs,1975

*: Species reported from India; +++: Highly destructive; ++: Pathogenic; +: Found to multiply on mushroom; –: Pathogenecity not studied.

Saprophagus nematodes seem to cause indirect damage only when present in enormous number. The only nematode order that represent saprophages in mushroom cultivation is Rhabditida. The genera of common prevalence in the beds are *Acrobeloids, Caenorhabditis, Cephalobus, Diplogaster, Penagrolaimus* and *Rhabditis*. Since these nematodes do not feed upon the mycelium directly they were early thought to be harmless. But whenever their population is enormous, even in absence of myceliophages, the mushroom beds are severely damaged. Also certain excretory metabolities released by them may be playing a negative role. So the aspect of their exact role in mushroom industry definitely needs to be exploited.

Predatory nematodes, generally encountered in small numbers in mushroom beds belong to order Mononchida. Since they feed upon other forms of nematodes including myceliophagous forms, they have beneficial role in mushroom cultivation. However, their role as biocontrol agent against harmful nematodes has remained a possible but unattended management strategy till date.

Feeding Behaviour

The fungal feeding nematodes pierce the hyphal cell wall (mycelium) and suck the cell contents by to the fro movement of their hollow needle like stylet, thus leaving the cell devitalized. After feeding upon one cell, the nematode shifts to another with the help of a moisture film present in the compost medium. The optimum temperature for feeding and multiplication ranges between 22 to 28°C.

Symptoms

Compost trays/bags are usually not disturbed after spawning, making the growers unaware of nematode entry. As a result, the symptoms are overlooked. When whiteness of the spawn run starts vanishing suddenly, first symptom of nematode attack is evident. If observed carefully, the following symptoms appear in namic infestation.

1. Mycelial growth is sparse and slow. Mycelium develops in patches due to uneven initial distribution of nematodes.
2. Mycelium turns stingy.
3. The compost surface sinks and it has a typical nemic odour.
4. Button production is delayed and reduced.
5. Alternate medium to very poor sporophore yields are obtained in succession of flushes.
6. The number of flushes and duration of crop is reduced.
7. Total crop failures are common if nematode finds entry into the beds of spawning time in big numbers.

The discovery and subsequent use of DDT and other pestisides by the agricultural community initiated what is commonly called the "Pestiside age".

Such a dependence on chemicals led to undesired effects, and alternative were sought. About 1956, a new concept of pest control termed as "Integrated Pest Management" (IPM) appeared. According to 'Insect Pest Management and Control' from the National Academy of Sciences, IPM is "applied pest control that combines and integrates biological and chemical measures into a single unified pest control programme. Chemical control is used only where and when necessary, and in a manner that is least disruptive to beneficial regulating factors of the environment. It may make use of naturally occurring insect-parasites, predators and pathogens, as well as those biological agents artificially increased or introduced."

Integrated Pest Management

The mushroom IPM programme can be described in terms of the following five steps.

Acquiring Pest Information

Knowledge of biology and behaviour of insects and the conditions which favour the colonization and spread of disease pathogen is essential. *Verticillum*, for example, has a sticky spore. Controlling the spore will control the disease. Spores are spread by attaching themselves to dust particles, insects, water droplets or harvesting equipment etc. Even a grower's hand can spread these adhesive spores. Virus is spread through mushroom spores and mycelium. Similarly, insect and other pest have to be studied with respect to their biology, behaviour and requirements before adopting IPM programme.

Monitoring the Farm Environment

When the colonization of pest occurs, the size of initial population and prediction on the future size of the pest population are an integral part of any management programme. Monitoring pest population provides this information. Disease can be monitored either by spore traps, percent area of fruiting body damaged or bed/bag area covered. In addition to the monitoring of a farm climate and pest, growth particles are also required to be monitored.

Decision Making Based on the Determination of Economic Levels of Damage

Economic threshold have to be worked out for each pest and disease. Below a certain pest population, the return will not equal or exceed the investment on pesticide or other management activity. Even if economic threshold levels can not be computed, action levels, can be decided upon by previous experience with pest. Action levels refer to doing something when pest levels reach a predetermined size or incidence known to be deleterious to the crop.

Techniques to Manipulate Pest Population

The fourth procedure involved in the execution of the mushroom IPM is not actually a step, but rather the selection, integration and implementation of several individual control techniques into a whole system. There are many types of control techniques used in IPM. These take various forms, including exclusion, cultural, chemical and biological.

Cultural Practices and Exclusion

Composting is actually a kind of cultural control. Ideally, properly composted materials are selective for the growth of *Agaricus*. If phase-I and II are done properly, including an effective pasteurization, most weed moulds, nematodes and mites that can feed upon them will be eliminated and will not able to grow. High compost temperature after spawn introduction and cropping may encourage several serious pests. Similarly, high humidity free water on cap may also favour some parasitic moulds and bacterial diseases.

Sanitation is an important cultural practice which covers all the measures which are necessary to give pest and disease as little chance as possible of developing and spreading.

Biological Control

Basically, the term biological control is used to describe pest control by the use of predators, parasites and pathogens, however, other means of control which can be termed 'biological' have been included within this general expression. These include pheromones, kairomones, repellents, antifeedents and genetic control.

Advantages

A number of advantages are claimed for biological as opposed to chemical method of control. These include: (i) **pest specificity**–affecting only the target insects; (ii) **safety**–biological control methods rarely leave toxic residues; (iii) **availability**–the beneficial organisms usually already exist; (iv) **proactivity**–some beneficial organisms can seek out and find the host; (v) **dynamism**–some beneficial organisms can increase in number and spread from initial point to introduction; (vi) **continuation of efficacy**–the pest will be slow or sometimes only one introduction is necessary and (vii) **registration costs**–some biological control agents are exempt from some of the costly registration procedures that chemicals have to go through.

Disadvantages

As with any method of control, there are a number of disadvantages to the use of biological agents, these include: (i) **the speed of control**–by their very nature some biocontrol agents can take time to show an effect; (ii) the degree of control–unless used in the manner of an 'insecticide' they may not act as an eradicant; (iii) **unpredictability**–due to conditions 'invisible' to the growers, they can often be variable in their efficacy; (iv) **cost**–because they are biological in nature, they may be expensive and difficult to develop; (v) **ease of application**–they may require more expert supervision in their use; and (vi) **shelf life**–some biological control agents, especially those that are animal in nature are temperature sensitive and need to be applied soon after purchase.

Biological control agents often cannot be used in isolation. They must mostly be used in conjunction with other control measures within an integrated pest management system. The effect of these other control systems invariably insecticides or fungicides based on the various biological control agents must be assessed before considering the use of such programme.

Methods

Predators

predators are organisms that consume their prey, either partially or entirely. The typical insect predators include midges, beetles, bugs and mites and for mushroom insect pests, mites hold most promise. Mites, who are known to be predacious on Diptera, Acarina and Nematode are obliviously of most interest in mushroom pest control. Various mites which are predatory of different pests of mushroom are summarised in Table 4.

Table 4: Predatory Mites on Different Pests of Mushrooms

Predators	*Host*	*References*
Hypoaspis miles	*Bradysia* spp. *Lycoriella auripila*	Wright and Chambers (1994) Lind (1993)
Geolaelaps aculeifer	*Bradysia spp.*	Gillespie and Quiring (1990)
Digamasellus fallax	Nematodes	Binns (1973a, b)
Parasitus fimetorum	Sciarids Phorids Cecids Nematodes	Binns (1973a, b)
Arctoseius cetratus	Sciarid larvae and eggs Cecid larvae, Tarsonemid mite, Nematodes	Binns (1973a, b)
Linopodes antennacpes	Tarsonemus myceliophagus	Hussey *et al.* (1969)
Parasitus bituberosis	Sciarid larvae and eggs Cecid larvae, Tarsonemid mites, Nematodes, Spring tails	Al-Amidi and Downes (1990)
Macrocheles merdarious	Dipteran pests Tyroglyphid mites	Binns (1973a)

Parasites

These are the animal which normally develops within their host over a period of time. An obligate parasite lives within its host without killing it whereas a parasitoid lives and develops within its host for a period of time before eventually killing it. Three nematode genera namely, *Howardula, Steinernema* and *Heterorhabditis* have been reported parasitic on a large number of insect-pests (Richardson and Chanter, 1979; Richardson, 1987; Richardson and Grawal, 1991). *S. feltiae* and *H. bacteriophora* have been shown parasitic on sciarids, phorids and cecids. *S. carpocapsae* alone can infect more than 250 species of insects from over 75 families in 11 orders (Poinar, 1979). A large number of commercial products have been developed (Table 3) and some of them are being used by the mushroom growers to control the mushroom pests.

Pathogens

This group includes bacteria, fungi and viruses that can selectively infect specific insects and cause their subsequent death. Bacteria and fungi are of potential significance for the biocontrol of mushroom pests and viruses are yet to be exploited.

Bacteria

The most common bacterium used for pest control around the world is *Bacillus thuringiensis* (*Bt*). After a period of active growth, this bacterium produces a protein crystal and a spore, both are having an insecticidal action, although it is the crystal that is generally most active. Mushroom pests which have been controlled to the level at par with pesticides by using *B. thuringiensis* subsp. *isaraelensis,* its strains and formulations are summarized in Table 5. Another disease of worldwide distribution and significance, bacterial blotch, caused by *Pseudomonas tolaasii,* has also been controlled by the commercial product *victus;* containing *P. fluorescence* biotype G. (Miller, 1993).

Table 5: Some Commercial Products Containing *Steinernema* and *Heterohabditis* Nematodes

Formulation	*Nematode*	*Product*	*Company*
Alginate gel	*S. carpocapsae*	Mioplant	Dr. R. MAAG, Austria
	S. carpocapsae	Boden-Nutzlinge	Celaflor, Rhinepoulenc, Germany
Flowable	*S. carpocapsae*	Bio-safe	SDS-Biotech, Japan
	S. feltiae	Exhibit	Ciba Geigy, Europe
	S. feltiae	Stealth	Ciba Giegy, UK
Clay	*H. bacteriophora*	Cruiser	Ecogen, USA
	H. megidis	Nemasys-H	Agric. Generc. Co., UK
		Larvanem	Koppert B.V., Netherlands
	S. faltiae	Nemasys	Agric. Generc. Co., UK
		Entonme	Koppert B.V., Netherlands
	S. scapterisci	Proactant	SS Biocontrol Inc., USA
Sponge	*S. carpocapsae*	Gurdian	Hydro-Gardens, Inc. USA
	S. carpocapsae	Scanmask	Biologic, USA
	H. bacteriophora	Lawn Patrol	Hydro-Gardens, Inc. USA
Water dispersible granules	*S. carpocapsae*	Biosafe	Biosys, USA
		Biosafe-N	Biosys, USA
		Biovector	Biosys, USA
		Bio–Flea-halt	Farnam, USA
		Interrupt	Farnam, USA
		Sudbury-Nature Gard	Farnam, USA
		Vector TL	Lesco, USA
		Helix	Ciba Geigy, Canada
		Millennium	Van Water and Logers, USA
S. feltiae	X-GNAT	E.C. Geiger, USA	
		Magnet	Amycel-Spawn Mate, USA
	S. riobravis	Bio vector	Biosys, USA
		Vector MC	Vector MC Lesco USA

Source: Grewal and Smith (1995)

Fungi

A number of phatogenic fungi are capable of invading the bodies of insects, mites and nematodes. Different fungi which have been shown to be potential biocontrol agents against mushroom pests are given in Table 6.

Table 6: Pathogenic Bacteria and Fungi as Mushroom Pests

Sl.No.	Pathogens	Host	References
A.	**Bacteria**		
1.	*Bacillus thuringiensis* Subsp. *israclensis* (B.t.i.)	*Lycoriella mali*	Cantwell and Cantelo (1984)
2.	Formulation of B.t.i.	*L. mali* *M. halterata*	Keil (1991)
3.	Different straints of B.t.i. (Strains GC 315 and GC 327)	*L. auripila*	White and Jarrett (1990) Pethybridge (1991) Butt *et al.* (1994)
4.	*Pseudomonas fluorescence biotype* (Victus) G.	*Pseudomonas tolaasii*	Miller(1993)
B.	**Fungi**		
1.	*Verticillium lecanii*	*L. auripila*	White (1995)
2.	*Pandora gloeospora*	*L. mali*	White (1995)
3.	*Arthrobotrys robusta*	Nematodes	Hesling (1978)
	A. irregularis	Nematodes	Hesling (1978) Anonymous (1987)
	A. conoides	Nematodes	Grewal and Sohi (1988)
4.	*Candelalretta musiformis*	Nematodes	Anonymous (1987)
5.	*Pleurotus sajor-caju*	*Aphelenchoides composticola*	Sharma (1994)

Pheromones

Pheromones are volatile chemical substance which are secreted and released by the animals. Their function is usually to aid detection and/or to provoke a response by the same species and is often sex-oriented. In biological control regimes they can be either used "live" (*i.e.* using captured insects to attract others) or, more normally, a synthetic analogue of the natural pheromone is used for the same purpose. They can be used either for monitoring pest populations, so that other control methods can be used more effectively; or they can be used in mass-trapping exercises to remove one of the sexes from finding each other. This is aimed at limiting or entirely stopping as infestation by that pest. Pheromones are safe to use and are normally extremely species specific. These are more useful when other biological system is not to be disturbed. The detection and use of pheromones in pest management of mushroom is as under:

Sciarids

The existence of sex pheromones in *L. mali* was demonstrated by Girard *et al.* (1974) and pheromone constituents were determined by Kostelc (1977) where in heptadecane was the most active. However, Binns (1976) and White (1995) did not get encouraging results for trapping *L. auripila* with this compound.

Phorids

Burrage (1981) shows the existence of sex pheromones in *M. halterata* virgin females. Subsequent isolation, identification and synthesis of the presumed pheromones were carried out by Baker *et al.* (1982).

Cecids

Pheromones induced dispersive behaviour has great potential in the management of cecids which results only in the production and development of sterile males.

Kairomones

These are volatile chemical substances which are produced by the food source of animal and are used by that animal for its location. Kariomonal activities have been observed in exhaust gases of the compost for sciarid females *L. auripila* and in smell of growing mycelium for phorid females (*M. halterata*) but practical utility for managing these pests has so far been unsuccessful.

Plant Parts/Products

A good number of plants are known to possess pesticidal properties against mushroom nematodes and moulds and their use probably would be the safest if recommended as a part of compost ingredient. Fresh leaves of *karanj* (5 per cent) and *neem* leaf powder (2-5 per cent w/w) when mixed in compost reduce the nematode population and increase the yield of sporocarps (Rao and Pandey, 1991; Sharma and Khanna, 1992). Incorporation of oil cakes like neem (*Azadirachta indica*), karanj (*Pongamia pinata*), coconut (*Cocus nucifera*), castor (*Riccinus communis*) and ground nut (*Arachis hypogaea*) in compost before spawning with nematicidal application (Rao *et al.*, 1991) also proved benificial.

Genetic Control

At present there is no strain of mushroom which is resistant to pest and diseases. Selecting a strain that does not anastomose as readily with virus infected strain can help to reduce the virus inoculam at the farm. *Agaricus bitorquis* is reported to be tolerant to virus diseases. *Pleurotus sajor-caju* and *Stropharia rugoso-annulata* have been reported resistant against *Aphelenchida sacchari* and *Ditylenchus myceliophagus*. Strain K-30 of *A. bitorquis* has also been reported resistant against *A. sacchari* (Sharma, 1995).

Chemical Control

Dependence on chemical control has several major drawbacks, including resistance, resurgence, legal complications and high costs. There are a limited number of pesticides registered for use on mushrooms because mushrooms are grown indoor

Table 7: Recommended Chemical Pesticides for Mushrooms

Sl.No.	*Active Ingredient*	*Trade Name/Formulation*
1.	**Mushrooms (Mushroom Houses with Crop Present)**	
a.	**Insecticide**	
	Allethrin + Piperonyl Butoxide	Alleviate 25-5
	Azadirachtin	Azatin 3 EC and CWP
	Diazinon	Diazinon
	Dichlorvos	DDVP 5 EC
	Diflubenzuron	Dimilin 25 WP
	Malathion	Various trade names
	Methoxychlor	Various trade names
	Permethrin	Pounce 3.2 EC and 25 WP, Ambush 3 EC
	Pyrethrins + Piperonyl Butoxide	Pyrenone
b.	**Fungicides**	
	Benomyl	Benlate 50 DF
	Chlorothalonil	Bravo 500
	Thiabendazole	Mertect 340 F
c.	**Sanitizers**	
	Calcium Hypochlorite	Various trade names
	Chlorine Gas	Various trade names
	Chlorine Dioxide	Carnebon, Oxime
	Formaldehyde	Various trade names
	Sodium Hypochlorite	Various trade names
2.	**Mushroom house Soils (Compost and Casing)**	
a.	**Insecticides**	
	Azadirachtin	Azatin 3 C and WP
	Cyromazine	Armor 5 S
	Diazinon	Diazinon 7 E
	Diflubenzuron	Dimilin 25 WP and 4 L
	Malathion	Malathion 5 EC and 80 EC
	Methoprene	Apex 5 E
b.	**Fumigants**	
	Chloropicrin	Metapicrin, Pic-C100
	Formaldehyde	Various trade names
3.	**Mushroom Houses and Equipment (Empty)**	
a.	**Insecticides**	
	Allethrin + Piperonyl Butoxide	Alleviate 25-5
	Diazinon	Diazinon
	Dichlorvos	DDVP 5 EC

Contd...

Table 7–*Contd...*

Sl.No.	*Active Ingredient*	*Trade Name/Formulation*
	Malathion	Various trade names
	Methoxychlor	Various trade names
	Permethrin	Pounce 3.2 EC and 25 WP, Ambush 3 EC
	Pyrethrins + Piperonyl Butoxide	Various trade names
b.	**Sanitizers/Fumigants**	
	Chlorine Dioxide	Camebon, Oxime
	Formaldehyde	Various trade names
	Iodine + alcohol complex	Various trade names
	Methyl Bromide	Various trade names
		Pheno Cen
		Sectrol
		Stericide
		1–Stroke Environ

and have a very short shelf-life. Mushroom themselves are fungi and most of the pathogens are also fungi. Thereby making the choice of fungicides very difficult. Moreover, because of short cropping cycle, residual toxicity of different chemicals is of great concern and this must be below the tolerance limit. Mushrooms are very sensitive to fumes, toxic gases and several chemicals. This also limits the free and frequent use of chemicals in mushroom industry. Equally important factor which limit the use of chemical for the management of pest in mushroom is the problem of acquired resistance. Repeated and regular applications of any chemical greatly increase the chance of resistance developing. If equally effective alternate chemicals are available, the problem of pesticide resistance can be avoided. However, in mushroom production there are rarely suitable alternatives. There are, unfortunately few pests or diseases that can be controlled satisfactorily by environmental manipulation alone except bacterial blotch. Since pesticides appear to provide the most effective form of pest control, they should be most judiciously used in the IPM program as a last resort. They should be saved for when they are really needed. Pesticide application should be based on monitoring, established economic threshold and temperatures. Various chemical pesticides which have been recommended for use on mushrooms, compost and casing mixture, and empty mushroom house and equipments are given below:

Following are some of the recommended fungicides for the control of some of the main fungal pathogens of *A. bisporus* (Fletcher *et al.*, 1986).

1. Benomyl (Benlate 50wp) = *Dactylium, Mycogone, Trichoderma,* and *Verticillium* = Mix 240g/100 m^2 with casing or add in water at 240g/200 liters/100 m^2 during first watering.
2. Carbendazim (Bavistin) = same as for Benomyl.

3. Chlorothalonil (Bravo or Repulse) = *Mycogone* and *Verticillium* = Apply as spray 1 week after casing and repeat not less than two weeks laler at 200 ml in 100-200 liters/100 m^2.
4. Prochloroz manganese (Sporgon) = *Mycogone, Verticillium, Dactylium* = if single application, apply 300g/100 liters/100 m^2, 7-9 days after casing. For double applications, use 113g/100 liters/100 m^2, 7-9 days after casing then between second and third flushes. For triple application, use 57g/100 liters/100 m^2, 7-9 days after casing and after first and third flushes.
5. Thiabendazole (Tecto) = *Dactylium, Mycogone, Verticillium* = same application rate as Benomyl.
6. Zineb (Zineb, Tritofloral) = *Dactylium, Mycogone, Red Geotrichum* and *Verticillium* = For 7 per cent dust, apply at 350g/100 m^2 every week after casing or 140g/100 m^2 before watering. For wettable powder, 1 kg/1000 liters at the rate of 5 liters/10 m^2 after casing and between flushes. For tritofloral 5 kg/100 m^2 between flushes.

Chapter 10
Postharvest Management of Edible Mushrooms

Like all fleshy fruits and vegetables, mushrooms are highly perishable because of their high moisture content and delicate nature, and cannot be stored for more than 24 hours at ambient temperature. Once the fruiting body matures, degradation process starts and it becomes unconsumable after sometime. Development of brown colour is the first sign of deterioration and is a major factor contributing to quality losses. The enzyme, polyphenol oxidase, in the presence of oxygen and the substrate catalyses the oxidation of colourless phenolic compounds into quionones which combine with amino acid derivatives to form highly coloured complexes thus making them highly unacceptable and, therefore, should be disposed off as soon as possible.

There are two main problems in so far as the marketing of fresh mushrooms is concerned. The first one is that mushroom is delicate and highly perishable item. It gets bruised and damaged in transit unless packed with skill and handled carefully. Slightly high temperature (30°C or above) will turn it brownish in colour, in which case there will be no buyers. Another difficulty which is experienced only in our country is that a slightly open mushroom is rejected outright as unfit for human consumption. In Europe and America mushrooms are sold in all three different stages of their growth *viz.*, buttons, opens and cups; preference being given to the latter two categories for normal daily use.

Another constrain of mushroom market is its poor demand, although there is no dearth of market in India as it is a vast country and has a large population. If each individual in the country is to eat even 10 gm of mushroom in a year, we would need roughly 6000 tones for our home consumption alone. The average rate of per capita consumption in western countries is of course very high; the highest being that of

Germany with a figure of 2020 gm in1974 followed by Canada and France (Singh 1977).

The reason behind poor consumption of mushroom in India is primarily ignorance. Even educated class of people in India has little knowledge about the properties of cultivated mushroom. A number of individuals firmly believe these mushrooms are poisonous and cause hallucinations. Further that these mushrooms are eaten for their taste but have no food value. There is another category that enjoys eating but do not know how to cook.

To improve the mushroom market is to educate the public on the following points

1. Cultivated mushrooms are not poisonous.
2. The food values of mushroom vis-à-vis other vegetables, fruit and meat.
3. Mushrooms which have been slightly bruised or have turned brownish do not loose their face value or taste.
4. Open mushrooms also have the same food value and taste as the buttons. They can conveniently be used for making pickles and enhancing flavour in day to day preparations.

A concerted effort has to be made to drive these points home to begin with, in big cities and then gradually extending the efforts to smaller towns. This could be achieved by suitable articles in newspapers involving eminent doctors and scientists, arranging TV and Radio talks. Mushroom market can also be improved by cooking demonstration and issue of small booklets showing mushroom recipes. Inspite of witnessing high consumption and appreciation the western countries spent vast sums on publicity to make mushrooms popular. Unfortunately this aspect is completely ignored at present in India.

Price is another constrain in popularisation of mushroom. The current price at which mushrooms are available to the consumer is very high compared with other vegetables or even meat and chicken. Although production cost of mushroom is low but false hike in its price is due to the middle man who takes the maximum share of profit. If somehow the middle man is eliminated, the price of mushroom would come down to about Rs. 8/- per kg. and this would increase the number of buyers many folds.

A number of western countries produce surplus mushroom and they export canned mushrooms. But at present we cannot compete with them because of the aid given by the respective Government to the mushroom industries and the cost of production is low. Canning of mushrooms in India is expensive due to the cost of empty tins. Secondly due to lack of proper facilities and insufficient knowledge we have not been able to achieve the target of getting 20-25 kg of mushrooms per square meter of bed area. When this is achieved, the cost of production is automatically reduced by half or even more.

Government help is needed by the Mushroom Industry which is in its infancy, specially if we are to compete in the world market. It could be done in the following manner:

1. The tin plate for canning mushrooms for export only should be provided at international rates.
2. There should be no levy of excise and other duties on canned mushrooms meant for export.
3. Greater emphasis be given to research work to make better quality spawn which would give higher yield per sq. meter.
4. Special subsidy on export of mushrooms be given in its initial stages to be able to compete with countries like Taiwan and N. Korea.

In the peak period of harvesting due to gluts in the market and owing to highly perishable nature, its preservation into more stable products is of great importance. Development of appropriate storage and processing technology in order to extend their marketability and availability to consumers in fresh as well as processed form is of great significance. Short term preservation methods like pre-packaging coupled with low temperature storage, irradiation and steeping preservation help to prolong storage life from one to three weeks. Prolongation of stalks, opening of veils, toughening of texture and loss of whiteness are deteriorative changes observed during prolonged storage. Long-term preservation methods, such as canning, drying, pickling, etc., can make the availability of mushroom of a good quality throughout the year at reasonable cost. Freezing and freeze drying methods have also been recommended for long-term storage but these are energy intensive, sophisticated and cost prohibitive techniques to be adopted for local markets. However, such techniques may be used for the export of mushroom products.

White Button Mushroom (*Agaricus bisporus*)

Quality Parameters

Fresh mushrooms are white or light buff without dark marks on either cap (pileus) or stem (stipe). The veil is closed and the gills (lamellae) are not visible. The upper surface of the cap should be strongly convex and stem should be plump, rather than elongated. To maintain whiteness, dipping of mushrooms in dilute solution of hydrogen peroxide (1:3) for half and hour and then steeping in 0.25 per cent citric acid solution containing 500 ppm sulphur dioxide had significant effect (Pruthi *et al.*, 1984).

The white button mushroom should be harvested when cap size is 30-45 mm in diameter. In general, button mushroom should be harvested at a stage when cap diameter is twice the length of stipe/stem (Kohlil, 1984). The shelf-life is very short extending upto one or two days under ambient conditions. The post-harvest maturation of the fruit body during storage causes deterioration in the quality. Discolouration is caused by drying, damage and enzymatic browning. Barwal (1992) observed that pre-harvest spray of ascorbic acid (2 per cent) improve the colour by inhibiting the polyphenol-oxidase activity, and responsible for browning.

Packing

Before packing in suitable containers, the mushrooms immediately after harvest are trimmed and washed to remove casing soil, compost and other foreign material.

Mainh *et al*. (1983) have reported that washing of mushroom prior to packing is very important for enhancing the shelf-life and thus extending the marketing period. Besides moisture, mushrooms are also sensitive to desiccation and drought, consequently a suitable package is very important during storage. Mushrooms are usually packed in polypropylene bags of 250-500 gm capacities. Quantities more than this have a tendency to loose acceptability. Mushroom can also be packed in perforated polythene pouches (Seth, 1980). Chopra *et al*. (1985) recommended 100 gauge polythene bags with 0.5 per cent venting area for packing mushroom for the refrigerated storage. On the other hand, studies have shown that polythene bags should not be perforated and be carefully sealed air-tight. Saxena and Rai (1988) used non-perforated polythene bags for the storage of button mushrooms for 4 days at 5°C. For transporting mushroom to long distance, polythene bags cannot be used for packing. Polystyrene or pulp-board punnets should be used for packing large quantities of mushrooms to distant markets. The punnets are over-wrapped with differentially permeable polyvinyl chloride (PVC) or polyacetate films. These over-wrappings create modified atmosphere in punnets producing an atmosphere at about 10 per cent CO_2 and 2 per cent O_2.

Transportation

The precooled mushrooms should be packed in insulated ice containers for 8-9 hours for distant transport. The container is usually lined internally with 4" thermocole which is covered with tin sheet on both sides to give it strength. Ice is usually placed at the bottom with a shelf put on top in which the mushrooms are preserved. In this way, mushroom remain healthy and white in colour (Seth, 1980).

Storage

Short-term Storage

After harvesting, the quality of mushrooms deteriorates quickly at ambient temperature, the stalks grow longer and the caps grow larger, and at last they open, exposing the gills. Further, the product becomes brown and tough, and is unsaleable. Treatment with solution of salt (2 per cent) and potassium metabisulphite (0.15 per cent) has been effective in reducing discolouration. Sethi and Anand (1978) reported that mushroom cannot be stored for more than 24 hours at room temperature or one to two weeks at refrigerated temperature. Mushrooms can be stored in non-perforated polythene bags for 4 days at 5°C without much deterioration, and the washing of mushrooms in 0.5 per cent potassium metabisulphite improves the whiteness which eventually deteriorates slowly during storage (Saxena and Rai, 1988). Maini *et al*. (1987) reported that treating mushrooms at lower concentration of potassium metabisulphite was found suitable for short time storage (24 hours). However, for long time storage, citric acid (0.1 per cent) and potassium metabisulphite (0.03 per cent) or the combination of citric acid (01 per cent) and tartaric acid (0.3 per cent) were found most suitable and white colour was retained upto a storage period of 48 hours In the treatment, where potassium metabisulphite was added to the citric acid and tartaric acid combination, there was a development of yellowish colour at the end of storage period of 36 hours. Rai and Saxena (1989) observed decline in total sugars, soluble proteins and total phenol content during storage while there is an upsurge in the enzyme (polyphenoloxidase) activity. These changes are aggravated with increase in storage temperature. Rai and Saxena (1988) studied the effect of storage temperature on vitamin C content of *Agaricus bisporus*. They stored mushrooms in perforated polythene packets at different temperature and relative humidities upto 4 days. After 4 days of storage, the mushrooms lost 12-25 per cent vitamin C with loss being highest at 15°C and 55-60 per cent relative humidity, and lowest at 5°C and 85-90 per cent relative humidity.

Chopra *et al*. (1985) studied the effect of pre-harvest aqueous sprays of honey, citric acid and *Euphorbia royleana* latex in different combinations, on the storage life of *Agaricus bisporus*. The study revealed that mushrooms treated with honey (0.5 and 1– per cent) 18 ± 2 hours before harvest, air-dried to remove surface moisture, packed in 100 gauge polythene pouches with 0.5 per cent venting area and stored at 3 to 5°C, reduced the physiological weight loss and polyphenol oxidase activity. Veil opening was delayed and shriveling was negligible even after 21 days of storage. The shelf-life was increased by more than a week over control at 3-5°C and 2-3 days at ambient temperature. Combination treatment of honey + citric acid + latex (0.5 per cent + 0.5 per cent + 0.5 per cent) was also fairly effective. Pruthi *et al*. (1984) reported that the shelf-life and whiteness of mushrooms can be maintained by preserving in steeping solutions. Various concentrations of acetic acid, citric acid, ascorbic acid and potassium metabisulphite were utilized for steeping preservation of mushrooms. Water blanched mushrooms steeped in 0.5 per cent citric acid, was the best treatment. Mushrooms can be steeped in this solution for 8-10 days without loosing much flavour and texture. Irradiation with gamma rays (250 krad) helps in increasing the storage life for 9-10 days at 15°C (Roy and Bahl, 1984). Prolongation of stalks, toughening of texture, opening of veils and loss of whiteness are deteriorative changes observed at prolonged storage.

Long-term Storage

Freshly harvested mushrooms are highly perishable and require processing for stable preservation. Steeping solution preservation (Bano and Singh, 1972; Adsule *et al*., 1981; Pruthi *et al*., 1984; Sethi *et al*., 1989) helps to extend storage life of *Agaricus bisporus* from 2 to 5 weeks. Long-term preservation methods which have been employed for mushroom include canning, freezing and freeze drying (Kaul, 1983). Pickling and sun-drying are economically viable methods of preserving mushroom during the glut periods in country like India.

Drying

Drying refers the removal of water by heat to such a level that the biochemical and microbial activity is checked due to reduced water activity in the product. Mushrooms are dehydrated by air-drying as well as freeze drying techniques. Freeze drying is a novel development in food dehydration, but the cost of removal of water is reported to be ten times higher than the conventional air-drying (Ramachandra and Ramanathan, 1978). Dehydration by sun-drying has been advocated by Anand (1975) to be an appropriate technology for our country because of the inexhaustible free source of energy and minimal investment in capital cost. However, dehydration in a cabinet drier equipped with hot air circulation was found to be superior to sun-drying (Mudahar and Bains, 1982). General procedure recommended for mushroom drying includes washing, slicing, blanching, drying, packaging and storage. Whole or sliced mushrooms blanched in steam or boiling water for 2-5 minutes are dried to a moisture content of about 5 per cent using a finishing temperature of not more than 65.5°C.

Slicing

Pruthi *et al.* (1984) recommended longitudinal slicing as well as cross slitting while Mudahar and Bains (1982) sliced *Agaricus bisporus* mushrooms lengthwise into four portions and Singh *et al.* (1984) used 3/8 inch thick slices for drying.

Blanching

Blanching is done to control the activity of enzymes but under and over blanching can cause poor quality of colour and texture with a loss in free amino acids and sugars. Mudahar and Bains (1982) reported 25.2 to 26.9 per cent loss in weight during blanching while Pruthi *et al.* (1984) reported it to be 33.3 per cent. Katiyar (1985) noted that blanching did not help in improving the dried product. The blanching time for button mushrooms ranges between 4 and 6 minutes (Tanga, 1974). Sharma *et al.* (1991a) also reported the effect of blanching media on the weight loss and final quality of button mushrooms.

Chemical Treatment

Sulphuring or sulphiting are known to prevent enzyme catalysed oxidative changes, inhibit microbial deterioration and facilitate drying by plasmolysing the cells. Mudahar and Bains (1982) have recommended the immersion of sliced and blanched *A. bisporus* in a solution of 0.5 per cent sodium metabisuiphite + 0.25 per cent citric acid as a predrying treatment for obtaining a quality product. Soaking of unblanched mushroom slices in 3 per cent salt + 6 per cent sugar solution for 12-16 hours before drying yielded product with better colour, shape, texture and flavour (Singh *et al.*, 1984). Bano and Singh (1972) reported that soaking mushrooms in salt concentration higher than 2.5 per cent reduced protein content. Lal *et al.* (1990) reported that a 16 hours soaking pre-treatment in 1 per cent potassium metabisulphite + 0.2 per cent citric acid + 6 per cent sugar + 3 per cent salt solution followed by drying at 60 ± 2°C for 8.5 hours gave the best product with higher yield as well as shelf-life. Pruthi *et al.* (1984) observed that the use of hydrogen peroxide helped in maintaining the whiteness of *Agaricus bisporus* mushrooms during steeping solution preservation upto 3-4 weeks storage. However, Lal *et al.* (1990) reported that soak treatment of hydrogen peroxide (30 per cent) for 30 minutes improved the whiteness of dehydrated *Agaricus bisporus* but adversely affected the texture, taste and flavour.

Drying Time

Mudahar and Bains (1982) dehydrated sliced *A. bisporus* at 45°C for the initial 6 hours followed by raising the temperature to 60°C for another 4-5 hours so as to reach the moisture level of 5 per cent in final product. Various treatments like sulphitation and Ethylene Diamine Tetra Acetic Acid (EDTA) soak did not appreciably affect the drying time sun-drying took 33-39 hours with maximum temperature of 22°C to reach moisture content of 5-6.5 per cent. Drying in mechanical dehydrator was reported to be faster by Katiyar (1985) because of higher air temperature and forced air circulation as compared to sun-drying. The use of black polythene surface helps to raise sun-drying temperature by 10°C. Mean dehydration time was 8.4 hours whereas 16.8 sun hours were needed in sun-drying. However, Kumar, Ashwani (1992) reported dehydration time of about 9 hours at 60 ± 2°C to a constant weight.

Yield and Quality

Yield and quality of dried product is influenced by factors, such as initial moisture content, drying temperature and time, susceptibility of the material to heat damage, pre-drying treatment and moisture content of the finished product. Dehydration ratio is an indicator of yield while rehydration ratio is an indicator of quality.

Mudahar and Bains (1982) observed that the dehydration ratio of *A. bisporus* was 9.9-10.9 and it was not affected by pre-treatments, while the rehydration ratio was 2.6 for blanched and 3.6 for blanched as well as sulphited air-dried mushrooms against 2.4 for the sun-dried. The rehydration ratio slightly increased with the storage period of 100 days. The rehydration time in boiling water for the blanched mushrooms was 37 minutes as against 21 minutes for the sulphited one. Pruthi *et al.* (1984) reported that the longitudinally sliced and blanched mushrooms when dried at 60°C for 5 hours, had drying ratio of 10.8:1 and rehydration ratio of 2.78 as against cross slit mushrooms with drying time of 8 hours which gave drying ratio of 10.9:1 and rehydration ratio 2.80.

Packaging and Storage

Pruthi *et al.* (1984) reported that the dehydrated button mushrooms packed in friction top tins had storage life of 9 months and those packed in polythene bags turned pale yellow during storage and could hardly be stored for 5-6 months. The dried mushrooms packed in polythene bags picked up 1.5-2.0 per cent moisture. Singh et a!. (1984) reported the storage life of sugar and salt treated dried *A. bisporus*, packed in flexible laminated pouches upto 6 months at room temperature (25-30°C) and 3 months at 37°C while Kumar, Ashwani (1992) reported storage life of pretreated unblanched mushrooms (0.5 per cent potassium metabisulphite + 0.2 per cent citric acid for 12 hours) by more than 6 months at ambient temperature. No change in total protein content in dried *A. bisporus* during storage but a sharp decrease in the sulphur dioxide content has been observed by Mudahar and Bains (1982). According to Sethi and Anand (1983), both dried and fresh mushrooms are used as a delicacy in a variety of food products like stew, baked, fried, boiled, pickled, in soup, sauce and pulao.

Canning

Canning is a well practised technique all over the world. White button mushrooms are canned either whole, sliced or in smaller pieces. Generally, for canning purpose only, very small buttons are required without stem attached. The efficiency of canning is highly affected by loss in weight of raw material during processing. This is known as shrinkage which is caused by the removal of water as well as solids from the mushroom tissues during processing operations. Dang *et al*. (1978) conducted experiment on the influence of pre-treatment on the yield and quality of canned mushrooms and reported soaking of mushrooms in water before blanching, and combination of soaking and storage have proved useful in reducing the shrinkage losses thereby increasing the canned product yield. Tanga (1974) reported that mushroom strain D-9 white had low shrinkage as compared to other three strains, D-4 white, D-7 cream and D-1 white. Rough handling of mushroom during blanching has a positive effect on the shrinkage and increases losses. He reported that if the mushrooms are blanched properly and subjected to gentle treatment after blanching, the loss in weight can be checked by about 4-8 per cent and thus bringing the loss to around 34-35 per cent. Hence, mushroom should be treated gently while handling and processing. Dang and Singh (1978) reported ascorbic acid, ethylene diamine tetra acetic acid (EDTA), sulphur dioxide and citric acid as useful adjunct in canning of mushroom for improving the colour of canned mushroom.

Singh *et al*. (1982) tried various water binding agents to minimise shrinkage losses in canned mushrooms. Steeping the buttons in 0.5 per cent solution of methyl cellulose and carboxy methyl cellulose helped in increasing the yield by 4.9 and 9.0 per cent respectively without any deleterious effect on the quality. Further, increase in yield upto 10.3, 11.7 and 15.0 per cent was recorded with the pectin-calcium combination, methyl cellulose and carboxy methyl cellulose at 1.5 per cent concentration level of each but the organoleptic quality was affected. Food additives like agar at 0.125 per cent and pectin at 0.5 per cent level helped in improving the flavour of the canned product. Srivastava *et al*. (1987) reported the effect of soaking and various chemical additives on yields and quality of canned mushroom. They tried twenty two treatments. These treatments included soaking, blanching and resoaking in different solutions followed by varying composition of brine to improve the canning quality of mushrooms. Among these treatments, pre- and post-blanch soaking in 0.1 per cent ascorbic acid solution followed by canning with 0.1 per cent ascorbic acid in covering liquid was found fairly acceptable and scored 88 per cent marks on organoleptic evaluation. The treated canned mushrooms also showed satisfactory drained weight (60.66 per cent) and better colour (0.006 O.D.).

Blanched mushrooms are filled in the cans containing brine solution of 2.5 per cent salt and 0.25-0.5 per cent citric acid, exhausted until the temperature in the centre of the can is 85°C, sterilized for 25-30 minutes under 15 lb/sq inch pressure, cooled and stored. Adsule *et al*. (1983) reported tomato juice medium to be better for retaining the mushroom quality than the brine solution. Unlike using brine solution, there is no need to add citric acid to tomato juice for lowering the pH of the filling medium. Further, the nutrients of the mushroom could be trapped in the tomato juice for human consumption.

Oyster Mushroom (*Pleurotus* spp.)

Quality Parameters

Oyster mushrooms are harvested when the fruit body has curled under edges and well formed gills (*i.e.*, wrinkled stage of umbrella) because fully mature fruit body is fragile and difficult to handle. As the maturity advances, the fruit body loses water and thus shrinkage occurs. The loss in turgidity is followed by darkening of colour and liquescent margins. Oyster mushroom has a longer shelf-life in comparison to others at ambient temperature. After 3-4 days at ambient temperature, the texture of fruit bodies also becomes slimy and spores are shed. This leads to a foul smell and abnoxious appearance.

Packing

Fruit bodies harvested at wringled stage of umbrella are cleaned and packed in the polythene bags having perforations to check condensation, slimness and texture. Sealed polythene bags should be stored at a temperature not less than 5°C. Bags can be successfully stored in refrigerators. For large scale production, the cooling of sealed bags is done with positive ventilation and the temperature in cold room should be maintained at 8-10°C.

Transport

Trays or baskets are used for transporting oyster mushrooms. Polythene pouches containing crushed ice and over wrapped in paper are also put alongwith. The trays baskets should be covered with perforated polythene sheets. The prepacked perforated polythene packs containing 250-500 g of fruit bodies can also be transported in this way.

Storage

Short-term Storage

Changes in post-harvest storage of *Pleurotus flabellatus* at ambient temperatures include decrease in respiratory rate, soluble carbohydrate and water content. During storage, the mushrooms show an increase in the activities of odiphenol oxidases and proteases accompanied by a fall in total phenols and an increase in free amino acids. The degree of discolouration increases with storage time. Fresh mushrooms (200 g) packed in 25 mm polythene bags with one pin hole on either side store well upto 24 hours at ambient temperature and upto 6 days at 5 ± 2°C in intact bags (Rajarathnam *et al.*, 1983).

Jandaik and Sharma (1987) studied the effect of different storage conditions on shelf-life and reported that perforated polypropylene/polythene are effective in reducing moisture loss as compared to unpacked conditions in which the loss of moisture was about 32-35 per cent at 15-18°C and 6.0 per cent at 6-8°C after 72 hours. There was no loss of moisture in packed bags but the fruiting bodies were slimy in appearance due to accumulation of excess of moisture inside the bags which resulted in liquefaction and unacceptability of fruit bodies.

Freshly harvested fruit bodies of *Pleurotus sapidus* can be easily stored in completely sealed non-perforated polythene bags upto 72 hours at room temperature (20-30°C) and at low temperature (0-5°C) upto 15 days (Mehta and Jandaik, 1989).

Adsule *et al.* (1981) reported the preservation of *Pleurotus sajor-caju* in steeping solution. They used preserving solution containing 5 per cent salt, 0.2 per cent citric acid and 0.1 per cent potassium metabisulphite in glass bottles for storing mushrooms and observed that the fruit bodies could be kept upto 3 months without losing much of its organoleptic quality.

Long-term Storage

Pleurotus flabellatus and *Pleurotus eous* species of oyster mushrooms are found to be quite suitable for dehydration and pickle preparation (Srivastava and Bano, 1970; Singh and Rajarathnam, 1977).

Drying

Pleurotus spp. could easily be dried in the sun or mechanical dehydrators at 50°C and rehydrated without loss of flavour (Jandaik, 1978). Jandaik and Sharma (1987) studied effect of different drying methods and reported that the sun-dried fruit bodies have 3-4 per cent moisture in comparison to 2.0 per cent in fruit bodies dried at 40-45°C. The change in colour was also slight in sun-drying as compared to hot air-drying (55-60°C) which resulted in dark brown colour. The fruit bodies dried either in the sun or mechanically at 45°C were acceptable organoleptically, whereas those dried at 55-60°C resulted in burnt taste. The time required for rehydration was 12-15 minutes. The product can be stored for 120 days in sealed polythene bags. Mehta and Jandaik (1989), in a similar type of study, reported rehydration ratio 1:8.

Pandey and Aich (1989) reported equilibrium moisture content (EMC) of dehydrated *Pleurotus sajor-caju* mushrooms determined at five levels of temperature ranging from 0- 50°C and relative humidity ranging from 30-95 per cent using static desiccator technique. Free energy change and heat of vaporization of mushroom decreased as the moisture and temperature increased. Mould growth was observed at relative humidity greater than 80 percent.

Kumar *et al.* (1980) observed that the blanched and sugar treated dehydrated *Pleurotus flabellatus* had no visible browning upto 6 months, whereas the unblanched dehydrated mushrooms had shelf-life of one month only, having turned distinctly dark in colour, leathery in texture and slightly bitter in taste. Further, they reported that the dried product picked up 6.7 per cent moisture at higher storage temperature (38°C and 90 per cent relative humidity) in 4 months and only 1.7 per cent at moderate storage conditions (27°C and 60 per cent relative humidity) during 6 months storage.

Paddy Straw Mushroom (*Volvariella volvacea*)

Quality Parameters

Paddy straw mushroom, which can be extensively grown in tropical regions of the country, is highly perishable having a shelf life of less than a day. These should be harvested at button or egg stage. The button stage occurs when the mushroom has begun to differentiate into the cap and stem inside the universal veil which continues until the differentiation is complete and the universal veil is broken. Once the stipe is visible, the elongation starts which is considered undesirable. Ramasamy and Kandaswamy (1978) reported possible causes of quick deterioration of paddy straw mushroom. They observed that deterioration was less in the mushrooms harvested at egg stage because the polyphenol oxidase activity was much less. Due to the fast post-harvest maturity, many changes occur as the caps open and it gets the appearance of flattened umbrella. This is accompanied by loss in weight due to loss of moisture and release of spores. These spores cause pink colouration of fruit bodies and thus causing off flavour and rendering the produce unmarketable.

Packaging and Transportation

Procedure are same as in case of white button and oyster mushrooms.

Storage

Short-term Storage

Low temperature storage causes frost injury and quality is adversely affected. The best way to maintain the freshness is to store in the perforated polythene bags at 10-15°C for about 4 days. Ramasamy and Kandaswamy (1978) reported that deterioration in the quality of stored mushroom is mainly due to polyphenol oxidase activity. They recommended the use of sodium chloride or ascorbic acid or citric acid solutions for enhancing the storage life.

Long-term Storage

Like oyster mushrooms, these are also dried or pickled during the gluts in the market. The process of dehydration involves the pre-treatments like blanching and steeping in chemical solutions before drying to a constant weight.

Pruthi *et al.* (1978) reported that grading of mushrooms prior to dehydration is absolutely necessary to attain uniform heating. Blanching is important to maintain the colour of dehydrated mushrooms during storage. Water blanching for 3-4 minutes or steam blanching for 4-5 minutes are optimum. Dehydration of sliced mushrooms in a phased manner at 70-65-60°C give better colour to dried product. Mushroom steeped in 0.5-1.0 per cent citric acid for 16 hours gave best dried product in respect of colour, appearance, rehydration properties and better keeping quality. Organoleptic evaluation of mushrooms stored in friction top tins (wax sealed) revealed that they could easily be kept well for one year under ambient conditions.

Deshpande and Tamhane (1981) reported that 2000 ppm of sulphur dioxide inactivated the polyphenolase oxidase enzyme activity and sulphited fruit bodies having higher drying rate at the optimum tray load of 6.25 per square metre. They further reported drying ratio of 12:1 for unblanched and sulphited, and 11.6:1 for steam blanched, whereas rehydration ratio for unblanched, sulphited and blanched mushooms were 4.9, 4.7 and 4.4, and the organoleptic score 13,22 and 20, respectively. Observations on storage studies revealed that the unblanched sulphited mushrooms stored at 25°C with 60 per cent relative humidity showed increased browning index during entire period of 6.0 days storage while texture remained crisp and flavour decreased.

Pruthi *et al.* (1984) reported 18.2 per cent increase in weight during blanching and decrease in protein content from 37.0 to 27.0 per cent in water blanched while in steam blanched, decrease in protein content was marginal.

Pickling

Mushrooms grown in the country can be processed into more durable products like pickle, chutney, ketchup, etc. (Sethi and Anand, 1978, 1982; Dang and Singh, 1978; Kumar and Nagia, 1989). Singh and Bano (1977) studied the suitability of *Pleurotus* spp. for pickle preparation. They reported that the product could be stored for a minimum period of 6 months at ambient temperature (22-34°C) without any off flavour. Joshi *et al.* (1991) reported a sweet chutney from edible mushrooms having a shelf-life of over a year with better sensory qualities. They further observed that pre-cooking of mushrooms is necessary for absolute microbiological safety and to increase the shelf-life of the product. Khader and Pandya (1981) prepared a pickle from paddy straw mushroom having good keeping quality.

Other Products of Mushroom

In addition to canned, dehydrated and pickled products, mushrooms can also be utilized in the preparation of weaning foods (Khader and Pandya, 1981), biscuits (Sharma *et al.*, 1991 b) and soup powder (Jandaik *et at*, 1989). Dhar (1978) has given the recipes for the preparation of mushroom dishes like mushroom vegetable, mushroom kidney fry, mushroom curry, mushroom salads, cheese sandwiches, mushroom stuffed capsicum, stuffed morels, mushroom fritters, meat stuffed mushroom, mushroom hot dog, mushroom burgers, mushroom omlette, mushroom and pouched eggs, stuffed potato, etc.

Chapter 11

Conservation of Mushroom Biodiversity

The presence of biosphere with its wealth of life forms distinguishes earth from other eight planets of solar system. The varied life forms coexisting on earth is commonly referred to as biodiversity. All life forms contributing to the biodiversity are intricately woven together in the web of nature constituting an ecosystem. Each component of the ecosystem, small or large plays a well defined role in maintaining its equilibrium. Loss of any life component in the ecosystem upset its equilibrium posing a far reaching effect. Therefore checks over any loss of species or in other words conservation of species deserve special attention.

The earth biodiversity is the product of more than 3.5 billion years of evolution including evolution, migration and extinction. Natural background of the rate of extinction during past 600 million years has been estimated of the order of only one species every year or so. Lately due to human interferences, the rate of extinction has tremendously increased. All over the world increasing concern is being shown for the conservation of natural resources facing threat of man made extinction. With this aim and objective, Earth summit at Rio De Janeiro was organized at Brazil, in the year 1992. Conservation has been defined by International Union for Conservation of Nature and Natural Resources (IUCN) as the management of human use of biosphere, so that it may yield the greatest sustainable benefit to present generation while maintaining its potential to meets the needs and aspiration of the future. This generated awareness amongst the environmentalist and encouraging results of conservation of biological biodiversity have started to increase fastly since over two decades. Limited attention has, however, been given to the role and conservation of the Fungi in general and Mushroom biodiversity in particular. The groups were almost ignored even during the Rio meeting in 1992, especially convened to control and preserve the biodiversity of our planet during the 21st century through "Agenda-21" adopted at the meeting. Reason behind such ignorance may be due to the fact that active part of the fungal mycelium is usually inconspicuous and in the case of mushroom below the ground. Mycology and nature conservation have developed separately for a long time, as mycologist were not aware of possible threat to mycoflora and conservationists believe that special measures for this group were unnecessary. In the recent past, however, mycologist especially in western and central Europe have noticed alarming changes in mycoflora–mainly in larger fungi or what are commonly called mushrooms.

Consequently, some activity in the field of mushroom conservation has been taken up in very limited places world over (*vide* Kaul, 1993). In Netherlands, all the methods of investigations have revealed a strong decrease in abundance and species diversity of ectomycorrhizal fungi, a strong increase in wood inhabiting fungi and relatively few changes in saprotrophic soil inhabiting spices. Derbsch and Schinff (*vide* Kaul, 2002) investigated the spices diversity of microfungi in Saarland, Germany and concluded that atleast 40 per cent of mycoflora has become extinct within last 35 years. Watling and Arnolds, discussed about the changes in species diversity and Red data list of many other countries (*vide* Kaul, 1993).

Red data book is the record of threatened species of a particular country and region and hence it differs from country to country. It contains the name of threatened and extinct species, accompanied by details of distribution, ecology and resource of threat. Red data book is published by international union for nature and natural recourses (IUCN). The species in red data book are usually assigned to four classes ranging from (presumably) extinct to potentially threatened. Occasionally a fifth category is established to accommodate those taxa in which the degree of threat is unknown (indeterminate).

Species of mushroom may be considered as threatened when their fruiting population are actually decreasing in number and (or) productivity over large area. Reduction in population of fruiting body does not always imply disappearance of mycelia, it may indicate anyhow reduced dispersal capacity of a species. Indication exists that the fruiting process is more sensitive to unfavourable external factors than in mycelial survival. In addition, rare species with local distributions and specific habitat requirements may also be considered as (potentially) threatened, even if there are no identification of a decline.

The number of threatened edible mushrooms varies strongly in different countries. It has been assessed 20 in USSR, 132 in Norway, 1208 in Saarland. In Netherlands, this number has been assessed 944 which alone constitute 28 per cent of the flora of larger fungi. A similar decline may be occurring in parts of North America and Asia as well, but data on long term changes in these regions are not available. India had compiled red data books having information of endangered plant species but not such information is there about Indian fungi in general and Indian mushrooms in particular.

The wild mushrooms are one of the minor forest products and in some instances; the commercial value of the forest mushrooms may surpass the value of timer. Increased collection of wild mushrooms may have adverse effect on following counts:

1. It can directly effect production of fruiting bodies in subsequent years.
2. Indiscriminate collection can make scientific recording much more difficult.
3. It may deplete the food source of other organisms which rely heavily on mushrooms.
4. There will be aesthetic effect of depriving many amateur naturalists the pleasant sight of fungi in autumn woods.

Increasing collection pressure of wild mushrooms also poses threat to its global biodiversity in many other ways. Loss of global mushroom biodiversity may also be ascribed to the deterioration and for destruction of habitats due to the interferences of ambitious human bodies. Man made threats to the biodiversity of wild mushrooms identified so far may be read as follows:

Harvesting of wild mushrooms may strictly effect production of fruiting bodies in subsequent years by damaging or exhausting, mycelia shifting competitive relations with other species or by causing reproductive failure due to decreased spore production. In addition, harvesting may indirectly effect productivity owing to compaction of upper soil layer by heavy trampling or by raking forest floor in search of truffles and primordia of epigeous mushrooms. The mycelia of most saprotropic and ectomycorrhizal fungi are mainly found in the little and upper soil layer and fragmentation of mycelia may cause their destruction.

Among forest fungi, ectomycorrhizal mushrooms are most threatened group in densely populated or heavily industrialized regions. Various authors have reported a correlation between a decline of fruiting bodies of ectomycorrhizal fungi and ambient concentrations of the air pollutants such as SO_2 and NH_2 (Gulden *et al.*, 1992; Jansen and Van Doben, 1987; Fellnur, 1989; Termorshuizen and Schaffers, 1991). It is supposed that acidification of forest soil and (or) increased nitrogen concentrations are primarily responsible for the observed decrease in mushrooms population. However, the exact mechanisms are not yet clear.

Other strongly endangered groups are lignicolous mushrooms of larger logs in old growth forests and soil inhabiting mushrooms of peat bogs, wetlands and low productivity grasslands. Their decline is primarily attributable by change in land use. However, the edible mushroom falling under these category are very few and unpopular.

Habitat destruction due to forest fires is even more serious as it disrupts the inoculums completely. This has been a regular problem in every summer in the North Western Himalayas. Forest fires have been reported to cause disappearance of ectomycorrhizal mushrooms (Egli *et al.*, 1990) but some specialist from such as *Morchella* species, may flourish during a short period after forest fire (Moser, 1949). Mushrooms growth on wooden logs and humus are lost when these are used as fuel by the local population. This is yet another reason for the loss of habitat and mushroom mycelium.

Some ectomycorrhizal mushrooms *e.g. Cantharellus cibarius* and *Boletus edulis* are symbionts of a wide variety of trees, and also members of the necrotrophic *Armillaria mellea* complex can be found on many different hosts. Many other symbionts of a number of trees, such as the ectomycorrhizal *Tuber aestivum* with *Fagus* and *Quercus* and *Suillus luteus* and *Lactarius deliciosus* with two needled *Pinus* species. The wood inhabiting *Kuehneromyces mutabilis* and *Flammulina velutipes* are only found on woods of deciduous trees, whereas *Hypholoma capnoides* grows exclusively on coniferous woods. Although no decline in their population has been recorded, yet they are considered threatened due to their very selective growth conditions and they may be facilitated by specific plantations–potentially favourable habitats for these symbiotic mushrooms.

Also many group of mushroom species are threatened due to their restricted preferences to certain soil type. Chanterelles are mainly found on acidic sandy soil with a low nutrient status and thin litter layer, whereas truffles prefer shallow, basic, mineral soil above calcareous bedrock. On the other hand, some saprotrophic species *e.g. Lepista nuda, Macrolapiota rachodes* thrive in nutrient rich, deep litter layer. Consequently, application of fertilizers in forests effect species in different ways. In general, nitrogen fertilizers depress the yield of edible mushrooms except on extremely poor soils (Ohenoja, 1988) as does the nitrogen deposition from air pollution.

One of the proximate cause of the loss of biodiversity is indiscriminate hunting. In Tripura, centpercent population have been reported to possess food habits of mushrooms. The local mushroom hunters have been reported to sell there booty in the local market. *Termitomyces heimii* is one of the wild mushroom, which has been reported to be under immense collection pressure in the village of Madhya Pradesh. Similarly *Podaxis pistillaris* and *Phellorina inquinans* were also reported to be sold in plenty during rainy season on the roadside market of various districts of Rajasthan (Singh, 1994). In the North Western Himalayas, the edible species of *Russula* and *Lactarius* which are under continuous pressure of being picked up in bulk by the local people and the species whose population seems to be on decline include *Russula virescens, R.cyanoxantha, R.olivacea, R.paludosa, R.mustalina, Lactarius volemus* and *L. deliciosus.* Some other species which are being collected in bulk regularly during rainy season in North Western Himalayas include *Boletus edulis, Amanita caesaria, Amanita vaginata, Lepiota procera, Morchella* spp., *Lycoperdon* spp. and *Ramaria* spp.

One of the most important causes for the loss of mushroom biodiversity, however, is the destruction of their habitats and deforestation for human settlement,. for the purpose of agriculture, industries, road construction, dam building, drainage, grazing ground etc. In India deforestation is estimated as 13000 sq. km. of forest area annually (Agarwal, 1999). If this rate continues, the bioresources, including mushrooms will keep on shrinking and ultimately may disappear.

To put a rider over rapidly receding mushroom biodiversity immediate step is required towards its conservation. A conservation plan has been designed by WRI (World Research Institute) and IUCN (International Union for Nature and Natural Resource) which can be implemented through two complementary approaches *in-situ* and *ex-situ*.

For *in-situ* conservation, the different habitats of mushroom should be preserved by demarcating protected area, sanctuaries, biosphere reserve, etc. as out of bound for indiscriminate unlicensed picking of mushrooms, trampling and raking of forest floor, grazing and any others such activities which contribute to the factors responsible for the dwindling of the mushroom population in their natural habitats.

Selective plantation of certain specific trees in the bare/afforested areas should be taken up on war footing which may generate potentially favorable habitats for the specific mushrooms.

Maintenance and in-nature cultivation of threatened species is a ready solution in which semicultures of young trees with inoculated microbial partners should first be raised in the nurseries and after establishment of mycorrhiza, such seedlings should be shifted to the barren/afforested areas for plantation. This two pronged strategy is of immense significance for management and conservation of threatened mushrooms. This venture can be made effective by involving local population for upbringing the seedling with suitable mycorrhizal partners in the afforested area (Lakhanpal and Atri, 2000). Even an open canopy of mature trees can be maintained by allowing selective felling and initiating replantation along with cultivation with suitable mycorrhizal partner. In this way conservation can be achieved by reclamation of barren areas by reviving its biological potentialities.

A system of permit should be evolved for commercial exploitation of mushrooms. Over exploitation of existing bioresources should be checked by enforcement of regulations available under CITES (Conservation on International Trade in Endangered Species of wild fauna and flora) and making use of other enforcement agencies. For example in Himachal Pradesh for morel collection forests are leased out to license holders but in practice there is pilferage as many other having no authorized contract also do handsome business (Shad and Lakhanpal, 1991). This is where strict enforcement is required to regulate mushroom collection in general and morel collection in particular from the natural habitats.

The *in-situ* conservation of fungi is hampered by the lack of information as to the species present in particular sites, the length of time and labour intensiveness of producing tests, knowledge of the rarity of individual species and in most cases a lack of understanding of the precise ecological requirements of species. Integrating this activity with the demarcated protected area networks, sanctuaries and biosphere reserve of the country will be fruitful exercise. Besinger (1990), however, maintains that "*in-situ*" conservation will be effective if the local population reaps some short-term benefits from it.

Ex-situ conservation of plant and animal species is achieved by the establishment of the botanical gardens, seed banks and geological gardens. However, establishing culture collection is *ex-situ* procedure for microorganisms. Hawksworth (1991) records that 2,54,000 strains representing 11,500 species are held in culture collection over the world at present (up till 1989) representing 17 per cent of the known and just under 1 per cent of the estimated world fungi. It is essential that much larger number of strains and species should be preserved in the culture collection to conserve the genetic resource of fungi. It would, however, need international collaboration and massive resources.

In the Indian context, even those species which are of high food value and have attained commercial status have not received comprehensive attention on their cultivation and conservation. Studies carried out by the Indian mycologist during 19th and most part of 20th century were restricted to their description and identification only without any attempt to isolate them in cultures. The first national effort comprising both assessment and conservation of fungi including mushroom were initiated in Indian Agriculture Research Institute (IARI)–New Delhi with the establishment of the advanced centre for biosystetematics (Microbes and Invertibrates) in June, 1993 (Varma and Sarbhoy, 1996), in collaboration with the Commonwealth Agricultural Bureau International (CABI). Although assessment part of wild fungal resources of a biodiversity rich country like India still need a lot of efforts, yet the conservation measures can not wait any longer due to fast depletion of forest resources and consequent destruction of the natural habitat/substrate of macrofungi in the country. Verma and Upadhyay (1998) after updating the mushroom wealth of India have emphasized over the need for urgent collection of mushroom germplasm of the country and their conservation in the national bank being established in 1983 at the National Research Centre for Mushroom (NRCM), Solan. At the global level also, limited attention has been given to the role and conservation of the mushroom germplasm. There are but a few Culture Collection and Conservation Centres specializing in mushrooms.

Mushroom cultures, particularly those of commercial importance are invariably stored in vegetative mycelial form to preserve their vigor and the desirable genetic characteristics. In contrast to spores, which are generally hardy structure with thick or double wall, the vegetative mycelia are much more susceptible to sudden changes in temperature and pressure (Franks, 1981) and therefore they need to be handled with special care. Safe and effective protocols for conservation of mycellial culture of mushroom species of threat but of commercial importance have been developed in NRCM gene bank. The NRCM gene bank for mushroom is gradually becoming the most sought after source for pure authentic cultures of cultivated edible fungi within the country as well as some near by SAARC countries.

There are but a few culture collections and conservation centers specialized in mushroom across the world, particularly those registered with the World Federation for Culture Collection (WFCC), which coordinates, supports and promotes for their networking and exchange of microbial genetic resource amongst them. It also helps to built the database through World Database Centre on Microorganism (WDCM) with valuable information on mushrooms also. A few such centers having culture collection of mushroom including commercial species are as under:

Major culture collection and conservation centre of mushroom germplasm:

1. American Type Culture Collection (ATCC): It is private non-profit making organization recognized as the first International Depository Authority (IDA) and it has more than 27000 fungal cultures including mushrooms species preserved in lyophilized and cry preserved states.
2. International Mycologial Institute (IMI): Situated in Egham(UK), the genetic resource collection of IMI has both fungal herbaria and fresh cultures in large numbers. It holds 3,60,000 dried fungal specimens and about 18000

live cultures, with the help of which it provides training as well as authentic identification services.

3. Central Bureau Voor Scimmelcultures (CBS): Situated in Baarn, The Netherlands, it has a rich collection (29000) of fungal cultures including mushrooms. These are preserved as slànts in mineral oil, as lyophilized or as cry preserved cultures at ultra low temperature depending upon the species requirements.
4. Mycothecque de l'universite'catholique de Louvain (MUCL): Situated in Belgium, MUCL holds both fungal harberia and pure cultures from all over the world. It has about 24800 fungal collection including yeast, filamentous and fleshy fungi preserved as fresh, submerged in mineral oil, freeze-dried and frozen cultures at -80°C to -140°C.
5. Fungal Cultures of University of Goteborg (FCUG): Situated in Sewden, this collection holds about 10000 fungal culture including mushroom specimens obtained from Europe, Canada, Argentina and Turkey.
6. International Bank of Edible Saprophytic Mushrooms (IBESM): Established in Italy, under the auspices of the FAO global network of mushrooms, IBESM has about 434 commercial and wild strains of mushrooms especially of *Pleurotus spp.*
7. The Instituto Polytechnico Nacional: Situated in Mexico City, it has more number (60, 000) of fungal herbaria (including 35000 mushroom specimens) than living cultures. About 135 cultures of edible mushrooms are preserved in the Instituto de Ecologia (Xalapa) comprising mostly tropical and subtropical species.
8. Global Bank of Mushroom Genetic Resource: Situated in Aquitaine (Bordeaux), France, this bank holds about 3000 strains and 80 species of edible mushrooms from wild and commercial collection from Europe, Asia, Latin America and Ivory coast.

Although importance of mushroom in fighting against increasing malnutrition and receding land resource has been well realized, corresponding initiative of its conservation has not been attempted so far. For implementation of proposed tentative plan for conservation of macrofungi (Kaul, 1992) there is need for:

1. A nationwide coordinated programme for compilation of existing mushroom wealth of India. As a step forward the conservation and monitoring of mushroom flora of National Parks and Biosphere Reserves etc. can be taken up.
2. Evolving a system of permits and registration for commercial mushroom harvesters from the natural habitats. Indiscriminate picking of mushrooms should be banned in the National Parks and Biosphere Reserves. Only licensed commercial harvests should be permitted as has been done in the Netherlands and the other European countries.

3. Raising semicultures of mycorrhizal fungi. It should be encouraged in afforested areas.
4. Persuing researches on population ecology of mushrooms. This should be initiated with the ideas to generate awareness about the need for regulated protection and conservation of existing mushroom wealth.
5. Preparation of Red Lists of larger fungi on priority basis.

The preparation of a project report on conservation of mushroom resources in India in 1991-92 (Kaul, 1992) which includes action plans suggested for the country may be read in brief as follows:

Action Plan

At a meeting of leading mushroom research workers of India held on 31st July, 1991 under the auspices of World Wide Fund for Nature–India, a tentative action plan for conservation of larger fungi in the country was formulated. Components of action plan are enumerated below:

1. A coordinated programme on compilation of information on the existing mushroom resources in the country, involving all the institutions active in this field be drawn up. This programme funded by some central agency can compile information on resources of larger fungi in India.
2. Red Lists of larger fungi for the entire country and region-wise can be prepared under carefully formulated scheme. This is a long term activity but is essential for conservationa work.
3. A programme for conserving and monitoring mushroom flora in the national parks or biosphere reserves be taken up. Committee selected following areas for conservation studies: (a) Silent valley (Kerala), (b) Nilgiris and Thanjavur district (Tamil Nadu), (c) Narkanda and Manikaran valley (Himachal Pradesh), (d) Katra (Jammu), (e) Assam hills and Assam/West Bengal plains (North–East), (f) Tribal areas (Orissa and M.P.), (g) Thar desert (Rajasthan) and (h) Terai belt (U.P.).
4. Germplasm collection of some prominent edible species *i.e. Pleurotus spp., Agaricus spp., Volvariella spp. and Morchella spp.* be taken up with involvement of culture collection centers in the country.
5. Involvement of tribal communities in conservation.
6. Involvement of rural education institutions in mushroom conservation.
7. Organizing training courses on mushroom taxonomy.
8. Preparation of mushroom field guides.
9. Establishment of mushroom conservation information service.

Efficient contribution to the field of mushroom conservation can be made by individual workers and small institutions spread over the country.

Chapter 12

Genetics of Mushroom and Perspectives of its Strain Improvement

Over the centuries of mushroom cultivation throughout the world, greater attention has been paid to the method of cultivation than to the breeding of new strains to improve the quality and productivity of the mushroom crop. The yield and the quality of the mushroom is determined by the two factors, the genotype of mushroom strain, and the environmental conditions in which the strain is grown. The character, behaviour and over all performance of the strain is the outcome of the interaction between genotype and environment. The main hurdle in serious scientific work on the improvement of existing available strain and breeding of new strains has been the mystery about the life cycle of fungi as a whole. With the progress made in understanding the life cycle and sexuality in cultivated edible fungi, some works have started in breeding. In *Agaricus bisporus* the life cycle was clearly elucidated (Elliott, 1972; Miller, 1971; Raper *et al.*, 1972) and this resulted into the release of first hybrids (Dutch hybrids) in 1981 by Dr. Gerda Fritsche. This opened up new vista of hope that it was possible to generate genetic variation by cross breeding strains in the predominatly inbred button mushroom.

The primary aim in most of the breeding attempt is to increase productivity and quality. However, there can be other objectives also. *i.e.* obtaining sporeless strain, evolving a high temperature or low temperature stock, resistance to diseases etc.

To develop a breeding programme on a rational scale we must know the sexual behaviour of the species. Majority of the cultivated edible species belong to Basidiomycetes. The life cycle, pattern of sexuality and sexual mechanism are the three parameters of sexual behaviour. Life cycle has already been discussed in preceding section of the book.

Pattern of Sexuality

Sexuality is the efficient means for generating the genetic variation, both for genetic reassortment and generation of additional diversity by mutations following the interaction between genotypes. The pattern of sexuality is controlled by two mechanisms (1) the distribution of four post meiotic nuclei to the basidiospores and (2) genetic factors of a mating system known as incompatibility system which may be either unifactorial or bifactorial. This is the only differentiating factor in most of the higher Basidiomycetes because of the absence of sexual differentiation on morphological basis. On the basis of mating system, edible mushrooms have been grouped into two types *viz.*, Heterothallic and Homothallic.

Heterothallism

Heterothallism was first described by Bensaude (1918) in *Coprinus fimetarious* and Kneip (1920) in *Schizophyllum commune*. Ninety per cent of the basidiomycetes studied so far are heterothallic in which single basidiospore germinates to produce monokaryons mycelium having the nuclei of same genotype. Hyphal fusion takes place between compatible monokaryns to produce dikaryon characterised by binucleate cells and clamp connections. The interacting monokaryotic mycelia usually exchange nuclei by migration through dolipore septum. The formation of clamp connections is the indication of a dikaryotic hyphae. The paired nuclei of dikaryon compose the two parental genotypes and this dikaryotic mycelium initiates fruit body formation. In some species (*Agaricus bitorquis*) formation of clamp connections is absent and it appears that no nuclear migration has occurred by mating of two compatible homokaryons. In this case heterokaryosis has been identified by fluffy growth and by complimentation auxotrophic strains (Raper, 1976). In the ascomycete *Morchella* (moral mushroom) similar type of sparse heterokaryotic growth occurs at the junction of two homokaryotic mycelia (Harvey *et al.*, 1978). Heterokaryons are also confirmed by using markers of mutants resistant to chemical inhibitors of growth on selective media. In *Armillaria melea*, no evidence of interaction between mycelia could be detected by a large number of workers, rather regular sectoring of the dikaryon into monokaryotic mycelium with diploid cells and without clamp connections was studied.

Genetic Regulation in Heterothallic Species

The compatibility of monokaryotic mycelia is regulated genetically by mating type factors. Two distinct, but surely related, incompatibility systems are responsible for obligatory cross mating in the Hymenomycetes and Gasteromycetes:

1. Unifactorial, and
2. Bifactorial

About 25 per cent of all investigated species are unifactorial in which single factor termed A is responsible for diakaryon formation while 65 per cent have bifactorial system of control with factors termed A and B. In both systems there are multiple alleles for each factor. A full interaction between vegetative mycelia leading to fertility occurs only if different alleles of each factor are present.

Unifactorial Heterothallism or Bipolar Incompatibility

In this case, dikaryon formation is regulated by a single mating factor with heteroallelic condition at that locus. Thus, mycelium bearing different mating type alleles are brought together *i.e.* AX compatible with AY, not with AX. for development of fruit body. After nuclear fusion, the two different A factors segregate in meiosis and one meiotic nucleus passes into each spore. Thus, the tetrad of spore formed in each basidium will contain parental factors in equal frequencies, further developing into two different mating types each cross fertile in interclass matings and self sterile in inter class matings. *Agaricus bitorquis, Auricularia auriculla* and *Pholiota nameko* are common examples having unifactorial system of heterothailism.

Bifactorial Heterothallism or Tetrapolar Incompatibility

In this case, dikaryon formation is regulated by two different mating type factors which are heterothallic at both loci. The mating type, loci A and B for dikaryon formation designated as AX, AY + BX BY, the diploid nucleus AX AY BX BY and meiospore as AX BX, AY BY, AY BX. Since A and B factors are unlinked, the meiospores occur in equal frequencies 1:1:1:1 and after germination give rise to homokaryotic mycelia.

Homothallism

In homothallic species individual uninucleate haploid spores germinate to produce mycelia which are capable of producing fruit bodies and each spore is self fertile. Three types of homothallism are found among self fertile species.

Primary Homothallism

Homokaryotic mycelium, established from a single meiotic nucleus has the potentiality to progress through dikaryosis to the completion of sexual cycle as in the case of *Volvariella volvacea.*

Secondary Homothallism

A fertile dikaryotic mycelium is established from a single basidiospore carrying two meiotic nuclei of different mating types due to the formation of 2 spored basidia and *Agaricus bisporus* is the best authentic example of this system. This mating type system in this mushroom is similar to unifactorial heterothallism but the two post meiotic nuclei migrate into each spore. Each spore on germination will establish a dikaryotic mycelium capable of forming fruit bodies. Although majority of the basidia in *A. bisporus* are bispored, but sometimes 3 and 4 spored basidia are also formed. Among the 2 spored basidia, not all germinating spores will produce fruit bodies. This can be explained by lavelling four post meiotic nuclei in unifactorial method of AX_1, AX_2, AY_1 and AY_2. Only those spores which are carrying both AX and AY alleles, will give rise to fertile spores. Hence two third spores of bispored basidia will be fertile and one third self sterile (Langton and Elliott, 1990).

Unclassified Homothallism

This situation remains unresolved in detail but differ from either of the two. These distinctions, while helpful in an appreciation of broader aspects of sexuality in the higher fungi, are often of little practical value, since to distinguish among them involves a very considerable amount of labour. There are a very few species that have been reported as "homothallic, without clamp connections" *viz.*, *Calocera cornea*, *Ephemeroides*, *Coprinus*, *Octajuae pleurotelloides* and *O. pseudopinsitus*. Whether this homothallism of these species is basically different from the first two categories above remains uncertain.

Heterothallism promotes outbreeding while homothallism promotes inbreeding. Thus hetcrothallic forms permit exploitation of heterosis or combination of desired traites while the task of achieving these results in homothallic group is little difficult.

Like other conventional crops in cultivated mushroom also, use of superior strains supported by improved crop management practices is the cheapest and surest way to achieve high productivity per unit area/time.

Although, a large number of basidiomycetes still remain to be studied on these aspects, yet an insight into the biology and reproduction of commercially important mushrooms have opened new vistas for their genetic improvement to meet the requirements of both the producers and the consumers alike. What is needed is a well planned strategy to bring in the desired traits like (*i*) high yield potential, (*ii*) greater yield stability and (*iii*) higher production efficiency in a strain having the other desirable quality traits.

Enhancement of Yield Potential

Any crop improvement programme lays primary emphasis on identifying or incorporating genetic potentials for higher yield, since the production performance of any crop variety is the outcome of its genotype and the grows environment-generally manipulated by way of crop management practices. Attempts to identify/incorporate the genotypes with high yielding potentials were rather restricted till the intricacies of the life cycles/sexuality of the edible mushrooms were worked out in the seventies (Elliott, 1972). However, once these mysteries were unfolded, significant break through were made on this front *e.g.* release of two high yielding hybrids U1 and U3 of *Agaricus bisporus* by Fritsche in the year 1981. Besides, there is a definite spurt in activities related to strainal improvement of various cultivated mushrooms, and much more is expected in the years to come with more and more application of molecular techniques for breaking the yield barriers. Nevertheless, the following breeding approaches are considered useful for improving the genetic yield potential of edible mushrooms.

Conventional Techniques

There has been a continuing attempt to identify and use improved strains of cultivated mushrooms particularly in the button mushroom right from the day when its indoor cultivation became a commercial enterprise. Due to limited knowledge of biology and reproduction of these fungi as well as for want of improved techniques, the earlier workers relied mostly on traditional breeding strategies of (*i*) introduction and selection, (*ii*) strainal improvement, (*iii*) mutation and only at a later stage on (iv) hybridization.

Introduction and Selection

Although, procedures of introduction and selection have yielded rich dividends in other crops, it could not prove much successful in edible mushrooms particularly in button mushroom–the most widely cultivated and commercialised mushroom during the last 2 centuries. This is because of the fact that any successful selection procedure primarily requires the availability of genetic variability, and only then individuals with desirable traits like high yield potential, growth rate, climatic adaptation, resistance to biotic/abiotic stresses and other quality parameters are selected. However, due to predominance of inbreeding and poor collection/ conservation of wild germplasm of *A. bisporus*, there is very limited genetic variability available in this taxon and hence these methods could not succeed and will not be able to meet the future challenges of the button mushroom industry.

Strainal Improvement

With the perfection of sterile techniques for spawn production and spore germination, a new possibility was opened to obtain superior strains by raising tissue culture and multispore/single spore isolates.

Tissue Culture

Raising of tissue cultures from the sporophores of wild collection of edible mushrooms and their use for germplasm conservation and spawn production has been in vogue for many years. However, their use to obtain improved strains has not been found very rewarding. Although, tissues from different parts of the sporophore have been found to differ in their growth rates and even yield performance, yet these have not been considered very dependable and are also known to carry some undesirable parental traits. Recently, Mehta (1991) reported that only 18 per cent of tissue cultures isolated from pileus could not yield the parent strain. The performance of isolates taken from stipe and the gills were still inferior. Obviously, tissue isolates can hardly be useful for getting improved strains but its usefulness for raising and conservation of cultures of wild germplasm can not be overemphasised, since this technique ensures their conservation in an unchanged form.

Multispore Culture

Although, raising of multispore isolates may yield some variability due to anastomosis of the hyphae originating from different spores, which may prove useful for further breeding purposes, but this technique is more in use for rejuvenating the old and degenerated cultures due to repeated sub-culturing, particularly by spawn makers. As regards the yield potentials of multispore cultures, there are already reports that they may perform better, worse or at par with the parent strain. To add to this, Mehta (1991) has reported that although some multispore isolates exceeded the yield of the parent strain, but the former degenerated faster than the latter. It could thus be concluded that instead of using this technique directly for breeding purposes, it could be more useful for rejuvenation of old cultures and isolation and conservation of wild germplasm.

Strain Mixing

Inter-strainal fusion of hyphae has been tried by earlier workers by allowing the mycelia of different strains to intermingle and anastomose. Even exchange and fusion of nuclei of different strains have been claimed due to such intermixing and the isolations made from the meeting zone have been supposed to perform superiorily due to complementary/supplementary action of the nuclei derived from the two parent strains. However, there are either contradictory reports of such possibility, or examples of such an improvement are too few to be of any practical consequence.

Single Spore Isolations

Strainal improvement directly by isolating single spore isolates (SSIs) is possible only in homothallic mushroom species like *Agaricus bisporus* and *Volvarieila volvacea* and not in heterothallic species. In both these mushrooms SSIs are self-fertile, although basidiospores of *A. bisporus* derived from 2-spored basidia contain two nuclei of different parental types and those of *Volvariella volvacea* are uninucleate receiving only one of the four meiotic nuclei in each basidiospore. Monosporous cultures of both these mushrooms exhibit variability in terms of growth-rate, morphology, and yield-potential and provide ample opportunity to select improved strains of both these homothallic mushrooms through this technique. Even after the release of hybrids of *A. bisporus*, by Fritsche in 1981, the single spore isolation technique is being used by several workers due to its simplicity and being quite rewarding as well. Recently, out of four promising single spore isolates developed at NRCM by Bhandal and Mehta (1989) and after multilocational testing two single spore isolates of *A. bisporus* have been released namely NCS-100 and NCS-1O1 as highyielding cultivars (Anonymous, 1997). Similarly, Kalra and Phutela (1991) could raise and evaluate several SSIs of *Volvariella volvacea* and *V. diplasia* and recorded upto 29 per cent higher yield in one of the isolates.

Hybridization

Hybridizaton, *i.e.* combining the gene pools of two different strains or stocks has proved a very effective means to exploit the advantage of heterosis in a good number of crops. In mushroom also, breeding of hybrids became a reality when the Dutch hybrids U1 and U3 of *A. bisporus* were released in by Fritsche the year 1981. This was made possible primarily because by that time the life-cycle and sexuality pattern of *A. bisporus* was almost fully understood and hence a right action plan for producing the hybrid was used. With more information accumulating in this respect (Table 8), the prospects and problems inherent in this mode of strainal improvement may be examined as under:

It is established that greater the genetic diversity between the parental strains, higher is the magnitude of heterosis. Hence, a good species to attempt heterosis would be the one wherein ample variability is available. However, this is not the case at least with *A. bisporus* since it is neither abundant in the wild nor it exhibits sufficient range of variations in genetic traits. Further more, due to predominance of inbreeding in this mushroom homogeneity is promoted.

Table 8: Pattern of Sexuality and Compatibility Systems in Cultivated Edible Fungi

Species	*Sexuality*	*Incompatibility*	*Fertility*	*No. of Spores per Basidium*	*No. of Nuclei per Spore*
Agaricus bisporus	Secondary Homothallism	Unifactorial	Self-fertile (Heterokaryon)	2	2
Agaricus bitorquis	Heterothallism	Unifactorial	Cross-fertile	4	1
Pholiota narneko	Heterothallism	Unifactorial	Cross-fertile	4	1
Auricularia auricula	Heterothallism	Unifactorial	Cross-fertile(?)	4	1
Auricularia polytricha	Heterothallism	Bifactorial	Cross-fertile(?)	4	1
Lentinula edodes	Heterothallism	Bifactorial	Cross-fertile	4	1
Pleurotus ostreatus	Heterothallism	Bifactorial	Cross-fertile	4	1
Flammulina velutipes	Heterothallism	Bifactorial	Cross-fertile	4	1
Coprinus fimetarius	Heterothallism	Bifactorial	Cross-fertile	4	1
Volvariella volvacea	Primary Homothallism	None	Self fertile (Homokaryon)	4	1

For breeding hybrids, heterothallic fungi are the ideal candidates, where each basidiospore is uninucleate and produces a self-sterile homokaryotic mycelium, which favours out-breeding and thus promotes heterogeneity. Obviously, mushroom species like *Agaricus bitorquis, Lentinula edodes, Pleurotus ostreatus, Auricularia auricula* etc. are better materials for hybrid-breeding programmes. However, mushrooms like *Volvariella volvacea* and *Agaricus bisporus*, which are primary and secondary homothallic species respectively, are much more difficult materials to handle. Of course, of the two, *A. bisporus* is comparatively less difficult to produce hybrids. Yet due to some peculiar features, *A.bisporus* also poses many problems to the mushroom breeders. These include:

1. Most basidia produce only 2 basidiospores, each having 2 non-sister nuclei derived from meiosis and produce a self-fertile heterokaryotic mycelium. Only a negligible proportions of basidia of *A. bisporus* produce 3 or 4 basidiospores each having only one daughter nucleus and hence germinate to produce a homokaryotic self sterile mycelium, a pre-requisite for hybrid breeding.
2. The identification of such homokaryons either as basidiospores or as mycelia is a very difficult and challenging task because (i) the number of such basidiospores is too small and can be identified only on the basis of slow growth and non-fruiting of its germinating mycelium/and because (ii) there is no morphological markers *e.g.* clamp connection in the hetorokaryotic mycelium of this fungus, without which the heterokaryon is difficult to distinguish from the homokaryon, and thus can be identified only on the basis of fruiting test. Both these operations are evidently highly laborious and time consuming.

3. Since the slow-growing and non-fruiting criteria are less reliable and more time consuming, for successful breeding in a difficult species like *A. bisporus*, some cultural markers like auxotrophs showing nutritional deficiencies in respect of some amino-acids and (ii) fungicidal resistance have been used. The auxotrophic markers are generally obtained through induced mutation and are recessive to the corresponding wild alleles and, therefore, they express their inability to grow on minimal media in the haploid or homoallelic condition.

However, if two such auxotrophs deficient for two different amino acids, *e.g.* proline and methionine, are crossed they complement each other for the deficiency and hence the hybrid so produced does not exhibit any nutritional deficiency and can be easily identified owing to its growth on a minimal medium. However, use of such markers in a breeding programme has its own limitations due to the fact that such mutations are very rare and difficult to induce. Although occurrence of naturally occurring auxotrophs has been reported, particularly in mushroom spawn, but they go undetected since they are masked by multinucleate nature of the mycelium of *A. bisporus*. Another cultural marker used in a similar way is the use of fungicidal resistance markers, which are somewhat easier to obtain, since mutants with such markers are readily identified by their growth on a medium supplemented with the fungicide.

Mutation

Although, both spontaneous and induced mutation have been reported in *Agaricus bisporus* but it has not yet found any major application in the genetic improvement of cultivated mushroom, except the chance spotting of the natural mutant pure white and smooth capped individuals in a bed of cream strain of *A. bisporus* more than 7 decades ago since then the pure white strain has attained the status of the most popular commercial culture of button mushroom. Both physical and chemical mutagens like, X-ray, UV-rays and Nitrosomethyl urea, Methyl uracil, Thio-uracil, Chloroethyl phosphoric acid and Uranium nitrate have been used for obtaining induced mutants of mushroom.

Molecular Techniques

With the recent advancements in biotechnology, several molecular techniques have become available to fungal breeders also and during the last one decade or so, intense activities have started in the areas related to genetic improvement of mushrooms. Some of the major applications of these techniques are in (1) Genome charactarization, (2) Molecular markers, (3) Genetic transformation and (4) Protoplast fusion.

Genome Characterisation

Before any attempt is made to improve a particular strain, an understanding of the nuclear genomes and any extra chromosomal genetic elements is essential. With the availability of various molecular techniques, more and more information on genetic characterization and gene mapping of mushrooms have been pouring, with maximum emphasis on the commercial mushroom *A. bisporus*. The *A. bisporus* genome size has been estimated to be 34 Mb (Mega base) pairs (Arthur *et.al*, 1982). The karyotype of *A. bisporus* using CHEF/Orthogonal field electrophoresis has resolved the haploid genome into 13 chromosomal bands ranging between 3,5 to 1.2Mb (Royer and Horgen, 1991; Royer *et. al*. 1992). Similarly, the mitochondrial DNA of *A. bisporus* has been estimated to be 136 Kb (circular). Recently, 3 double stranded (ds) RNA of sizes> 13.1 Kb, 2.4 Kb and 5.2 Kb have been isolated from *A. bisporus* (Romaine *et al.*, 1994). In addition, two mitochondrial plasmids namely pEM (7.4 Kb) and pMPJ (3.9 Kb) have been recovered from this mushroom by Mohan *et al.* (1984).

Molecular Markers

For effective exploitation of hetersosis in mushrooms, the genetic variability of parental strains need to be authentically defined, particularly in those mushroom species where phenotypic variability is not well explicit, *e.g. A. bisporus*. Allozyme markers, Restriction Fragment Length Polymorphism (RFLP) and Randomly Amplified Polymorphic DNAs (RAPD) analyses have proved very fast and unambiguous methods to assess the genetic diversity in a number of mushrooms:

a) RFLPs: RFLP has been the other molecular tool, which is proving a dependable and rapid analytical technique for detecting intra and inter-specific variations as well as for confirming the success. The genetic diversity in *A. bisporus* was assessed by employing allozyme markers by a host of workers (Callac *et al.*, 1993; Kerrigan et. al,1993,1996; Kerrigan Rose,1986). In fact, more than a dozen allozyme markers have so far been identified in *A. bisporus* alone. These naturally occurring markers are easily identified by specific staining for particular enzymic activity. Similar studies have also been made in *A. bitorquis, Lentinula edodes* (Itvaara, 1988) and *Pleurotus spp* (Magae *et al.*, 1990, Zervakis and Labarere, 1992) Recently, Royse and May (1993) have found extensive allozyme variation in all mushrooms they studied, including *A. bisporus, A. campestris, L.edodes, Morehella spp., Pleurotus spp.* and *V.volvacea.*

b) Hybridization: RFLP markers were employed to establish inter specific genetic diversity in *Agaricus* spp. (Castle *et al.*, 1987) and in *Laccaria spp*. (Gardes *et al.*, 1990). However, Iracabal and Labarere (1993), and Labarere *et al.* (1993) compared the levels of variability detectable with isozyme and RFLP analyses in case of *Pleurotus cornucopiae* (Iracabal and Labarere, 1993 and 1994 Labarere *et al.*, 1993) Similarly, RFLP of ribosomal DNA (Iracabal *et al;* 1995; Vilgalys *et. al*, 1996) mitochondrial DNA and total genomic DNA (Sagawal *et. al*. (1992)) have been found very handy and useful for understanding the inter-specific diversity and molecular systematics of the genus *Pleurotus.* Although RFLP markers have continued to be used in researches related to breeding of several crop varieties due to their codominance, (*i.e.* with their help heterokaryon can, be distinguished from either homokaryons and they provide complete genetic information at a single locus), but they still have several disadvantages. These include, high cost, too much laborious, time consuming, requirement of large quantity of DNA and till recently, it required radioactive probes.

c) RAPDs: Random Amplification of Polymorphic DNA based technique has become very popular because of its simplicity and low cost and because it provides automated genotype assay with very modest laboratory facilities. It provides quick and efficient screen for DNA sequence based polymorphism at many loci. Moreover, it requires a very small amount of DNA and a non-radioactive assay system which can be completed in hours. DNA amplification and sequencing has already been used in several edible fungi. This method was first reported in *A. bisporus* mushroom by Khush *et al.* (1992). Hibbett (1992) found a portion of nuclear encoded large sub-unit (LSU) RNA gene useful for assessing the phylogenetic relationship at genus level. Vilgalys and Sun (1994) has described procedure for DNA extraction, PCR amplification and manual sequencing. Recently, Vilgalys *et al.* (1996) have used nuclear ribosomal DNA corresponding to large Subunit (LSU) RNA coding region and its corresponding Internal Transcribed Spacer (ITS) region for getting an insight into molecular systematics of *Pleurotus* spp. Similar studies have been made by Neda and Nakal (1995) using 18s RNA and ITS sequences from several *Pleurotus* spp. and found this technique useful in molecular systematics. Moncalvo (1995) however, confined his studies based on sequence data to establish a close relationship between a newly described taxon *Pleurotus cystidiosus* var. *formosensis* and other members of the *P. cystidiosus* complex.

Genetic Transformation

Genetic transformation of a strain enables us to introduce dominant single genes that can selectively modify the yield determining factors, etc. without other changes in the genetic constitution of the recipient strain. During the last one decade or so, transformation systems for a variety of Basidiomycetes have been developed including *Schizophyllum commune* Munoz (Munoz-Riwas *et al.*, 1986), *Coprinus cinereus* (*Binninger et al.*, 1987), *Agrocybe aegerita* (Labarere *et al.*, 1993). However in the most widely cultivated mushroom *Agaricus bisporus* no transformation system proved successful till recently despite a series of attempts (Challon and Elliott, 1994). However, reports of generation of an effective transformation system for homokaryotic *A. bisporus* have come later and mating of such compatible homokaryons results in the formation of transgenic fruit bodies. Further this system has been used successfully by Mooibroek *et. al.*(1996) to investigate the browning phenomenon and to find out ways for gene silencing in a heterokaryotic organism with one transformed nucleus per cell.

Another technique for transformation already proved successful in other plant species is the ballistic delivery of DNA into the recipient organisms, which normally do not respond to protoplast mediated transformation. To carry out such an operation, a gene gun has been developed and tested in some edible mushrooms like *Coprinus* spp. (Moore *et al.*, 1995). For obtaining transformants by this technique, auxotrophic host cells are bombarded with tungsten microcarriers coated with plasmid DNAs carrying the requisite wild genes. By this method, any kind of tissue like mycelium, sexual and asexual spores and sporophores, can be bombarded and hence it is possible to apply this method to any mushroom species at any stage of their life-cycle.

Protoplast Fusion

Among the biotechnological techniques, cell fusion is becoming a handy method to overcome the natural barrier existing between strains/species genera to mate and exchange the genes with each other. To achieve such cell fusion, however protoplasts are required to be isolated, tested for reversion and then made to fuse with another type (both monokaryotic types) to produce a somatic hybrid protoplast. The parental types to be fused are carefully selected and are fully characterized by creating a selection marker, which helps to identify the fusants. The fusants are later put to identical tests to compare them amongst themselves and with the parent strains which together help to identify and evaluate the fusants. Further, the fusants are cultured in pairs with the parent monokaryons to test for antagonistic reactions-a barrage like phenomenon, which suggests that the fusants are different from the original monokaryons. The fusants are also put to fruiting test to identify those having vigour and high yielding capacity.

The protoplast isolation and reversion has been attempted in a number of Basidiomycetes including *Agaricus bisporus, Armillaria mellea, Coprinus* spp., *Ganoderma* spp., *Hericium coralloides, Lentinula edodes, Morchella* spp., *Pholiota* spp., *Pleurotus* spp., *Schizophyllum commune* and *Volvariella* spp., *Auricularia* spp. and *Lentinula edodes*, Also protoplast fusion and intra and inter-specific heterokaryons have been obtained in several edible mushrooms like *Auricularia auricula, A. polytricha* (Sucnegawa, 1992), *Pleurotus* spp. (Toyomas, 1991), *Lentinula edodes* (Zhang *et al.*, 1996) and intergeneric protoplast fusion between *Lentinula edodes* and *Pleurotus* spp. (Zang *et al.*, 1996).

Increasing the Yield Stability

The full yield potential of a crop species is often difficult to realise due to losses caused by biotic and abiotic stresses. In mushrooms also, a number of extremely harmful pests and diseases cause losses both in quality and quantity of the produce. Mushrooms themselves being fungi pose special problems in adopting chemical control measures, particularly against the diseases. Hence, strains with genetic resistance or tolerance to the biotic and abiotic stresses should be the preferred strategy to impart yield stability as well as quality assurance. However, very little attention has been paid on breeding such strains, even in the most commercially exploited mushrooms like *A. bisporus, Pleurotus spp., Lentinula edodes* etc. due to which there is a definite lack of sustainability in mushroom production even in high tech-mushroom farms, what to talk of low cost seasonal farms. In future, therefore breeding-strategy must include the following parameters to ensure yield stability in this high income generating cash crop.

1. Improvement for resistance to insects
2. Improvement for resistance to diseases
3. Improvement for competitiveness against moulds/weed mushrooms
4. Improvement for tolerance to abiotic stresses like low oxygen, high temperature, low humidity, excess watering, tolerance to pollutants and fumes of machines etc.

Increasing Production Efficiency

A perfect breeding-strategy must not confine to yield potential and yield stability only, if the ultimate aim is to obtain higher production per unit time/cost. Many other attributes like nutrient use-efficiency, and crop-duration parameter carry a lot of meaning in a crop like mushroom, if the above two objectives are to be met by developing efficient crop cultivars. Similarly, for getting higher profit, strains capable to suffer least post-harvest losses and possessing high nutritional and culinary qualities will be essential. Hence, future strategies for breeding mushroom strains will have to include:

1. Breeding for short-duration strains
2. Breeding for nutrient-use efficiency
3. Breeding for more nutritious and tasty strains
4. Breeding for increased shelf-life.

With molecular tools available-to-day and likely addition of more and more biotechnogical techniques in future, it would be very much possible to incorporate genes governing these traits from wild germplasm of the same species, different species or from a different genus, and such a possibility may not be far-off.

Chapter 13

Economic Importance of Mushroom

Mushrooms were introduced in the kitchen of aristocrats primarily for their taste and flavour. But increasing knowledge opened their various qualities including high nutritive value, value added products of mushroom, mushroom as a natural scavenger, varied utility of spent mushroom substrate, medicinal value, mushroom poisoning and psychoactive properties as well as other uses which may be read under respective heads as follows:

Nutritional Value of Mushrooms

Evaluation of nutritive quality of mushroom was initiated as easily as 1898 by Mendel and since then quite good number of literatures have accumulated on it which have been summarized by Purkayastha and Chandra (1985). Data collected so far clearly reveals that the nutritive quality of mushrooms not only varies from species to species but also from authors to authors probably due to two reasons: first, due to technological variations and second, due to the variations in age of fruiting body taken for evaluations as well as the variations in the nature of the substrates on which the mushroom in question was grown. Further, the nutritive value of the species of mushrooms worked out so far is far less than the number of edible mushrooms known. Therefore this is too early to authenticate the true nutritional status of mushroom.

Protein content in mushroom is highly variable. It has been estimated as low as 2.78 g/100g of dry wt in *Pleurotus* sp. and as highly as 47.93g/100g of dry wt in *Pleurotus sajor-caju* with common range variations between 25-35 g/100g dry wt (Purkayastha and Chandra, 1985). As regards the protein content, mushrooms can be ranked higher than several vegetables (viz. peas, cabbage, carrot, cauliflower potato, tomato) and fruits but lower than some of the cereals (rice, wheat) and animal products (viz. beef, egg, chicken, fish and cows milk). Data on production of Crude Protein and Gross Energy by Crops and Animals compared with the Edible mushrooms have been summarized in the Table 9. Variations in protein content is not limited to the species of mushroom but also with stages of development of its sporophore, parts of sporophore and nature of growing substrate. Young sporophores are usually richer than the old ones (Motskus, 1977). Pileus contains greater content of protein than that of stipe and volva (Chang and Chan, 1973). Difference in protein content was recorded in *Volvariella volvacea* when grown on different substrates (Chang, 1964). Mushrooms have unique efficiency of producing protein from a given quantity of carbohydrate which has been estimated 65 per cent against about 20 per cent in Pork, 15 per cent in Milk, 5 per cent in Poultry and 4 per cent in Beef. Further, the production capacity of an acre of land has been computed ten times greater for mushroom protein than for meat protein.

The quality of protein plays an important role rather than its quantity present in mushroom. The mushroom protein has a high digestibility and in its overall quality can be classed as being intermediate between low grade vegetable and high grade meat protein. The digestibility test of mushroom nitrogen in human has revealed that 69 per cent of total nitrogen of *Agaricus compestris* exists in the form of protein and 50 per cent of the total nitrogen was in the form of a digestible protein. The biological value of mushroom proteins is intermediate between vegetable and animal protein (Worgan, 1968 and Chang, 1980). Its digestibility varies from 72 to 83 per cent (Lintzel, 1941) and its caloric value ranges from 18 to 29k.cal./100g of mushrooms.

Amino Acid is the basic unit of protein. Therefore its composition plays a significant role in determining the nutritive value of the protein. Although qualitatively animal protein is ranked high from nutrition point of view, it lacks essential amino acids. The sole source of essential amino acids is vegetable protein. Since mushrooms are considered vegetable, their screening for essential amino acid profile is necessary. Considerable amount of work has been done on the essential amino acid composition of some common edible mushrooms (Table 10). It is obvious from the available data that the protein hydrolysate of some of mushroom contains all the amino acids including essential amino acids. Percentage of essential amino acid in several wild and cultivated edible mushrooms has been reported to vary from 2.6 to 7.6 per cent (Matti *et al.*, 1976). Usually 25-35 per cent of total amino acid occur as free amino acids while the rest are present in bound form. Mushrooms in general are rich in lysine and tryptophan (cereals are deficient of it) but are deficient in sulphur containing essential amino acid. Primarily mushrooms are deficient in phenylalanine (except in *Volvariella diplasia*) and methionine when compared with egg protein. Supplementation of mushroom protein with phenylalanine and methionine would be necessary, when used as sole source of protein in diet. Further that lysine content is highest in white mushroom (*Lactarius*) and all mushrooms contain relatively high amount of aspartic acid, glycine and serine.

In India people are underfed and suffer from malnutrition especially for protein and amino acids. Cereals are supplemented with legumes for protein. The rate of pulse is touching sky and has gone beyond the reach of poors. Also its production

Table 9: Production of Crude Protein and Gross Energy by Crops and Animals compared with the edible Mushrooms.

Species	*Yield Component*	*Yield* (kg)*	*DM per cent*	*Yield DM (kg/ha)*	*Crude Protein (per cent)*	*Gross Energy (MJ/kg)*	*Crude Protein (kg/ha)*	*Gross Energy (MJ/ha)*
Grass (Perennial)	Total harvested	60,000	10.0	12,000	18.5	15.5	2,100	222,000
Cabbage	Total harvested	54,545	11.0	6,000	13.6	17.5	316	105,000
Peas	Seed	2,511	86.0	2,159	26.1	18.9	566	40,805
Potatoes (main crop)	Tuber	27,621	21.0	5,800	9.0	17.6	52.2	102,080
Wheat (winter)	Seed	4,394	86.0	3,779	12.4	18.4	469	69,534
Maize	Seed	4,654	86.0	3,995	9.8	19.0	392	75,905
Rice	Seed	5,670	86.0	4,876	7.7	18.0	375	87,768
Cattle, beef (suckler)	Carcass	–	–	360	14.8	10.9	53	3,924
Cattle (dairy)	Milk	–	–	3,386	3.5	2.6	118.5	8,770
Sheep	Carcass	–	–	462	14.0	16.2	65	7,486
Pigs (baconer)	Carcass	–	–	875	12.0	16.5	105	14,438
Hen	Egsg (edible portion)	–	–	624	11.9	6.6	74	4,118
Pleurotus sajor-caju	Total harvested	248,888	10	24,888	22.5	12.5	5,599	312,344
Pleurotus flabellatus	Total harvested	133,333	10	13,333	21.6	11.3	2,880	151,196
Agaricus bisporus	Total harvested	189,000	10	18,900	26.3	13.7	4,970	259,308
Volvariella volvacea (on straw)	Total harvested	10,864	10	1,086	22.5	12.7	244	13,792
V. volvacea (on cotton waste)	Total harvested	22,222	10	2,222	22.5	12.7	500	28,219

*: Fresh weight.

Source: Rajrathnam and Bano,1991.

Table 10: Comparison of Amino Acids Composition of Edible Mushrooms (g/100 g protein)

Amino Acids	*Agaricus bisoporus*	*Lentinus edodes*	*Volvariella volvacea*	*Pleurotus*
Essential				
Isoleucine	4.3	4.4	4.2	4.0
Leucine	7.2	7.0	5.5	7.6
Lysine	10.0	3.5	9.8	5.0
Methionine	Trace	1.8	1.6	1.7
Phenylalanine	4.4	5.3	4.1	4.2
Threonine	4.9	5.2	4.7	5.1
Valine	5.3	5.2	6.5	5.9
Tyrosine	2.2	3.5	5.7	3.5
Trypotophan	ND	ND	1.8	1.4
Total	38.3	35.9	43.9	39.3
Non-essential				
Alanine	9.6	6.1	6.3	8.0
Arginine	5.5	7.0	5.3	6.0
Aspartic acid	10.7	7.9	8.5	10.5
Cystine	Trace	ND	ND	0.6
Glutamic acid	17.2	27.2	17.6	18.0
Glycine	5.1	4.4	4.5	5.2
Histidine	2.2	1.8	4.1	1.8
Proline	6.1	4.4	5.5	5.2
Serine	5.2	5.2	4.3	5.4
Total	61.6	64.0	56.1	60.7

Source: Chang, 1991.

has fastly declined due to receding land resources. The edible mushrooms can only replace/substitute the legumes to a greater extent.

The digestibility of mushroom is also directly related to the cellular structure of their bodies. Mushrooms are less digestible than other plants due to the fact that their cell walls are largely composed of the polysaccharide chitin resistant to the action of starch content and are free of cholesterol. As such many of the edible mushrooms could be an ideal food for diabetic patients.

Mushroom is a low caloric food with little but showing wide range of fat content. It has been reported as low as 1.2 per cent in *Auricularia polytricha* (FAO, 1972) to as high as 20.6 per cent in *Volvariella esculenta* on dry wt basis (Adriano and Cruz, 1933). The constituents of crude fat include all classes of lipids (Hughes, 1962; Zhuk *et al.*, 1981) such as free fatty acids, various sterols, lecithin, sterol esters, phospholipids, mono, di and triglycerides often in the form of oil droplets within spores or in the tissue. Mushrooms have no starch and are very low in sugars.

Mushrooms are ranked high for their vitamin content. Both mycelia and sporophores contain significant amount of vitamins particularly the provitamin carotene, thiamine, riboflavin, biotin, ascorbic acid, niacin, pantothanic acid. Vitamin A, B or E are usually absent in edible mushrooms but some species contain vitamins C, K and B complex group. Thiamine, nicotinic acid, riboflavin, pantothenic acid, niacin and biotin are well retained even when the fresh mushrooms are cooked by various methods, canned, dried and freezed.

Mushrooms are also excellent source of folic acid, the blood building vitamin which counteracts pernicious anaemia. Folic acid was at one time thought to be confined to spinach, kidney and raw liver but doctors now place mushrooms in the same category. *Morchella*, which is one among the species of fleshy fungi which are consumed for their taste and flavour, contains all the essential vitamins including folic acid. Like protein, vitamin content do vary with the stages of development and part of the sporophore. Thus, thiamine is present in highest concentration in young sporophore of *Rozile caperata* but that of rioboflavin in its mature cap.

The ash analyses of mushroom have revealed that among the mineral constituents, phosphorus, sodium and potassium are predominant while low quantity of calcium and lowest amount of iron are also present. The iron content has been reported to vary from 70 to 1530 mg/kg of dry matter and that of manganese from 9 to 100 mg/kg of dry matter (Drabal Karel *et al.*, 1975). Potassium and Sodium ratio is very high which is desirable for patients of hypertension. Occurrence of copper and zinc has also been detected in a number of edible mushrooms including B*oletus edulis, Cantharellus cibarius, Armillaria mellea, Flemmulina velutipes* and *Pleurotus ostreatus* (Matti *et al.*, 1976). The capacity of accumulating these minerals especially the trace elements from the soil, are so important for the enzymatic processes taking place in the human system, presumably is highest in mushroom than that of green plants. The concentration of trace elements in the fruit bodies of mushroom depends on the habitat.

Mushrooms are free from cholesterol but ergosterol is present. They are richer than most vegetables in water soluble vitamins, besides being a good source of minerals.

Fibers are also part of healthy diet. In a modern society, some food stuffs are refined (for example white bread) and thus contain less fibers. Fresh mushrooms contain relatively large amount of fibers.

Desirable progress in commercial exploitation of the edible mushrooms has not been achieved so far mainly due to technological constrains in its cultivation. The growth and development of its sporophore is dependent upon natural conditions and therefore, yield is unpredictable. Such a problem has been solved by growing their mycelia in culture which are now being considered as a potential source of proteins, amino acids, vitamins and mineral (Wargel *et al.*, 1974; Jandaik, 1976 and Kaul, 1978a). It is also possible to enhance protein content of mycelia by cultural amendments. Since amino acid composition of protein in sporophores and mycelia are more or less similar (Worgan, 1968), large scale production of mushroom mycelia can provide promising source of proteinaceous food.

Among the mushrooms gathered in their natural state, high value is attached to those belonging to the Boletaceae family. In most cases they are easily recognized and endowed with a pleasant smell, taste and consistency. The most popular and generally known bolet is *Boletus edulis*. Otherwise it is recommended to eat mushrooms in mixtures in which specific taste qualities, consistency and smell of the single species come to the fore. Species with a less distinct taste are added to strongly aromatic or more acridly tasting species. Tougher species are cut into thin slices, dried mushrooms ground into powder can be used as spices, many spices can be preserved in salt or pickled in various solutions. The trimmings added to mushroom dishes should be easily digestible *e.g.*, potatoes and vegetable salads.

Unlike many other single call protein, mushrooms are directly edible and are praised as food delicacies because of their characteristic texture and flavours.

Value Added Mushroom Products

Button Mushroom (*Agaricus bisporus*) has a poor shelf life at the ambient temperature and is highly perishable because of high moisture content (90 per cent) and delicate texture. The shelf life of common fruits and vegetables can be taken care of harvesting them before ripening which is not applicable to the mushrooms. Even after harvesting mushrooms continue to grow, mature and respire resulting in weight loss, veil opening, browning and microbial spoilage (Rai *et al.*, 2003). Technology has been developed very recently by Arumuganathan *et al.* (2006) for retention of the fresh mushrooms at various marketing levels by conversion of mushrooms into the value added products which not only reduce the post harvest losses but also enhance the income and marketing. Some value added products, are mushroom nuggets, mushroom biscuits, mushroom–soup powder, mushroom ketch up and pickles from the dried button mushroom. Among different bakery products, biscuits are more amenable to variations in the formulation to meet a wide range of consumer demands with respect to taste and nutritional requirements. Dried mushroom powder is used to prepare mushroom cookies, nuggets are generally prepared from the ground pulses (*dhal*) namely black gram powder, soybean powder, urad *dhal* powder etc. and are used in the preparation of vegetable curries in North India. The nuggets add taste as well as nutrients to the meal, since it is prepared from *dhal* powder rice in protein. Though mushroom soup powder as well as pickles are already acceptable by the consumer and freely available in mushroom can also be utilized in the preparation of weaning foods. The recipes for preparation of dishes like mushroom kidney fry, mushroom salads, cheese sandwiches, stuffed capsicum, stuffed morel, mushroom fritters, meat stuffed mushroom, mushroom hot dog, mushroom burgers, mushroom omlette, mushroom and pouched eggs and stuffed potato are also available in the market. Two novel products namely mushroom biscuits and mushroom nuggets need a bit introduction and promotion.

Preparation of Mushroom Soup Powder

Button mushroom slices are dried at 55°C for 8 hours in a cabinet–air drier and pulverized in a pulverizer to pass through 0.25 mm sieve and mushroom powder is produced. Mushroom soup powder is prepared by blending as per the following formulation which has shown excellent acceptability in preliminary trials: Mushroom powder (16 per cent), corn flour (5 per cent), milk powder (50 per cent), refined oil (4 per cent), salt (10 per cent), cumin powder (2 per cent), black pepper (2 per cent), sugar (10 per cent), ajinomoto (2 per cent).

Ingredients are mixed with desired quantity of water for the preparation of good quality mushroom soup with characteristic aroma and taste. The flow chart for the preparation of mushroom soup powder is given in Figure 8.

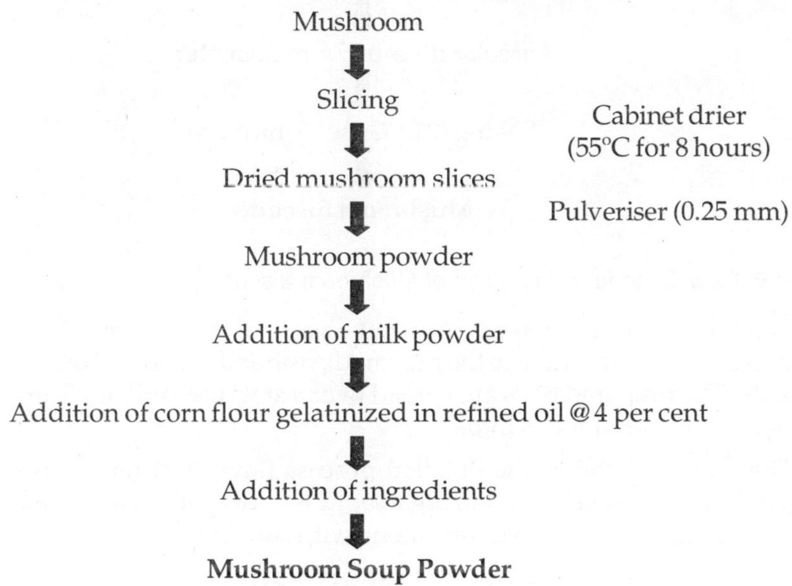

Figure 8: Flow Chart for Preparation of Mushroom Soup Powder

Preparation of Mushroom Biscuit

Refined wheat flour (Maida), refined oil (sunflower oil), sugar, milk powder (Everyday) glucose/fructose salt, ammonium bicarbonate and vanilla are purchased from the local market, while mushroom powder is prepared as described above. The detailed process flow chart for the preparation of mushroom biscuit is given in Figure 9.

The various ingredients required for the preparation of mushroom biscuit are maida (14 per cent), sugar (30 per cent), oil (40 per cent), baking powder (0.6 per cent), ammonium bicarbonate (0.2 per cent), salt (0.5 per cent), vanilla essence (0.02 per cent), milk powder (1.3 per cent), glucose (1.2 per cent), water (12.22 per cent) and mushroom powder (8 per cent).

Preparation of Mushroom Nugget

The various ingredients used in the preparation of nuggets are mushroom powder (10 per cent), urad dhal powder (80 per cent), salt (2 per cent), red chilly powder (0.09 per cent), sodium bicarbonate (0.01 per cent) and water (7 per cent). Mushroom powder

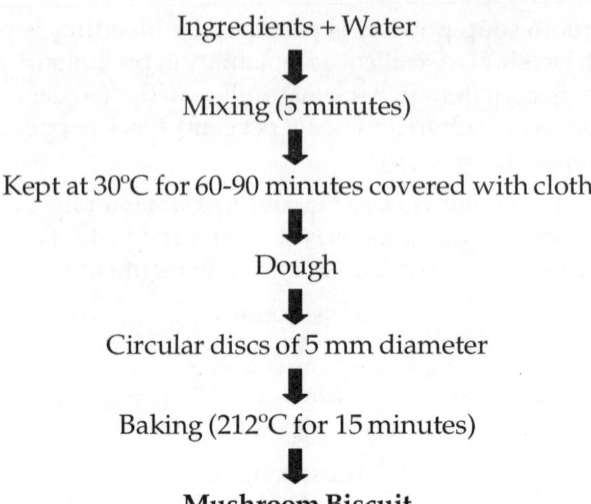

Figure 9: Flow Chart for Preparation of Mushroom Biscuit

is mixed with the urad dhal powder and a paste or dough is prepared by adding water. Ingredients are added to the prepared paste and round boll of 3-4 cm diameter are made. The prepared balls are spread over a steel tray and are dried by the sun-drying method as is done in India.

The Figure 10 shows the detailed process flow chart for the preparation of nuggets. Nuggets are subjected to organleptic test: deep–fried and used as snacks or used in vegetable curry preparation along with suitable vegetables.

Figure 10: Flow Chart for Preparation of Mushroom Nuggets

Preparation of Mushroom Pickles from Dried Mushroom

Pickles are normally prepared from fresh mushroom but method has been devised by Arumuganatham *et al*. (2006) to prepare the pickles from the dried mushrooms to diversify its uses. In this process, dried mushroom are dehydrated first and then pickled. Dried button mushroom slices with 9 per cent moisture content are taken and soaked in hot water at 80°C for 5 hours to dehydrate the mushroom for pickle preparation. The ingredients used in the dried mushroom pickle preparation are salt (9 per cent), acetic acid (1 per cent), sodium benzoate (650 ppm), rai powder (3 per cent), turmeric powder (2 per cent), red chilli powder (1.5 per cent), cumin powder (1 per cent), aniseed powder (1 per cent) black pepper powder (1 per cent), kalonji (1 per cent), ajwain (1 per cent), sugar (10 per cent) and oil (15 per cent). The mushroom pickle is prepared by mixing the ingredient with the rehydrated mushroom. Mushroom pickles can also be prepared from fresh mushroom as shown in Figure 11.

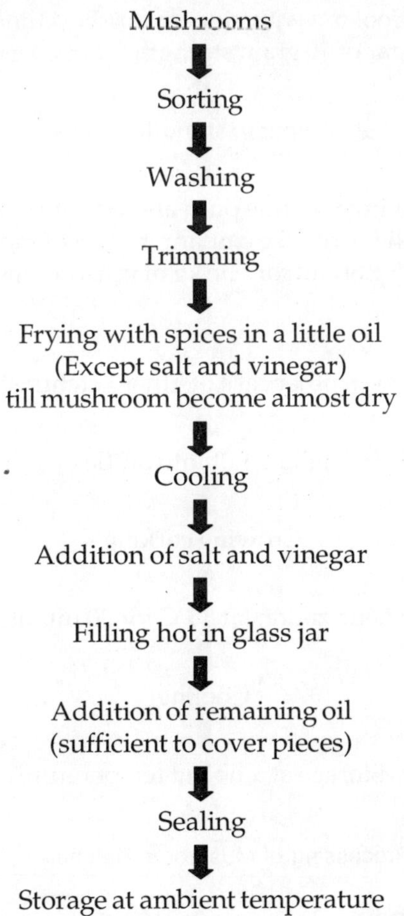

Figure 11: Processing Flow-chart for Mushroom Pickle

The organoleptic evaluation of all the four value added products for their taste, texture, colour and flavour by panel of ten judges on the nine point Hedonic Scale has rated very high of liking (rating as 9) and very low of disliking (rating as 1).

Mushroom Ketchup from Fresh Mushroom

It can be prepared adhering to and using the ingredients shown in Figure 12.

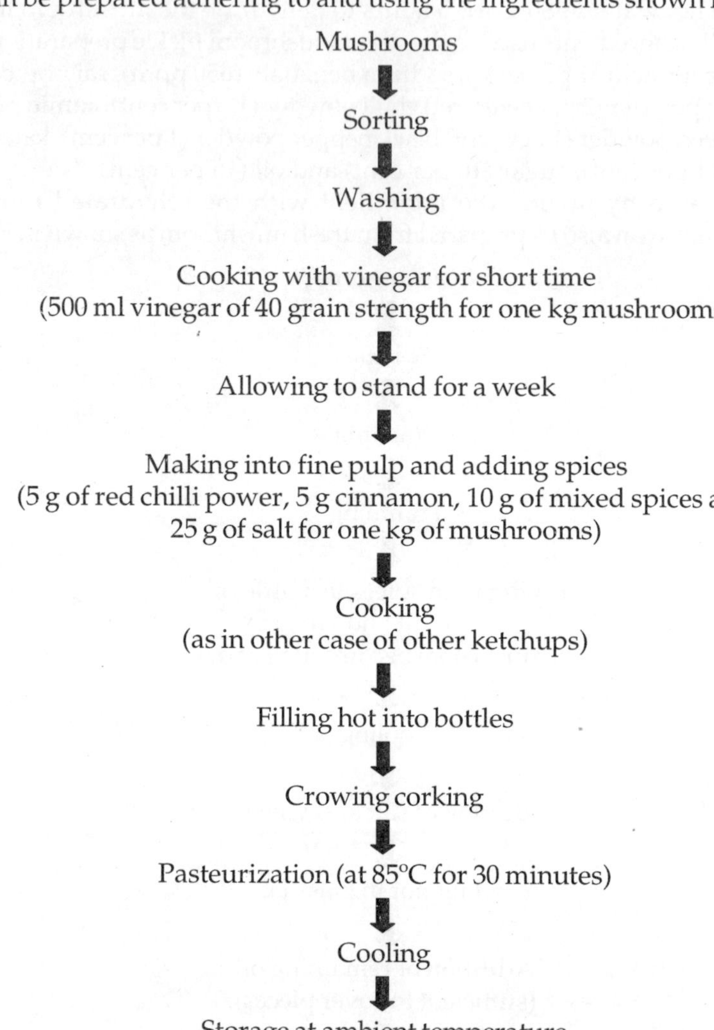

Figure 12: Flow-chart for Processing of Mushroom Ketchup

Pickle of Milky Mushroom

Milky mushroom (*Calocybe indica*) is third commercially grown mushroom in India after button and Oyster mushrooms. Use of this mushroom in the form of pickle and chutney is a new concept due to short shelf life of mushrooms owing to high (90 per cent) moisture content (Jalali *et al.*, 2003).

For preparing pickle, mushrooms are thoroughly washed in running water to remove adhered dust particles and are cut into one inch cube or triangular shapes alongwith stalk. For pickle, the excessive water from cut pieces of mushroom is removed by heating them in little quantity of refined oil and half of the quantity of required table salt. The heating is continued till all excessive moisture evaporated from the vessel. It is important to remove excess water from fresh mushroom to extend shelf-life of the product (Singh and Bano Zakia, 1977).

The remaining quantity of the refined oil is heated and asafetida (0.2 per cent) is added followed by required quantity of red chilli powder. The chilli powder is heated in oil. To stop cracking, a dash of fresh water is given to the material. The spices are heated and powdered individually. One lot of spices is powdered and used in one treatment and for other treatment coarsely grounded spices are used. Ratio and quantity of spices used in different treatments is given in Table 9. In all the treatments cinnamon and cloves are added due to their inhibitory action on the germination of mould *Aspergillus niger* spores (Anand and Johar, 1957). Spices in required quantity (Table 11) are added to all the treatments as material is mixed with remaining quantity of salt while heating. After cooling varying quantity of Acetic Acid is added in different treatments. Material from all treatments are bottled and left for balancing of salt and acidity.

Chutney of Milky Mushroom (*Calocybe indica*)

Chutney is prepared using 1 kg of milky mushroom, 1.5 kg of red, ripe and peeled tomatoes, 750 mg of vinegar and 35 g of garam masala. Material is concentrated till the volume reduced to half while heating. On concentration, the mushroom paste get blended with tomato paste. The material is then bottled hot. The products are evaluated for sensory attributes like colour, flavour, texture and taste by a panel of 10 judges on a scale of 10 points for each parameter and 30 point scale for over all (Table 12).

As shown in Table 12, sensory score for flavour declared best in T3 followed by T1 due to the presence of cumin and coriander powder since the cinnamon and clove were absent, methi and cumin flavour was dominant. T5 was graded with lowest flavour showing 4 per cent of *saunf* is not desirable. For taste sensory score was highest for T4 (6.3) followed by T3 (Zero storage, Table 10) Highest test scoring for T4 can be imparted to the presence of ginger extract (20 per cent) and garlic paste (5 per cent).

After three months of storage at room temperature (Table 12) same panel of judges who evaluated the pickles again preferred and graded T1 as best with overall score of 16.2 (after 3 months storages, Table 12) followed by T3 with score of 14.9 and T4 with score of 14.0. Presence of ginger and garlic extract resulted in browning of T4 sample and high acidity in T6 due to less quantity of spices and absence of red chilli powder.

Table 11: Quantity of Different Materials Used in Mushroom Pickle

Treatment	*Quantity Used per kg. of Fresh Mushroom*																
	Oil	*Salt*	*Red Chilli*	*Mustard Seeds*	*Mustard Powder*	*Methi Seeds*	*Coriander Powder*	*Cumin*	*Cloves*	*Cinnamon*	*Pepper*	*Asafetida*	*Saunf*	*Ginger Extract*	*Garlic*	*Acetic Acid Glacial*	*Dye Edible Orange Colour*	*Sugar*
T1	10	3.0	2.5	3.0	0.5	1.6	Nil	1.5	0.3	0.2	0.6	Nil	Nil	Nil	Nil	2.0	Nil	Nil
T2	12	3.0	2.0	2.0	0.5	2.0	Nil	1.5	0.2	0.2	1.0	Nil	2.5	Nil	Nil	2.0	Nil	Nil
T3	10	3.0	3.0	3.0	0.6	1.0	0.3	1.5	Nil	Nil	1.0	Nil	Nil	Nil	Nil	2.5	Nil	Nil
T4	13	4.0	2.0	Nil	0.5	Nil	Nil	0.7	Nil	Nil	Nil	0.2	Nil	20.0	5.0	3.0	Nil	1.0
T5	12	3.0	2.5	3.0	0.6	2.5	Nil	1.5	0.3	0.2	1.5	0.2	4.0	Nil	Nil	2.5	Nil	Nil
T6	10	3.0	Nil	3.0	0.6	2.0	0.25	1.5	0.3	0.2	2.5	Nil	Nil	Nil	Nil	2.5	Traces	Nil

Oyster Mushroom Powder: As Effective Supplement for Daily Bread of Indians

In order to popularise consumption of Oyster mushroom in daily diet particularly for extending shelf life efforts were made in the department of Home Science, Amravati University, Amravati by Mane *et al.* (2003) to incorporate Oyster mushroom powder in daily bread known as Roti 5 per cent and 10 per cent respective percentage of mushroom powder was incorporated in wheat, sorghum, maize, pearls and finger millet flour. The ingredients and their proportions, procedures were standardised subjected to sensory evaluation till get constant result. The standardized and breads were evaluated for sensory attributes by a panel of 8-judges on Hedonic scale on three consecutive days. The evaluation scores revealed that all experimental breads were acceptable. This finding was then extended to growers and consumers by organizing nutrition education programme on "Importance of Oyster Mushroom in Daily Diet".

Table 12: Sensory Score Sheet for Pickle

Treatment	*Sensory Score (Zero Storage)*				*Sensory Score (After 3 months Storage at room temperature)*			
	Colour and Texture	*Flavour*	*Taste*	*Total*	*Colour and Texture*	*Flavour*	*Taste*	*Total*
T1	7.0	5.3	5.3	17.6	6.5	4.2	5.0	16.2
T2	5.8	5.2	4.1	15.1	4.9	4.5	4.0	13.4
T3	5.0	5.5	5.4	15.9	5.8	4.6	4.5	14.9
T4	5.8	5.2	6.3	17.3	6.0	5.0	3.0	14.0
T5	6.0	4.4	3.5	13.9	5.0	5.0	3.0	13.0
T6	4.4	4.8	3.8	13.0	4.0	3.5	2.0	9.5

Besides the products described above, mushrooms blend well with most of the vegetables and spices to form delicious item of food. In India based on the spices and oil used mushroom dishes can be divided into north Indian dishes and south Indian dishes. About 60 different recipes have been standardised at the Department of Plant Pathology, College of Agriculture, Vellayani (Das Lulu, 2003).

Mushroom as Natural Scavanger

In addition to its high rate of productivity per unit area and per unit time, high nutritive as well as medicinal values, mushroom has proved it potency to escape organic pollution and save environment from the hazards of pulp and effluent of paper industries.

Recycling of Inedible Plant Wastes

On the surface of our planet, about 155.2 billion tons of organic matter is synthesized through photosynthesis every year: while only a portion of this organic matter is directly edible in the form of cereal grains, fruits, vegetables etc., much of it taking varying forms such as cereal straws, dried leaves, twigs, sawdust, maize cabs, cotton wastes and so on are inedible organic wastes have to be disposed off by burning or some wasteful methods, which causes environmental pollution.

Mushroom are biotechnologically endowed with the ability to secrete a variety of hydorolyzing and oxidizing enzymes that aid in the degradation of these plants wastes (which are chemically lignocellulosics); they then use some of the degraded products for their growth and fructification and leave behind the rest in the form of spent substrates which are good manure and are used in many ways. Thus mushrooms are potent biological agent, given to us by nature to convert inedible organic wastes directly into palatable human food.

Establishment of industries in and around big cities like Delhi have tremendously increased the problem of pollution particularly that of agro industrial waste and its bioproducts. Air pollution among many other social concerns is becoming alarming. Shutting down all polluters will shatter all over economy and disrupt our way of life. Thus mushroom cultivation is one of the viable answers to reduce the pollutants to some extent by utilizing the agro industrial wastes and its biproducts in a nutritionally and economically beneficial manner.

Besides cleaning the environment, mushroom, cultivation also attracts the attention of the farmers as cash crop as well as remunerative supplementary enterprises which can give them additional income with requirement of less land, labour and time having low pay back period. Also protein conversion efficiency and productivity of mushrooms per unit of land and per unit of time is far superior compared to any plant and animal sources.

Biological Treatment of a Pulp and Paper Industry Effluent

The pulp and paper industry produces effluents with large BODs and CODs. One of the specific problems that has not yet been solved is the strong black/brown color of the effluent, which is primarily due to lignin and its derivatives released from the substrate and discharged in the effluents, mainly from the pulping, bleaching and chemical recovery stages. The brown colour of the effluent may increase water temperature and decrease photosynthesis, both of which may lead to decreased concentration of dissolved oxygen (Kingstad and Lindstrom, 1984). Lignin and its derivatives are difficult to degrade because of the linkages within the molecule, especially the biphenyl type carbon-to-carbon linkages. Because chlorolignins are not easily biodegradable, conventional biological treatment methods cannot efficiently decolorize these effluents (Eaton *et al.*, 1982). In India 34 large scale paper mills account for 51 per cent of total capacity and 271 small paper mills account for the remaining 49 per cent. Chemical recovery is not carried out in small paper mills due to economic reasons. The pollution load in terms of BOD from small paper mills is 2.5 times higher than that of large paper mills, which employ the soda recovery process (Sastry, 1986). Several methods have been tried for the removal of chloro-organic materials, but all are quite expensive and are not feasible in common practice.

The ability of *Pleurotus* spp., *P. sajor-caju*; *P. platypus* and *P. citrinopileatus* to treat pulp and paper mill effluent on a laboratory and pilot scale were studied by Rangnathan and Swaminathan (2004). On the laboratory scale treatment, *P. sajor-*

caju decolorized the effluent by 66.7 per cent on day 6 of incubation. Inorganic chloride liberated by *P. sajor-caju* was 230.9 per cent (814.0 mg/dl) and the COD was reduced by 61.3 per cent (1302.0 mg/dl) on day 10 of treatment. In the pilot scale treatment maximum decolorization was obtained by *P. sajor-caju* (60.1 per cent) on day 6 of the incubation. Inorganic chloride content was increased by 524.0 mg/dl (113.0 per cent.) and the COD was reduced by 1442.0 mg/dl (57.2 per cent) by *P. sajor-caju* on day 7 of incubation. These results revealed that the treatment of pulp and paper mill effluent by *P. sajor-caju* proved better for the purpose than *P. platypus* and *P. citrinopileatus*.

Spent Substrate: Additional Benefit of Mushroom Cultivation

Mushrooms are biotechnologically endowed with the ability to secrete a variety of hydrolising and oxidising enzymes that aid in the degradation of agricultural wastes which are chemically lignocellulosics. They then use some of the degraded products for their growth and fructification and leave behind the rest in the form of spent substrate.

The mushrooms like *Pleurotus* and *Agaricus* during their growth in lignocellulosic waste, degrads lignin preferentially so the degraded substrate becomes more bright (white) in colour due to the exposed cellulose, such fungal degradation is called *"white rot"* decay. This characteristic has been studied in three categories (i) abitity of fungal cultures to colour phenolic media, (ii) secretions of phenoloxidases and (iii) capacity to dergade lignin by monocultures.

Mushroom production is increasing at the rate of 4 per cent annually and so is the case with the spent substrate. Usually this is dumped in the nearby areas but now technologies have been evolved to utilize them in many ways:

1. *As animal feed*: Mushroom degrades the agrowaste increasing its nutritional quality and digestibility (Diaz–Godinez and Sanchez, 2002 and Braun *et al.*, 2000). It has been found that when maize straw generated after mushroom cultivation was addded to the diets of sheep, the weight gain of the sheep increased, as did the efficiency of the straw feed conversion.
2. *For mushroom recultivation*: Several studies have reported the use of spent mushroom straw for mushroom recultivation. The spent mushroom compost enriched with cotton seed meal and soya meal can be used for *Agaricus* cultivation, spent *Pleurotus* substrate for king *Stropharia* cultivation and spent *Agaricus* compost with added cotton waste for *Volvariella* production, spent *Volvariella* substrate for production of *Pleurotus sajor-caju*, spent straw of *Ganoderma lucidum* and *Flamulina vellutipes* for production of other mushroom such as *Agaricus bisporus* and *Coprinus comatus*.
3. *For enrichment of soil*: It is beneficial for enrichment of soil, restoring areas that have been destroyed through development, deforestations or environmental contamination. Some studies have been done on the use of spent mushroom substrate in the field of vegetable and fruit and flower greenhouse in nursery and landscape gardening. There are some industries that manufacture and sell different kinds of compost based on spent mushroom substrate.

4 *For degrading environmental pollutants*: The potential of spent mushroom substrate has been reported for degrading organopollutants and in the environmental bioremidiation (Law *et al.*, 2003; Xawek *et al.*, 2003).

5. *As casing soil*: Casing is an important event in the production of white button mushroom. It is referred to the covering of processed compost following inoculation of spawn. It has been found that 4-5 years old spent mushroom substrate has the right property and consistency of a reasonable casing soil.

6. *As a substrate for biogas production*: One of the biotechnologies which is getting increased attention is decomposition of organic compounds in microbial chain to volatile fatty acids as the first stage. In the second stage acids are decomposed to methane and carbon dioxide. Keeping this in view experiments were conducted with spent mushroom substrate and increase in biogas production was found to be 85 per cent to 102 per cent from the spent rice straw and *Cyamopsis tetragonoloba*. Increased nitrogen contents of the spent residues as a result of mushroom growth seem to be likely factors responsible for enhanced production of biogas.

7. *As substrate for vermiculture and vermicomposting*: Spent mushroom substrate are mixed with garbage or organic wastes produced from home or hotels, gardens, municipality, slaughter house which today is causing environmental problem. Vermicomposting can convert all such wastes into wealth. The term vermicomposting means use of earthworms for digesting organic residues in vermicompost. The worm cast or vermicompost is a dark coloured grannular aggregate, the stability of which is due to the coating of mucopolysachrides of microbes and earthworm. This is best organic fertilizer being rich in nitrates, P, K, Ca and Mg, Vitamins, natural phytoregulators and microflora in balanced form which helps in reestablishing the natural soil fertility.

Medicinal Value of Mushrooms

In addition to their nutritional value, many large edible mushrooms have long long been used in the orient for medicinal purposes justifying an old Chinese proverb "medicines and food have a common origin". Many non-edible mushrooms have also gained important medicinal uses.

The use of fungi for medicinal purpose has rather amious history. The fungi have only a very restricted application in modern 'Materia medica' but in olden days they were very much in use. Discorides (about 200 A. D.) and Hipocrates also recommended that cauterization should be made by means of fungus in order to cure certain complaints. Gerard, one of the famous herbalists gave too much importance to the curative property of *Agaricus* stating that it could cure asthama, jaundice etc. In the old herbal, larch fungus is called "Agaricus" and is found in many prescriptions upto later eighteenth century. Either agarick or any of the large fleshy puff balls were extensively used in the treatment of wounds to stop bleeding. Gentleman magazine for 1756 refers to the 'Agarick' sent from France and applied as styptic after amputation

Its mode of action was described by Todal as Agarick sponge entangled the blood and retained a coagulum on the spot.

Medicinal mushrooms have become even more widely used as traditional medicinal ingredients for the treatment of various diseases and related health problems largely due to the increased ability to produce the mushrooms by artificial methods. As a result of large numbers of scientific studies on medicinal mushrooms especially in Japan, China and Korea, over the past three decades, many of the traditional uses have been confirmed and new applications have been developed. Thus, in addition to various immunilogical and anti-cancer properties of the mushrooms many other potentially important therapeutic properties including antioxidents, anti-hypertensive, cholesterol–lowering, liver protection, anti-fibrotic, anti-inflammatory, anti-diabetic, anti-viral and anti-microbial have been explored. Many pharmaceutical companies in the Far East are viewing the medicinal mushrooms as a rich source of innovative biomedical molecules. A recent general paper by Wasser and Weis (1999 a, b) dealing with general mycological information on several of the most important medicinally valuable mushrooms, may be summarised as follows:

Ganoderma lucidum and *Ganoderma tsugae*

G. lucidum and related species have the longest historical usage for medicinal purposes. In Japan it is called Reishi and in China and Korea it is variously called. Reishi is also widely used in the orient as a talisman to protect a person or home against evil. The fungus grows in many parts of the world and in Japan is to be found mainly on old plum trees. Originally, rare and expensive it can not be artificially cultivated, which makes it more accessible and affordable. The mushroom and mycelium contain steroids, lactones, alkaloids, polyssacharides and triterpenes.

Pharamologically a number of the water-soluble polysaccharides have demonstrated antitumour and immunostimulating activities. At least 100 different alcohol-soluble triterpenes have been identified including highly oxidised lanostane-type triterpenoids such as agnoderic, ganoderenic, lucidenic, and ganolucidic acids. These triterpenoids have been shown to possess adaptogenic and antihypertensive as well as anti-allergic properties.

This mushroom possesses many different medicinal properties dependent on the stage and environment of its growth (Jong and Birmingham, 1992; Liu, 1999). Traditionally, it has been widely used in the treatment of hepatopathy, chronic hepatitis, nephritis, hypertension, arthritis, neurastheine, insomnia, bronchitis, asthma and gastric ulcers. Scientific studies have confirmed that substances extracted from the mushroom can reduce blood pressure, blood cholesterol and blood sugar levels as well as inhibit platelet aggregations.

Reishi extracts have been highly effective in alleviating altitude sickness and also in treating myotonia dystrophica. Several major biochemicals such as polysaccharides, proteins and triterpenoids with potent immuno-modulating action have been isolated from *Ganoderma* spp. The major immuno-modulating effects of these active substances include mitogenicity and activation of immune effector cells such as T cells, macrophages and natural killer cells resulting in the production of cytokines, including interleukins, tumour necrosis factor-a and interferons.

The therapeutic action of *G. lucidum* as an anti-cancer and anti-inflammatory agent has been associated with its immuno-modulating properties (Wang *et al.*, 1997). The extensive range of traditional medical treatment with this mushroom have not yet been fully substantiated by modern scientific standards. They are being extensively scrutinised in the far East and the USA (Chang, 1995, 1999; Chen and Miles, 1996). In view of its bitter taste and indigestible structure (often similar to varnished wood in appearance) this is not an edible mushroom but, in hot water extracted form it is available worldwide in tablet and liquid products (Stamets, 1999).

Key Active Constituents

Beta and hetero-Beta-glucans (antitumour, immunostimulating).

Ling Zhi-8 protein (anti-allergenic, immuno-modulating) Gandomermic acids–triterpenes (anti-allergenic agents, cholesterol and blood pressure reducing).

Lentinus edodes

This fungus is indigenous to Japan, China and other Asian countries with temperate climates. It is to be found in wild on fallen deciduous trees special chestnut, beech, oak, shia, alder etc. However, it has been grown artificially for centuries on logs which can support seasonal fruiting. In more recent times, mass cultivation has predominantly been achieved by the sawdust culture technology. It is undoubtedly the leading mushroom worldwide that can be used both as a nutritious and tasty food and a highly effective medicine.

In oriental medicine it has been used for a wide range of health problems and its curative properties are well attested to in folk medicine (Hobbs, 1995). It has been particularly valuable in treating high blood pressure and lowering blood cholesterol. The fungus is the source of two well-studied polysaccharide preparations, *viz.* Lentinan–a cell wall polysaccharide extractable from both the fruit-body and mycelium and LEM–a protein-bound polysaccharide derived only from the mycelium (Chihara, 1992). Both compounds have demonstrated anti-cancer activity. It is believed that such compounds function by enhancing the immune system rather than attacking the cancer cells directly. Such compounds are increasingly used in Japan as adjuvants to help support immune function in cancer patients during radio- and chemo-therapy and can prolong survival times in some types of cancer (Mizuno, 1995).

Key active constituents

Beta-D-glucan (Lentinan) and heteroglucan-protein (LEM) (anti-tumour, immunostimulating)

Eritadenine (cholesterol-reducing)

Ergosterol (provitamin D-2)

RNA fractions (antiviral nucleic acids)

Phellinus linteus

The fruiting-bodies of this fungus are called 'song gen' in Chinese medicine and 'meshimakobu' in Japanese. The fungus grows as a parasite mostly on living deciduous trees (but occasionally on *Pinus* spp.) in Japan and Korea but also in other parts of the world (Teng, 1996). The fungus can now be cultured by log and sawdust technology. However, extensive studies are underway for mycelium culture in fermentors.

In traditional Chinese medicine, hot water extracts of the inedible furit-bodies have been used for an extensive range of ailments and it is believed to work as a 'miracle medicine' refreshing the human body and prolonging longevity (Ying *et al.*, 1987). Recent studies have compared to water extracts of *Phellinus* with the other main anti-cancer medicinal mushrooms when tested against xenographs. The *Phellinus* extract showed the strongest evidence of tumour proliferation suppression (Mizuno, 2000). It has been extensively studied for effects on digestive system cancers eg. gastric, duodenal, colon and rectal and also liver cancer. Special attention has been given to the beneficial effects of such extracts before and after cancer operations or in the adjuvant setting (Aziawa, 1998; Mizuno, 2000).

In recent years, using cultured mycelium of selected strains of *Phellinus*, compounds have been extracted and used in Korea as medicine–especially for cancer treatment.

Key Active Constituents

Beta-glucans (anti-tumour and immunostimulating)

Marketed as Meshima capsules

Poria cocos

This fungus, sometimes referred to as Wolfiporia cocos, is the most commonly used of all Chinese medicinal fungi. It is called Hoeleu of Fu Ling in Chinese which refers to the hard sclerotium of the fungus. This is a mycorrhizal fungus grows in association with the roots of various conifers, especially Chinese red pine. Large tuberculiform structures–the sclerotia–are formed underground and can be collected all year round but especially early autumn (Liu and Bau, 1980). Apart from the medicinal uses, the large scleratial structures have been used as food in Nigeria and in parts of eastern and southern US–"Indian bread".

Pharmacologically, the polysaccharides–spachyman and pachymaran exhibit strong anti-cancer and immuno-modulatory activities. Low-molecular weight tetracyclic triterpenes have considerable immuno-stimulating and antiviral activities (Hobbs, 1995). The extensive range of traditional medicinal uses are documented in Hobbs (1995) and Ying *et al.* (1987).

The main clinical trials have been concerned with the treatment of viral hepatitis in which Poria was one of several herbal ingredients. Hoeleu (*Poria cocos*) is widely available in Chinese herbal shops in bulk, and is also included in many commercial preparations.

Key Active Constituents

Polysaccharides-spachyman and packymaran (anti-tumour and immunomodulat- ing.)

Tetracyclic triterpenes (immunornodulation and antiviral).

Auricularia auricula

This fungus is widely known as Jew's ear (a contraction of Judas' ear), wooden ear, or tree ear and in Japan, Kikurage. They are facultative parasites on trunks of many broadleaf trees or on dead wood (Hobbs, 1995). The fruiting body is gelatinous, elastic, rubber in texture. It is widespread in U.S., Europe and Asia. Historically, it has been used in China both as a food and medicine. It is particularly useful for stopping pain and bleeding, and is regularly prescribed in traditional Chinese medicine to treat haemorrhoids and excessive uterine bleeding (Ying *et al.*, 1987). It has a high content of undigestible polysaccharides or dietary fiber.

Pharmacologically, the polysaccharides have been used as immune toxins, anticoagulants and to lower cholesterol. Extracts of *Auricularia* prevent egg implantation in animals terminating early and mid-pregnancy (Ho and Chen, 1991). Owing to this possible teratogenicity, it is recommended that *Auricularia* extracts should not be taken by pregnant or lactating women and those planning to conceive.

Key Active Constituents

Polysaccharides (immune stimulation, anticoagulant, lowering cholesterol).

Hericium erinaceus

This is an edible mushroom occurring widely in Japan and China, growing on standing and decayed broadleaf trees such as oak, walnut and beech. It can also cause heart rot in standing trees. Originally collected from the wild, it is now extensively grown artificially on logs and sawdust mixtures making this mushroom available all the year round. It is known in the West as the hedgehog or monkey head fungus and in China as Shishigashida because the fruiting body looks like the head of a lion. When air dried and extracted with hot water it is used extensively in traditional Chinese medicine (Houtou), to promote digestion and general vigour, strength and general nutrition. The polysaccharide from this mushroom have cyto-static effects on gastric, oesophageal, hepatic and skin cancers (Mizuno, 1999; Mizuno *et al.*, 1992). Mycelium produced from several *Hericium* spp. and then extracted with hot water formed the basis of a sports drink named Houtou that was used in the 11th Asia Sports Festival (1990) and is believed to have contributed to the remarkable activities of Chinese players.

Key Active Constituents

Beta-D-glucans (antitumour)

Ergosterol (provitamin D)

Cyathane derivatives (nerve growth stimulators)

Grifola frondosa

It is often called Hen of the Woods' or 'Sheep's Head' while it is more often called in Japan, Maitake, which mean 'dancing nymph'. In the early stages it is sought after for its delicious taste and excellent aroma. Previously, it was only collected from the wild and, consequently, was highly prized and priced. Since the late 1970s, it can be artificially cultivated on logs or sawdust mixtures and now many thousands of tons are being grown in Asia and more recently in USA (Chen, 1999). In traditional Chinese medicine it has been used for improving spleen and stomach ailments, calming nerves and treating haemorrhoids (Hobbs, 1995). Maitake is a component of a wide range of Chinese medicines.

Recent studies have shown that polysaccharides and polysaccharide-protein complexes from this mushroom have significant anti-cancer activity (Hishida *et al.*, 1988, Kurashige *et al.*, 1997). A limited number of clinical studies in Japan and the USA have shown that a purified fraction of polysaccharide is highly effective against cancers of the breast, lung, liver, prostate and brain. Other fractions from *G. frondosa* exhibit immunological enhancement together with properties of anti-HIV, anti-hypertension, antidiabetic, and antiobesity (Zhuang and Mizuno, 1999). It is interesting to note that the b-glucan fractions from this mushroom are now being used by over 3,000 health professionals in the US for the prevention and treatment of:

Flu and Common infection (bacteria and viruses)

AIDS (HIV)

Diabetes mellitus

Hypertension

Hypercholesterolemia

Urinary tract infections (particularly for women)

Key Active Constituents

1,3 and 1,6 Beta glucans (antitumour and immunomodulating).

Commercial product "Grifolan".

Flammulina velutipes

This is a major edible mushroom. It can be slightly salty and bitter in taste and is used in traditional Chinese medicine to treat liver diseases and gastric ulcers. Polysaccharides from this mushroom have been shown to inhibit the growth of cancers in a number of xenograph models. Flammulin, a basic simple protein from *F. velutipes* is able to markedly inhibit tumour cells (Karnatsu *et al.*, 1963). Flammulin has been purified to a crystalline state and clinical trials are now in progress (Zhang *et al.*, 1999). The first scientific paper stating that edible mushrooms were effective against a solid tumour was with *Flammulina*.

A new antitumour glycoprotein has been isolated from cultured mycelium of this fungus–Proflamin. It is useful in combination therapy with other chemotherapy agents (Ikekawa, 1995). Furthermore, an epidemiological study in Nagano Prefecture,

Japan showed that the cancer death rate among farmers producing *F. velutipes* was remarkably lower than that of other people in the Prefecture and in Japan overall (Ikekawa, 2001).

Key Active Constituents

Beta-glucan-protein (antitumour and immunomodulating)

Beta-glycoprotein-Proflamin (antitumour)

Pleurotus ostreatus

The medicinally beneficial effects of *Pleurotus* spp. were discovered independently on different continents *viz.* Asia, the folklore of central Europe, South America and Africa (Gunde-Cimerman, 1999). It was first artificially cultivated in USA, and now it is cultivated worldwide. There have been a number of studies suggesting a role in numerous diseases with its anti-cancer activity, immunomodulating effects, and antiviral, antibiotic and anti-inflammatory activities The major cause of death in the Western hemisphere is coronary artery disease with hypercholesterolemia as a primary risk factor. Drug therapy for lowering cholesterol has made considerable use of the pharmacologic agent lovastatin (mevinolin) and its analogues.

Species of the genus *Pleurotus* are excellent producers of lovastatin and as such, *Pleurotus* could be considered as a functional food with natural cholesterol-lowering ability (Gunde-Cimerman, 1999). However, large scale production of lovastatin from fruit-bodies is not deemed commercially viable because of variability in fruit-body composition. Lovastatin is normally found only in the lamella and basidiospores and not in the stipe and cap. Mycelial cultivation could be the way ahead.

Key Active Constituents

Beta-glucans (antitumour, immunomodulation)

Lovastatin (cholesterol-lowering)

Trametes (*Coriolus*) *versicolor*

This fungus has multicoloured cap resembling a 'turkey tail' and occurs as overlapping clusters on dead logs in most parts of the world. This is not an edible fungus but hot water extracts have been used in traditional Chinese medicine from historical times for a wide range of ailments (Ying *et al.*, 1987). Modern studies have produced two extremely important compounds, PSK or "Krestin", a water-soluble protein-bound polysaccharide and PSP, a polysaccharide-peptide both derived from mycelial cultures of this fungus. PSK has been shown to act directly on tumour cells (cytostatic and cytotoxic) as well as indirectly in the host to boost cellular immunity (Tsukagoshi, 1984). PSK also shows antiviral activity through stimulation of interferon production. PSP is a powerful immune-stimulant and anti-cancer agent (Yang, 1993; Ng,1998). There have been a wide range of clinical trials for a range of human cancers. In most cases when taken with traditional chemotherapy or radiotherapy there have been significant increases in patient longevity. A polysaccharopeptide isolated from this mushroom has been shown to inhibit the HIV-1 (Collins and Ng, 1997) while a polysaccharide showed chemopreventitiie activity in an *in vitro* model (Kim *et al.*, 1999).

Key Active Constituents

Beta-glucan-proteins (antitumour, antiviral, immunomodulating)

Ergopsterol (provitamin D2)

Tramella mesenterica and *T. fuciformis*

It has a long historical use in traditional Chinese medicine as an immune tonic and for treating debility and exhaustion together with many other ailments including skin-care. It contains acidic polysaccharides especially glucuronoxylomannan, readily extracted with hot water giving a smooth and stable solution used in Oriental cuisine. It can now be grown artificially and is being increasingly consumed in Asia.

The polysaccharides of this fungus show anti-cancer activity and can enhance immune functions (Hobbs, 1995). Clinical trials have shown it to be effective in treating radio- and chemo-therapyinduced leukopenia, boosting immunological functions and stimulating leukocyte activity (Hu and But, 1987). Med Myco Ltd. (Israel) have developed a submerged fermentation method to produce Tremellastin from *T. mesentrica* mycelium which contains 50 per cent glucuronoxylomannan, together with proteins rich in amino acids, dietary fibre and vitamins of the B group.

Key Active Constituents

Acidic polysaccharides (glucuronoxylomannan) (antitumour, immunostimulatory, antidiabetic, skin enhancing).

Cordyceps sinensis and *C. sobolifera*

The fungi grow as parasites in larvae of *Lepidoptera*, gradually taking over the entire larval body. The diseased larvae bury themselves in the soil and die. Later the fungal mass or stroma grows out of the pupa and can be identified and collected.

The caterpillar fungus or Tochukaso is not a mushroom type fungus and the fruiting structure cannot be cultivated or cultured. The complete structure can be used in many forms, whole, powdered or extracted and has many applications in Chinese medicine (Hobbs, 1995; Halpern, 1999). Anti-cancer polysaccharides have been isolated from several species of *Cordyceps* and some have been shown to have hypoglycaemic activity as well (Itami and Yahagi, 1996; Kim 1991). A major concern with herbal medicine using *Cordyceps* collected from nature is quality and safety. However, the pure mycelium of these parasitic fungi can now be easily cultivated in fermentors and is attracting considerable interest as an agent to treat fatigue and improve motor function (Mizuno, 1999).

Key Active Constituents

Galactomannans (antitumour, immunostimulating)

Cordycepin

Sterols

Schizophyllum commune

This is a small, whitish fungus with no stalk which grows on dead trees throughout the year. It is a very common fungus and has worldwide distribution

(Hobbs, 1995). Pharmacologically it is extremely important because it produces the polysaccharide schizophyllan which shows considerable anticancer activity in xenograph and clinical practice.

Key Active Constituents

Beta-glucons (antitumour and immunomodulation).

Agaricus blazel

This mushroom was discovered in the USA in the 1940s but its main commercial cultivation occur in Japan and Brazil. In Japan, it is called Himematsutake and is one of the most expensive medicinal mushrooms. A novel polysacchanride-protein complex obtained from it has been shown to be highly active against a variety of xenographs (Ito *et al.*, 1997).

Key Active Constituents

Beta (1,3) D-glucan, Beta (1-4) D-glucan,

Beta (1-6)–D–glucan (anti-tumorur and immune enhancing)

Proteoglucans (anti-tumour).

Amanita muscaria (Fly agaric)

It is used as intoxicant for killing flies. It has been used from the earlier time, therapeutically as a powder, a tincture for swollen glands, serious swelling, nervous diseases and epilepsy etc. For ailment of heart, rheumatoid arthritis and scrofulous inflammation of the eyes, the lotion made from *Amanita muscaria* was applied externally and internally, it is taken to remove ring worm and erruption. It has also been used as a narcotic. This fungus yields an uncrystallised alkaloid "muscarine', the nitrate of which was formerly used in medicine. *Amanita muscaria* is still used in homoeopathy under the name Agar, Agarus or Agaic. Recent studies on *Amanita muscaria* have led to the development of drugs that may treat diseases such as epilepsy and schizophorenia, thought to be characterised by malfunction of GABA, a transmitted system of brain, GABA is gamma aminobutyric acid which has been identified as "Brain break" or major inhibitory compound in central nervous system of mankind. Researches have found that the effect of barbiturates are mediated by GABA receptors. In addition, malfunction in the GABA system have linked with epilepsy schizophorenia and spasticity Krogagnard (1981) found that muscinol and ibotenic acid, compounds of *Amanita muscaria* interact with GABA system. These compounds are extremely toxic and are specific in action.

The other drugs which affect the mind have been isolated from *Psilocybe mexicana*, these are Psilocybin and Psilocin. The fungus is used by medical men to induce greater insight and to cause hallucination and other mental aberrations. *Amanita phalloides* was used against cholera, Brights disease and for intermittent fevers. *Auricularia auricula* frequently used as a poultice for inflammation of eyes and as gargles for inflammation of throats.

Boletus edulis

The fungus is dried and used as remedy for influxes from the bowels, which cause rheumatism and fleshy excrescenes of the annus. Use of *B. edulis* diminishes these ailments and in time these are cured. It removes blemishes from face and healing lotion is also made for some ailments of eyes. After soaking in water it is applied as a salve to foul ulcers and erruptions of the head and for dog bites.

Boletus satanus

It contains phosphoric acid, resins and fatty acids and is used in homoeopathy in very diluted form for dysentery for the diseases of liver and gall bladder.

Calvatia gigantia

It is still used for anaesthesia.

Formes officinalis

A resinous substance 'agaricin' found in its fruit body and is used for its diaphoretic and purgative action. *Fomes ignarius* and *F. fomentarius* were formerly known as "surgeon agarick". For the preparation of this, young specimens are collected, *rind* and tubes are removed. The remainder is softened by storing at a cool place for some time and is then cut into thin slices, and well beaten with a mallet, damped from time to time and beaten again. After further rubbing with the hands, the material acquire a certain softness and laxity and is suitable for styptic use. It can be impregnated with self petre to produce amode for cauterization or other purposes. The mechanism is that it entangle the blood and retains a coagulum, on the spot.

Lycoperdon gigantium and *L. bovista*

These are used as soft and comfortable surgical dressings. The dusty powder is a powerful haemostat. *L. sovista* and *L. gigantum* and many other species of puff balls preferably large size are used as a styptics. Dentists used it as an absorbent for drying cavities in teeth before filling them. The 'fumes' given off by burning large puff balls have the properties similar to chloroform; they are used in surgical operations but more frequently used by beekeepers to intoxicate the honey bees.

Polyporus officinalis

This fungus under the name 'agarick' is taken internally as a universal remedy for all complaints and disorders. The active principle is 'agaricic acid' or 'laricic acid'. This is obtained by extraction of fungus either by 90 per cent alcohol or dry ether. The alcoholic extraction is known as 'agaricin'. This fungus in powdered state is powerful purgative. In large quantity it causes diarrhoea, sickness and even death. In small dose it is used in the form of 'agaricin' to cure diarrhoea and to inhibit night sweats. This fungus is also used as 'amodou' and preparation is same as in case of '*Fomes officinalis*'.

In India, the young dried specimens of *P. officinalis* is known as 'Harkshai' and sold in the market as medicine. In Burma *Polyporus arthelminticus* grows at the roots of old bamboos and used as an 'anthelmintic'. In Malaya states *P. sacer* under the name of 'Susu Rimau' or 'Tiger's Milk' is used for treating colds, *P. sanguineus*, known as

'Chandawan Merah" or "Red fungus" is used for dysentery. Now in modern medicine few fungi are again gaining importance.

Antibacterial Activity

In higher fungi antibacterial activities has been reported from Basidiornycete and Ascornycete. Earlier workers reported antibacterial activity from *Fomes, Polyporus* and *Trametes* (Robbins *et al.*, 1945); mycorrhizal fungi, *Flammulina velutipes* (Sasek and Musilek, 1967); *Aleulrodiscus, Clitocybe, Daedalea, Marasmius, Merulius Pleurotus, Polyporus, Poria, Prathyrella and Tricholoma* (Benedict and Brady, 1972); *Agaricus bisporus* (Vogel *et al.*, 1974). These fungi contain polyacetylene phenolic componouds, *i.e.* purines, pyrimidines, quinones and terpenoides. Antibacterial properties of *Agaricus campestris, Flammulina mellae, F. odilpis, Polyporus schweinitzii* against *Staphyloccous aureus, Salmonella typhii* and *E. coli* are known (Bose, 1953).

Antifungal Activity

Antifungal activity is reported from *Sparasis ramosus* (Falck, 1923; Briain1951). *Lentinus edodes* and *Cortinellus shiitake* (Herrman, 1962). *Coprinus comatus* (Bohus *et al.*, 1961) and *Oudemansiella mucida* (Musilek *et al.*, 1969).

Antiprotozoal Effect

Illudin M., Illudin S. and terpenoids from *Clitocybe illudens,* (Mc Morris and Anchel, 1965) are reported to be active against *Plasmodium gallinaceum* (Coatney *et al.*, 1953). *Inpex flovus* is reported to be active against a protozoan (Gregory *et al.*, 1966).

Antitumour Effect

Calvacin an antitumour substance (Beneke, 1963), has been reported from giant puff ball *Calvatia gigantea*. Calvacin is a non diffusiable basic microprotein (Roland *et al.*, 1960; Gregory *et al.*, 1966) found near fifty basidiomycetes having antitumour activity, most among those are *Polyporus* spp., and *Merullius nivers*. Chemical compounds associated with antitumour activities are polysaccharides, basic protein acid protein, quinoid basic and these are found in *Lentinus edodes* (Ikekawa *et al.*, 1969), *Pleurotus* (Yoshika *et al.*, 1972), *Flammulina velutipes* (Komatsu *et al.*, 1963), *Poria corticola* (Ruelius *et al.*, 1968; Schilling and Ruelius, 1968), *Calvatia gigantea* (Rolande *et al.*, 1960), *Poria obliqua* (Kier, 1961), *Agaricus bisporus* (Vogel *et al.*, 1975). *Volvariella volvacea* and *Flammulina veltutipes* contain cardiotoxic proteins which cause swelling and inhibit the respiration of tumor cells.

Antiviral Activity

Antitumour and antiviral activities are believed to have at least one common focus in nucleic acid metabolism. *Boletus edulis, Calvatia gigantea, Suillus luteus* and *Lentinus edodes* (Cochran *et al.*, 1967) are found active against influenza *in vitro* and *in vivo*.

Hypollpiclemic Activity

Kameda and Coworkers (1964), Kanedà and Tokuda, (1966) reported the ability of lowering blood cholestrol in *Lentinus edodes, Auricularia polytricha, Flammulina velutipes* and *Agaricus bisporus*. Jiu (1966) reported hypocholestrodemic activity in *Fomes fomertariustoo*.

In addition to the medicinal properties of mushrooms referred above the extracts of *Agaricus bisoporus, Pholioto nameka, Pleurotus florida, Tricholoma nameka* and *Tricholoma matsutake* have been reported as the effective antitumor agent.

Compounds extracted from *Agaricus bisporus, Coprinus comatus, Oudemensiella mucida*, have been reported to possess antiviral and antibacterial properties.

Commercial cultivation of *Morchella esculenia* has not been successful till now and hence its mycelium is extensively used as flavouring agent. Ethanolic extract of its cultured mycelium has also been reported to process most effective anti inflammatory and antitumour properties.

Psychoactive Mushroom

Psylocybin mushrooms possess psychedetic properties. They are commonly known as "magic mushrooms" "mushies" or "shrooms" and are available in smart shops in many parts of the world, though some countries have outlawed their sale. They have recently been found to induce mystical experiences that give meaning to the subjects upto several months later (Griffiths *et. al*, 2008).

Psylocybin mushrooms are not the only psychoactive fungi. *Amamita muscaria* is also psychoative. The active constituents are libotenic acid and Musciml. The mushroom is traditionally dried in the sun to transform the libotenic acid in to the more psychoactive mushroom. Drying renders the mushrooms safer and less likely to cause negative side affects.

Other Uses of Mushroom

Mushrooms can be used for dying wood and other natural fibers. The chromoatophores of mushrooms are organic compounds and produce strong and vivid colour, and all colours of the spectrum can be achieved with mushroom dyes. Before the invention of synthetic dyes mushrooms were the source of many textile dyes.

Some polypores loosely called mushrooms, have been used as Fire starters (Known as tinder fungi). Otzi the Inceman was found carrying such fungi (http/cn.wikipedia.org/wiki/mushroom).

Mushroom Poisoning

In popular mushroom manual and mycological exhibitions, the described, illustrated and exhibited species are usually divided into poisonous, edible and indible ones. The category of inedible mushrooms is one which are neither edible nor poisonous, or whose edibility is still open to doubt due to a lack of exact information. They possess qualities making them unsuitable for table use: too small fruit bodies, unpleasant smell and taste, excessively tough flesh, and sometimes they inspire distaste by their appearance or place of occurrence (eg. coprophillic fungi). With the spreading of knowledge of individual mushroom species, some of them are gradually being released from the category of inedible mushroom into that of poisonous mushroom. In future, it might be expedient to omit the category of inedible mushrooms altogether, and to distinguish only mushrooms suitable for table use *i.e.* edible and mushroom detrimental to the human organism *i.e.* poisonous mushroom. The term toadstool has been used frequently to refer to poisonous mushrooms. Scientifically, the term has no meaning at all. It is suggested that the term toadstool be dropped altogether in order to avoid confusion and to use edible and poisonous mushroom. A third category-medicinal mushroom has been also proposed but again it creates confusion as the majority of mushrooms which possess tonic and medicinal qualities are primarily edible, again, some of the mushrooms are not edible but of medicinal importance.

There is no known test to tell if a mushroom is edible or not. An experienced worker can recognize the edible and the poisonous varieties. A number of tests however, have been referred. According to these tests, silver spoons or silver coins turn black when they are dipped in a dish of cooked poisonous mushrooms. The species which are brightly coloured; bitter in taste, or emit bad odour are usually treated as poisonous. A field mushroom is said to peel off readily. But *Amanita phailloides*, a dangerous fungus can also be peeled off easily as any other non-poisonous mushrooms. Exudation of milky substances from damaged fruit bodies, coagulation of milk, unusual change in colour of parsley or onion when cooked with mushrooms have often been considered as characteristics of poisonous species. The colour of gills of some *Agaricus* sp. is generally pink at first but turns black or brown with age. If the colour remains unchanged with age it may arouse suspicion. In some cases, it has been observed that the colour of the pileus or stipe is changed when it is slightly injured. For example, *Agaricus xanthodermus* changes saffron yellow colour when its stipe or pileus is rubbed or bruised. The gills of certain mushrooms are initially white or pale pink but later become black and produce drops of inky fluid. This autolytic stage of the fruit bodies is unsuitable for consumption although young fruit-bodies are some times recommended for edible purpose. Even though, poisonous mushrooms represent less than 1 per cent of the world's known mushrooms, we can not ignore the existence of the relatively few dangerous and some times fatal species.

The morphological characteristics used for the identification of poisonous mushrooms are, however, inconsistent and unstable criteria because they are strongly influenced by the environmental conditions. For this reason, poisonous wild mushrooms can be misidentified as edible mushrooms and they can be eaten by mushroom hunters. Furthermore, morphological characteristics of the mushroom samples brought for clinical diagnoses are time taking and are generally not well preserved and thus frequently unsuitable for a rapid identification. The analysis of cooked mushrooms or of gastric aspirates from poisoned patients is particularly very difficult because the spores are often few and the morphological features of the mushroom are generally altered. Therefore, techniques such as DNA based identification, which do not depend on morphology, are required. Identification of wild and cultivated mushrooms and determination of genetic differences among basidiomycetes by using DNA techniques are usually performed by comparing the nucleotide sequences of amplified DNA fragments from nuclear and mitochondrial ribosomal DNAs (r DNAs). Meeta *et al.* (2008) developed a rapid system of poisonous mushroom identification (within 1.5 hrs) using a real time PCR system. They proposed a new strategy for mushroom identification in medical facilities on the basis of species specific fluorescence signal detected among four poisonous mushrooms species that are frequently eaten by mistake in Japan and 27 popular mushroom species.

Apis *et al*. (2010) presented novel primer pairs and real time PCR method for the detection of four poisonous mushrooms that are a common cause of human intoxication in Italy *i.e. Amanita phalloides, Lepiota cristata, L. brunneoincarnata* and *Inocybe asterospora*. Identification technique of sample containing DNA from the aforesaid four mushroom texa was developed. To test specificity, all protocols were applied on DNA extracted from various mushroom species; sensitivity was assessed performing serial dilutions on all samples; versatility of the protocols was evaluated performing tests on DNA extracted from different matrices. The protocol showed high sensitivity (32 mg dried mushroom), high specificity and sensitive detection of DNA extracted from different samples, including pasta with mushroom, cooked mushroom and gastric aspirates.

The availability of molecular techniques for the identification of poisonous fungi in general including mushroom would support and integrate the work of mycologists in case of clinical poisoning.

The toxins contained in various species attributing poisonous nature of the mushrooms are very different in chemical composition. This difference is reflected in the effect of poisoning which too differ considerably according to the species involved. In any case, suspected mushrooms poisoning should never be taken lightly and medical assistance should be sought at once.

On the basis of the nature of toxicity and symptoms produced, mushrooms poisoning can be classified into following five categories.

Amanita-Type Poisoning

Unquestionably, the *Amanita phalloides* group causes the most dangerous type of mushroom poisoning. The toxins involved belong to the phallotoxin and amatoxin complexes. The phallotoxin phalloidin binds specifically to actin. While the phallotoins are not active following ingestion, they are potent when injected intravenously, they have proved useful in experimental studies. In such studies phalloidin, binding to actin, is coupled with fluorescent groups. By this means actin can be localized in the cells. It is the amatoxin such as α-amatine that is ivolved in amanita poinoning. α-Amatine is a specific inhibitor of RNA polymerase present in all eukaryotes. This blocking of the enzymes associated with the replication of RNA inhibits the formation of new cells. These toxins tend to accumulate in the liver and damage that organ severely. The RNA polymerase of the fungus is not affected. This group has caused the majority of recorded death from mushroom poisoning, especially in Europe. The general symptoms of this type of poisoning are severe abdominal pains, nausea, violent vomiting, diarrhoea, cold sweats, and excessive thirst. These may last for 48 hours with dehydration cramps, and anuria.

Muscarine-Type Poisoning

Two toxins, muscarine and ibotenic acid are involved. They occur in *Amanita muscaria*. *A pantherina* and in a number of *Inocybe* and *Clitocybe* species. Muscarine is known to be responsible for pupil contraction, blarred vision, lachrymation, salivation, perspiration, reduced heart rate, lowering of blood pressure, and asthmatic-like breathing. Ibotenic acid is responsible for the insecticidal properties of *A. muscaria*, the fly agaric. Both muscarine and ibotenic acids are intoxicants, and there is a long history of different cultures using these compounds from *A. muscaria* for this purpose and in religious rites. The symptoms usually appear soon after eating the mushrooms, with vomiting, diarrhoea, and salivation. The most characteristic symptoms are nervous excitement, difficulties in breathing, shivering and a tendency to collapse.

Psychotropic or Hallucinogenic Poisoning

Several different toxins are involved, including psilocybin, which are found in species of *Psilocybe*, *Conocybe* and *Stropharia*. These compounds are similar in their reaction to d-lysergic acid diethylamide (LSD). They act on the central nervous system, producing distortions in vision and of tactile sensations as well as mixed emotional feelings of happiness or depression. Other symptoms are varied, including vomiting, increased rate of heart beat, and hallucinations, which may last for various lengths of time.

Coprinus Poisoning

Several *Coprinus* species, such as *C. micaceus* and *C. atramentarius,* when consumed with an alcoholic drink, produce unpleasant but not dangerous symptoms. The symptoms include reddening of the face, increased rate of heart beat, and in some cases, vomiting and diarrhoea. The mode of action of the chemical in *C. atramentarius* mushrooms is similar to Antabuse, which is a drug used to induce nausea and vomiting in individuals who are trying to overcome an addiction to alcohol.

Poisoning from External Sources

The poisoning is not caused by mushrooms themselves but by toxic substances that have accumulated in the mushrooms. The principal causes are (1) heavy metals due to polluting environmental conditions from where the mushrooms are harvested that are far in excess of permissible levels, and (2) radioactive contaminants due to the pollution by contaminating radioactive materials in mushroom hunting areas and subsequent consumption of the collected mushrooms.

Chapter 14

Economics of Mushroom Cultivation

Mushroom is highly remunerative crop for Indian farmers. In India, four mushrooms namely button, oyster (dhingri), milky and paddy straw are under commercial cultivation. Button is the most popular mushroom among farmers and contribute almost 85 per cent of the total production of the country.

Economics of mushroom cultivation in Indian perspective has been analysed time to time by different workers but with changing price indices became outdated. Recently Kapoor (1989) published the economics of mushroom cultivation but only of button mushroom. Most recently in the year 2009, Shama and Dhar have projected the cost benefit ratio of button, oyster and paddy straw mushroom cultivations in details alongwith relevant dimensions of their economic sustainability which may be read as follows:

Mushroom cultivation is not land dependent and any type of land can be used for the construction of the mushroom house/hut for crop raising as mushroom is an indoor crop and does not require soil and sunlight. In the last 10-15 years, large numbers of commercial units have been built by the entrepreneurs/farmers throughout the country for export. The world's largest button mushroom growing unit is located in India (Punjab). The present production of all types of mushroom in India is about 1,00,000 tons per annum (2008) and the bulk of it is the white button mushroom. World's total production of mushroom is about 22 million tons with China contributing the lions share. Button mushroom is the most widely accepted mushroom in India both by urban and rural consumers. The second important mushroom is the oyster mushroom.

Mushroom cultivation requires cereal straws for substrate preparation and there is abundant cereal straw available in India for recycling to edible biomass of highly nutritional and medicinal value. India produces nearly 140 million tons of cereals and equal amount of straw is generated by the farmers, which can partly be utilized by the farmers for mushroom cultivation. Land requirement is a minimum and any spare room of the house can be converted into a mushroom growing room, or a hut built on a piece of land can also be used for the purpose. The raw materials required for crop raising are generated by the farmers on their own fields (paddy/wheat/or any other cereal straw).The family labour is used for different operations and the only input required from outside is the seed/spawn of mushrooms. This makes the farmer self confident in raising this crop with great remuneration. A farmer grows a crop or two in the season and his income subsisted to a greater extent. As most of the inputs are available to him from his farm only, this is one reason that mushroom cultivation is very remunerative for a farmer under seasonal growth conditions.

Present Status of Mushroom Cultivation in India

There are chiefly two types of mushroom growers in India. These are 1. Environment controlled mushroom growers (high cost technology) and 2. Seasonal mushroom growers (low cost technology). Both the technologies are relevant in Indian conditions. Large environment controlled units grow mushrooms mainly for export purpose only while seasonal growers take a few crops of mushrooms in winter months/all year round in hills. There are scores of spawn producing laboratories supplying mushroom seed/spawn to mushroom growers in India Mushroom cultivation has now become a household name in almost all regions in India. Button mushroom is grown in cool/hilly regions (17–18 °C) of the country like Himachal Pradesh, Kashmir, Uttrakhand, Ooty hills, Darjeeling hills and Gangtok (Sikkim) seasonally all the year round. Oyster mushroom is grown all over the country at temperature ranging from 20-30°C and paddy straw mushroom grown at temperatures around 30-35°C in hot and humid peninsular regions. Button mushroom gives an average yield of 150-200 kg per ton of compost on an average under controlled conditions. Oyster mushroom yield is around 300–350 kg per ton of wet substrate, while yield of paddy straw mushroom is 100-120 kg per ton of wet substrate.

The seasonal grower on an average earns a rupee over the investment of a rupee, with raw material (straw)/labour available from the farm. These growers raise a crop of button mushroom seasonally in winter and then go back to their cereal crop raising in their fields. This mushroom cultivation is primarily to subsist the income of a farmer, thereby making it very remunerative for the farmers.

Types of Mushrooms Grown in India

There are 5 -6 types of mushrooms grown in India presently. These are (i) white button mushroom (16-25 °C), (ii) oyster mushroom (20-30 °C), (iii) paddy straw mushroom (30-35 °C), (iv) milky mushroom (30-35 °C) (v) shiitake mushroom (17-23 °C) and other specialty mushrooms are grown on bench scale. These five types of mushroom are grown seasonally in different regions of the country by the farmers. The produce is sold fresh by them in the local markets/or in nearby cities. Some mushrooms like oyster and shiitake are dehydrated in absence of ready market. The per capita consumption of mushrooms in India is around 40-50 gm only, as compared to 4-5 kg in Germany/USA. All the above types of mushrooms find acceptability in local market and with increasing production and availability mushroom consumption is also increasing. No marriage party/get together/Kashmiri wazwan is complete without a dish of mushroom. The Chinese restaurants in India are using scores various of mushrooms and some of these are imported from China. The good news is that people have started developing taste for mushrooms slowly. If per capita consumption reaches 100 gm in India, 1.1 lakh ton will be sold fresh and then there will be no requirement of export, as the present day scenario stands. Use of dehydrated/preserved mushrooms is minimum as these are not liked by the consumers.

Raw Materials Available for the Mushroom Crop

The raw materials requirement for raising a crop of mushroom is recyclable cereal straw/organic waste/organic byproducts. The main byproducts used for substrate preparation for mushroom farming are:

1. Wheat/paddy straw
2. Sugarcane baggase
3. Saw dust
4. Cotton seed meal/soybean meal
5. Scores of the locally available agro byproducts and other agro waste materials like banana pseudo stem/corn cobs/groundnut hulls etc.

Besides above, poultry manure is also used as nitrogen rich supplement for compost making for button mushroom cultivation. These raw materials are locally available in all the rural areas of the country at very reasonable rates. The farm yard manure (FYM), spent mushroom substrate (SMS) and coir industry waste are three main ingredients used as casing materials after proper decomposition in button mushroom farming in India.

Bamboo and synthetic cloth are used for making temporary structures (mushroom growth huts), which are also available at very cheap rates all over the country. For seasonal growth no power is required for crop raising excepting the light. Good quality water which is good for human consumption is good enough for mushroom cultivation as well.

Seasonal Mushrooms for Different Regions in India

Seasonally, button mushrooms are grown all the year round on hills (2- 3 crops) with each crop lasting 4 months from start of composting to end of cropping. Seasonal mushroom growth is a long crop and the farmer has to depend on the season available and the farmers coincide various operations accordingly. Under hilly conditions three crops are raised. One from early spring to start of summer, second from early autumn to winter and third in winter months with some adjustments. In plains of NW India one seasonal crop is raised between October to February conveniently. In other parts of India, one crop with minor adjustments can he raised during winter months when temperatures are low. Oyster mushrooms can be grown round the year using different species for different seasons. Different *Pleurotus* spp grow at different temperature regimes–hot weather oyster, cold weather oyster and mid temperature loving oyster. These mushrooms grow directly on cereal straw without composting. Farmers enjoy growing oyster mushrooms but sometimes marketing of the oyster mushroom becomes a problem in some areas. Paddy straw mushrooms are principally grown in areas with high R.H. and temperature of about 30°C (Coastal Peninsular India). Milky mushrooms are grown all over on cereal straw substrates like oyster mushrooms but at 30–35°C.

Training Programme for the Farmers

Extensive training programmes are undertaken at various centers for mushroom training in the country. No crop is supported for such extensive training programme as mushrooms. The main centres where training is available are the Directorate of Mushroom Research, Solan; TART (Division of Plant Pathology), New Delhi: G.B. Pant Univ. of Agri. and Technology, Pantnagar; all agricultural universities; Deptts. of Agri./Hort. in states of India and various NGO's. Special trainings under Horticulture mission are organized under central schemes for inaccessible areas like North Eastern region, Kashmir and other distant states of the country. IGNOU, New Delhi also offers courses for the students and farmers. Special training programmes are organized at TART in response to demands of the farmers. There is no deficiency in this area and all the training programmes are well attended.

Cost Benefit Ratio of Mushroom Farming

The cost of cultivation is based on raw materials/labour costs prevailing in 2006 in rural India and may not be reflective of the present prevailing prices in 2009. The cereal straws cost between ₹ 3-4 per kg and labour cost is also escalated by about 50 per cent. The scenario is not effected much as selling price also has gone up by 50-60 per cent in the retail/wholesale market, thereby not effecting the profit margin.

Economics of Button Mushroom (*Agaricus bisporus*) Cultivation in India Under Seasonal Grows Conditions (Low-Cost Low Production Technology System)

A.	**Expenditure (production costs per crop of 20 tonnes compost)**	
I.	*Fixed cost (facilities used more than one time)*	₹
i	Land rent per crop season (land required–0.5 acres)	2,000
ii	Seasonal growth hut (bamboo/rawbricks/synthetic cloth thatched roof/bamboo shelves) 60 x 20 x 10-12 feet (h), 1,200 feet @ ₹ 25/squire foot = ₹ 30,000 life span of 3 crop seasons, 1 crop season cost	10,000
iii.	Spraying equipments, hoses, tubs, buckets, forks, spades, etc. (life span 5 seasons = ₹ 5000), cost for 1 season	1,000

II.	*Variable Costs (materials used one time only)*	
i.	Cereal straw–12 tonnes @ ₹ 1000/tonne	12,000
ii.	Supplement wheat bran 300 kg/tonne–3,600 kg @ ₹ 6/kg	21,600
iii.	Urea 20 kg/tonne–240 kg @ ₹ 5/kg	1,200
iv.	Cotton seed cake–60 kg/tonne, 720 kg @ ₹ 10/kg	7,200
v.	Gypsum–35 kg/tonne, 420 kg @ ₹ 2/kg	840
vi.	Spawn–5 kg/tonne of compost, 100 kg @ ₹ 50/kg	5,000
vii.	Casing material (decomposed FYM), 2 kg/bag of 10 kg compost–4 tonnes @ ₹ 250/tonne + Formalin treatment cost	2,000
viii.	Labour–composting 30 LD, spawning 10 LD, casing 10 LD, cropping 60 LD, Miscellaneous 10 LD = 120 LD © ₹ 1 00/LD	12,000
ix.	Water costs	1,000
x.	Power costs	500
xi.	Pesticides, disinfectants, chemicals	1,000
xii.	Polybags for packing	500
xiii.	Polybags for cropping	2,000
xiv.	Miscellaneous expenses	2,000
	Total of (I + II)	**81,840**
		Say ₹ 82,000

B.	**Returns**	
	Returns at 12-15 per cent of compost weight in 8 weeks of cropping at an average price of ₹ 500 kg	
	At 12 per cent, 20 tonnes x 120 kg per tonne compost, 2,400 kg	= ₹ 1,20,000
	At 15 per cent.-20 tonnes x 150 kg per tonne compost, 3,000 kg	= ₹ 1,50,000
	Profit at 12 per cent yield, ₹ 1,20,000–₹ 82,000	= ₹ 38,000
	Protit at 15 per cent yield, ₹ 1,50,000–₹ 82,000 (figure are calculated on prices prevailing in March' 06 in India)	= ₹ 68,000
	1 US $ = 45 Indian Rupees (approximately)	

Economics of Oyster Mushroom Cultivation (*Pleurotus* spp.) in India Under Seasonal Grows Conditions (Low-Cost Low Production Technology Systems)

A.	**Expenditure (production costs per crop of 20 tonnes wet substrate material)**	
I.	*Fixed costs (facilities used more than one time)*	₹
i.	Land rent per crop season (land required–0.5 acers)	1,000
ii.	Seasonal Grows hut (bamboo/raw bricks/synthetic cloth/thatched roof/bamboo shelves) 60 x 20 x 10-12 feet (h), 1,200 feet © ₹ 25.00/squire foot = ₹ 30,000, life span of 4 crop seasons, 1 crop season cost	75,000
iii.	Spraying equipments, hoses, tubs, buckets, forks, spades, etc. (life span 5 seasons = ₹ 5,000), cost per one crop season	1,000
II.	*Variable costs (materials used one time only)*	
i.	Cereal straw–5 tonnes at ₹ 1,000/tonne	5,000
ii.	Rice bran supplement @ 10 per cent of dry straw weight, 500 kg at ₹ 6/kg	3,000
iii.	Spawn 20 kg/tonne wet, 20 tonnes × 20Kg, 400 kg @ Rs.50 Kg	20,000
iv.	Labour–wetting 10 LD, hot water treatment 10 LD, spawning 10 LD, cropping 45 LU, miscellaneous 10 LD = 85 LD at ₹ 100/LD	8,500
v.	Water costs	500
vi.	Power costs for light/exhaust fans	500
vii.	Pesticides, disinfectants, chemicals	1,000
viii.	Polythene bags for packing mushrooms	500
ix.	Polythene bags for cultivation	2,000
x.	Miscellaneous expenses	2,000
	Total of (I + II)	**52,500**
		Say ₹ 53,000

B. Returns

5 tonnes of dry straw will turn to 20 tonnes of wet substrate on wetting, 10 kg per bag fill, average yield of 1.0-1.5 kg fresh mushrooms per bag (10-15 per cent) in 4 weeks of cropping, average selling price of ₹ 45.00*/kg.

Returns at 10-15 per cent of wet substrate weight in 4 weeks of cropping, at an average selling price of ₹ 45.00/kg.

At 10 per cent, 20 tonnes × 100 kg/tonne wet substrate, 2,000 kg	₹ 90,000
At 15 per cent, 20 tonnes × 150 kg/tonne wet substrate, 3,000 kg	₹ 1,35,000
Profit at 10 per cent, ₹ 90,000–53,000	₹ 37,000
Profit of 15 per cent, ₹ 1,35,000–53,000	₹ 82,000

(figures are calculated on prices prevailing in March' 06 in India)

1 US $ = 45 Indian Rupees (approximately)

* Though higher prices are prevalent for fresh oyster mushrooms, calculation has been done at minimum selling price.

Economics of Paddy Straw Mushroom Cultivation (*Volvariella* spp.) in India Under Seasonal Growth Conditions (Low-Cost Low Production Technology System)

A.	**Expenditure (production costs per crop of 20 tonnes wet substrate material)**	
I.	*Fixed costs (facilities used more than one time)*	₹
i.	Land rent per crop season (land required–0.5 acre)	1,000
ii.	Seasonal Grows hut (bamboo/rawbricks/synthetic cloth/ thatched roof/bamboo shelves) 60 x 20 x 10 feet (h), 1,200/feet @ 25.00/squire foot = ₹ 30,000, life span of 3 years, 5 crops in 1 year in tropical areas, cost per crop.	2,000
iii.	Spraying equipment, hoses, tubs, buckets, forks, spades, etc. (life span three years = ₹ 6,000), cost per one crop season	400
iv.	Cement tank for wetting 20′ x 10′ x 4′ (depth) of ₹ 100 quire foot, ₹ 20,000 life span of 100 years, cost per year	2,000
II.	*Variable costs (materials used one time only)*	
i.	Paddy straw 5 tonnes @ ₹ 1,000/tonne	5,000
ii.	Gram flour @ 5 kg/tonne of wet substrate, 20 tonnes wet substrate x 5kg = 100 kg @ ₹ 30/kg	3,000
iii.	Spawn (2 per cent wet weight of substrate), 2 kg/tonne, 20 tonnes × 2 kg, 40 kg @ ₹ 50/kg	20,000
iv.	Labour–welling 10 LD, steeping 10 LD, Hot water treatment 10 LD, spawning 10 LD, cropping 30 LD, miscell. 10 LD = 80 LD at ₹ 100/LD	8,000
v.	Water costs	500
vi.	Power costs for light/exhaust fans	200
vii.	Pesticides, disinfectants, chemicals	500
viii.	Polythene bags for packing mushrooms	500
ix.	Polythene sheets for covering beds for spawn run	20,00
x.	Miscellaneous expenses	20,00
	Total of (I + II)	**29,100**
		Say ₹ 29,000

B. Returns

5 tonnes of dry straw will turn to 20 tonnes of wet substrate on wetting, placed in shelves in small beds of 30 kg dry weight each, with mushroom yield of 1.25-1.5 kg/ bed (dry weight) in 20 days of cropping. This will correspond to 4-4.5 kg mushrooms/ 100kg wet substrate, 40-45 kg/tonne of wet substrate, average selling price of ₹ 50/ kg.

Returns at 4-4.5 per cent of wet substrate weight in 3 weeks of cropping, at an average selling price of ₹ 50.00/kg.

At 4 per cent, 20 tonnes x 40 kg/tonne wet substrate, 800 kg	₹ 40,000
At 4.5 per cent, 20 tonnes x 45 kg/tonne wet substrate, 900 kg	₹ 45,000
Profit at 4 per cent, ₹ 40,000-29,000	₹ 11,000
Profit at 4.5 per cent ₹ 45,000-29,000	₹ 16,000

(figures are calculated on prices prevailing in March' 06 in India) 1 US $ = 45 Indian Rupees (approximately)

Employment Opportunities Mushroom Cultivation Offers Ample

Opportunities for self employment, a new entrepreneur requires a basic training of 4-7 days on mushroom cultivation, some financial help from financial institutions, guidance from the nuclear centres and small piece of land for building infrastructure for raising mushroom crop indoors. Educated unemployed youth is a big class that can be benefited by adopting mushroom cultivation. Farmers can take a short crop in a lean season for additional income generation and women entrepreneurs can adopt mushroom cultivation on kitchen garden scale also to subsist their income for house hold expenses. Mushroom cultivation does not require very hard inputs from cities, and these are readily available in rural areas and hence it is an attractive way of generating employment at a very low cost. Mushroom growing units also provide employment opportunities for landless labourers, rural women and people from weaker sections and backward classes. Some entrepreneurs can establish mushroom seed/spawn production laboratory, especially the women entrepreneurs and educated unemployed youth. Finance is available from the banks now at an easy rate. So mushroom farming is a very lucrative and remunerative activity that a farmer can adopt. In fact it can play an important role in integrated rural development programme of India. Mushroom is a cash crop and the farmers earn cash in a short period of time. But one must plan properly for it and survey the markets before starting this activity.

Government Support for the Mushroom Crop

A full fledged institute was established on mushroom crop in India at Solan in 1983, which has now been given the status of Directorate of Mushroom Research by ICAR, Delhi very recently. It exclusively works on research and development of mushrooms in India. Various research projects are running at the centre on all aspects of mushrooms with extensive emphasis on training on mushroom cultivation. The centre trains the farmers, entrepreneurs and trainees from the industry. Mushroom breeding programme is also running at this centre for evolving high yielding strains with desired traits of different cultivated mushrooms. This centre is one of the best centres on mushroom in the world. The Government also supports mushroom farming in remote North Eastern states under special programme of horticulture mission. There is one institution namely "National Horticulture Board" which helps all horticulturists including mushroom growers in adopting large scale mushroom farming in the country. Nationalized banks extend financial assistance to entrepreneurs/farmers for starting mushroom cultivation units. NAB ARD (Refinance Bank) also finances mushroom projects all over the country at low interest rates. Moreover there are scores of spawn labs, demonstration centres and composting units functioning in the government sector in support of this crop in India.

Farmer's Awareness of Mushroom Cultivation in India

In recent time tremendous awareness is shown by the farmers for mushroom cultivation in India. They are ready to invest not only on larger units but also on smaller scale to adopt mushroom cultivation as a hobby/profession. The educated unemployeds are turning up in good numbers in response to training programmes for mushroom cultivation and the seriousness shown by the entrepreneurs is really encouraging. The farmers know that mushroom is a cash crop and they can earn easily with small investment primarily because raw materials are available at cheaper rates and there is a ready market for fresh mushrooms in the country. Both the seasonal growers and all the year round growers earn good income by adopting it as a business. More varieties of mushrooms grown by the farmers are readily accepted by the consumers now and there are no questions asked about its taste/edibility. Mushrooms are not only a delicacy but also as a health food for all with high medicinal value.

Mushroom Cultivation as a Means of Women Upliftment in Rural India

Women are now becoming more and more enterprising in all spheres of life. Rural women do require some means to be independent and earn some additional income to spend on themselves and their households. Mushroom cultivation on kitchen garden scale/small scale will help the rural women in generating some additional income and making themselves self confident and financially independent. This will also help in developing self satisfaction in them while contributing in meeting demands of their family. This self satisfaction will help in upliftment of their stature and self esteem, which is very important for developing a healthy society in rural India as more than 65 per cent of Indian population live in rural India. Women are the backbone of a family and a happy woman means a happy family.

Future Prospects and Problems

The mushroom industry is gradually taking root in India but the pace is rather slow because of insufficient scientific support and inadequate training programmes. With the exception of the Mushroom Research Centre, Solan, there is no other research laboratory or institute in India which has a proper mushroom house with necessary facilities. Pasteurization of compost is essential to get uniform higher yield and freedom to a large extent from pests and diseases and competitors but this practice is still to be adopted by the growers. Some of the problems of the pests, pathogens and competitors are exclusive to this country. Their control measures are not known and have to be worked out by Indian scientists. Nematode problems are also becoming a limiting factor in the successful production of crops as nematicides recommended abroad are not available in India. This problem needs special attention.

At present, there are no organized spawn-producing units and, therefore, spawn supply to growers is limited and uncertain. Also there is no advisory service. A few research workers engaged in this activity have to attend to research, training and advisory service, with the result that they are unable to do full justice.

The production of mushrooms world-wide is increasing at an annual rate of about 10 per cent. Hitherto considered a luxury, mushrooms are generally available now with the extension of their cultivation to almost all parts of the world. Unlike mushrooms gathered from nature, artificially cultivated mushrooms are truly edible.

Mushroom cultivation has a special relevance to India because straw and plant residues are abundantly available with our farmers and our environment is conducive to the cultivation of mushrooms. Hence the future of this industry is bright. The awareness for the mushrooms among the general consuming public has been created. The question at the moment is whether we should go for sophisticated, high-capital investment industrial production technology with capacity of thousands of tonnes per year or a large number of small farms distributed in the rural areas of the country. Swaminathan (1982) has advocated the adoption of the small-farm approach in developing agriculture production. This approach has been adopted for mushroom cultivation by China, Japan, South Korea and Taiwan with significant success. Mushroom cultivation as a cottage industry is quite valid for Indian growers also because majority of them cannot afford the expensive mushroom houses equipped with boiler and environment-controlling gadgets. However, a modest modernization of the existing technology is essential if India is to be counted among the mushroom-producing nations of the world. In this regard, we may follow Taiwan which has more or less similar climatic conditions. Mushrooms are grown in Taiwan during winter in improvised mushroom houses made of polythene supported on bamboo frames. Pasteurization of compost is accomplished by mobile boilers. With these modest improvements in India too, yields can be stabilized at 12–15 kg/m^2 as compared to 6–8 kg/m^{2}as at present.

Bibliography

Abraham, S.P. 1991. Kashmir agarics- an overview, *Adv. Mushroom Sci* P.6 (Abstr.).

Abraham, S. P. and Kaul, T.N. 1985. Larger fungi from Kashmir–111. *Kavaka* **13**(2): 77-81.

Abraham, S. P. Kaul T.N. and Kachroo, J. L.1984. Larger fungi from Kashmir–II. *Kavaka* **12**: 41-48.

Abraham, S.P. 1991. Kashmir Fungal Flora- An Overview. In: Indian Mushrooms, (M.C. Nair, Ed,). Kerala Agricultural University, Vellanikkara pp.13-24.

Abraham, S.P., J.L. Kachroo and T.N. Kaul. 1980. Fleshy fungi of Gulmarg forests-I. *Kavaka* 8: 29-39.

Abraham, S.P., T.N. Kaul and J.L. Kachroo. 1981. Larger fungi from Kashmir-l. *Kavaka* **9**: 35–43.

Abraham, T. K. Virnda, K. B. Pradeep, C.K. and Vijay Joseph A. 1995. Edible Cantherelle from Southern India. *Mushroom Res.* **4**: 73-76.

Adhikari, R. S. 1992. Some new records of Agarics from Kumaun Himalayas. *Geobios New Reports* **11**: 96-99.

Adriano, F.T. and R.A. Cruz. 1933. The chemical composition of Philippines mushrooms. *Philipp. J. Agric.* **4:** 1–11.

Adsule, P.G., Girija, V., Amba Dan and Tewari, R.P. 1981. A note of simple preservation of Oyster mushroom (*Pleurotus sajor-caju*). *Indian J. Mushrooms,* **7** (1 and 2): 2-5.

Adsule, P.G., Onkaraya, H., Tewari, R.P. and Girija, V. 1983. Tomato juice as a new canning medium for European mushroom (*Agaricus bisporus* (Lange) Sing.). *Mushroom J.,* **124**: 143-145.

Agarwal, K.C. 1999. Biodiversity. Agro Botanica Vyas Nagar, Bikaner.

Agha, Z.D. 1974. A Cultivated *Pleurotus* for the use in forest and uncultivable areas of the temperate zone. *The Mushroom Journal*. Jan: **28.**

Ahmed, S. 1939 Higher fungi in the Punjab plains. I. the Gasteromytes, *J. Indian Bot. Soc.* **18**: 47-58.

Ahmed, S. 1940 Higher fungi in the Punjab plains. II. the Gasteromytes, *J. Indian Bot. Soc.* **18**: 169-177.

Ahmad, S. 1941. Higher fungi of the Punjab plains-III. The Gasteromycetes. *J. Ind. Bot. Soc.* **20**: 135-143.

Ahmad, S. 1941. Gasteromycetes of the Western Himalayas-I. *J. Indian Bot.Soc*: **20**: 173-182.

Ahmad, S. 1942. Gasteromycetes of the North Western Himalayas II. *J. Indian Bot. Soc.* **21**: 283–293.

Ahmad, S. 1945. Higher fungi of the Punjab plains–IV. *Lloydia* **8**: 238-244.

Ahmad, S. 1949. A contribution to the fungal flora of Pakistan and India. *Indian Phytopath.* **29**: 11–16.

Ainsworth and Bisby's Dictionary of the fungi. 2001. Eds, P.M. Kirk, P.F. Cannon, J.C. David and J.A. Stalpers, 9th Edition, CABI Publ. Wallingtorol.

AI Amidi, A.H.K. and Downes, M.J. 1990. *Parasitus bituberosis* (Acari: Parasitidae), a possible agent for biological control of *Heteropeza pygmaea* (Diptera: Cecidomyiidae) in mushrom compost. *Exp. Appl. Acarol.* **8**: 13-25.

Alof, E. and Folaranmi, V. 1991. *J. Food Compos.* **4** (2): 167-174.

Anand, J.C. 1975. Development of appropriate technology for a breakthrough in fruit and vegetable processing industry. *Indian Food Packer* **29** (6): 31-35.

Anand, J. C. and Johar, D. S. 1957. *J. Sci. Ind. Res.* **16** (8): 370-373.

Anand Rao, T. 1964. Field notes on certain Gasteromycetes of Ahmedabad. *Sci. and Cult.* **30**: 555-556.

Anandh, K. and V. Prakashan 2008. Some cultivable mushroom flora from Western Ghats. In Current vistas in Mushroom Biology and Productions. Edo. R. C. Upadhyay. S. K. Singh and R.D. Rai, Mushroom Society of India.

Anjali Roy and De, A. B. 1977. A record of *Polyporus tricholoma* Mont. from India. Trans. *Brit. Mycol.Soc.***68**: 442-444.

Anonymous 1950. List of Common names of Indian plant diseases. *Indian J. Agric. Sci.* **20**: 107-142.

Anonymous 1952. Index of fungi. CMI, **2** (4): 51-70.

Anonymous 1956. Index of fungi. CMI, **2** (13): 257-302.

Anonymous 1987. Biennial Report of AICRP on nematode pests of crops and their control. Deptt. of Entomology, UHF, Solan (HP), India.

Anonymous 1997. Annual Report of National Research Centre for Mushroom, Solan. p–96.

Apis, Saza, Caterina Matinato, Gabriella Gentili, Fabia Varotto, Clandio, Bandi and Davide Sassera 2010. Molecular Detection of Poisonous mushrooms in different Matrices. *Mycologia* **102** (3): 747–754.

Arrold, N.P. and Blake, C.D. 1968. Some effects of the nematodes *Ditylenchus myceliophagus* and *Apheienchioides composticola* on the growth on agar plates of the cultivated mushroom *Agaricus bisporus*. *Neniatologica* **12**: 501-10.

Arrold, N.P. and Blake, C.D. 1968. Some effects of the nematodes *Ditylenchus myceliophagus* and *Aphelenchoides composticola* on the cultivated mushroom. *Ann. Appl. Biol*. **61**: 161-66.

Arthur, R., Herr, F., Straus, N., Anderson, J.B. and Horgen, P.A. 1982. *Exptl. Mycol* **7**: 127-132.

Arumuganathan, T., R. D. Rai, Anil Kumar Hemakar and Swet Kamal 2006. Mushroom soup powder, nuggets, biscuits and pickle from the dried white button mushroom (*Agaricus bisporus*) *Indian J. Mush*. **XXIV** (1 and 2): 40-43.

Atri, N. S. 1981 Study on the taxanomy of Agaricales, the M. Phil. Thesis submitted to Punjabi University, Patiala.

Atri, N.S. 2001. Mushroom Genetic Resources of Punjab. A Critical appraisal and prospects. National Symposium on Plant Biodiversity and its Conservation.Abstr. III.10, pp. 70-71, Feb.14-15, Botany Department, Punjabi University, Patiala.

Atri, N. S. and Amanjeet Kaur 2002. Mushroom Flora of Patiala The genus *Coprinus* Pers ex Gray. In Plant Diversity in India (Eds. J. S. Dargan and T. A. Sarma) Bishen Singh Mahendra Pal Singh, Dehra Dun, India pp 431–452.

Atri, N. S. and Amanjit Kaur 2004. Mushroom flora of Patiala–the genus *Coprinus* Pres. Ex Gray. In: Plant Diversity in India PP 427-448. (eds. J. S. Dargan and T. A. Sharma) Published by Bisen Singh and Mahendra Pal Singh, Dehradun, U. A., India.

Atri, N. S., Amanjeet Kaur and Saini, S.S. 2001. Mushroom Flora of Patiala–The genus *Agaricus* L. ex Fr. *Indian J. Mush*. **19**(1 and 2): 1–9.

Atri, N.S., Amanjit Kaur and Harvinder Kaur 2005. Systematic and Sociobiology of Termitophilous mushrooms from Punjab, In: The fungi-diversity and conservation in India. 159-182 (eds. J. S. Dargan, N. S. Atri and G. S. Dhingra) Published by Bisen Singh and Mahendra Pal Singh, Dehradun, U. A., India.

Atri, N. S. and Harvinder Kaur 2003. *Sinotermitomyces* zang–a new genus record from India. *Mushroom Res*. **12**(1): 15-16.

Atri, N. S. and Harvindar Kaur 2005. Some unrecorded wild mushrooms of Punjab, India. *Mushroom Res*. **14**(2): 56-59.

Atri, N.S., Kaur, A and Saini, S.S. 2000. Taxonomic studies on Agarics from Punjab Plains. *Indian. J.Mush*.: **18**(1 and 2): 6-14.

Atri N.S., Kaur, A. and Saini, S.S. 2001. Mushroom Flora of Patiala- The genus *Agaricus* L. ex. Fr. *Indian J. Mush.* **19**: (1 and 2): 1-9.

Atri, N.S. and Saini, S.S. 1986. Further contributions on the studies of North–West Himalayan Russulaceae. Geobios new Reports **5**(2): 100-105.

Atri, N.S. and S.S. Saini 1988 Studies on *Lacterius* Pers.–the subgenus *Piperites* (Fr.) Kauff. *Kavaka* **16**(1 and 2): 13-19.

Atri, N. S. and Saini, S. S 1989. Family Russulaceae Roza–A review- Plant Science Research in India: Present status and future challenges. Aspects of plant sciences II (M. L. Tewari *et al.*, eds.) pp. 115-128. Today and Tomorrows Printers and Publishers. New Delhi.

Atri, N.S. and Saini, S .S. 1990a. North Indian Agaricales–VIII. The section *Compactae* of *Russula* Pers. in India. *J. Indian Bot. Soc.* **69**: 343-346.

Atri, N.S. and Saini. S.S. 1990b. Studies on Russula Pers.–Section Decolorantes (Maire) Sing. Geohios New Reports **9**: 10-13.

Atri, N.S. and Saini, S.S. 1990c. North Indian Agaricales–X. *J. Indian Hort. Soc.* **69**: 425–429.

Atri, N.S. and Saini, S.S. 2000. Collection and study of Agarics–an interoduction. *Indian J. Mush.* **18**(1 and 2): 1-5.

Atri, N. S., S.S. Saini and A. K. Gupta 1992. *J. Indian Bot. Sec.* **71**(1 and 2): 119-121.

Atri, N.S. Saini S. S. and Gupta. A. K. 1991. Systematic studies on *Agricus campestris* (L.) Fr. Geobios new Reports **10**: 32-37.

Atri, N.S. Saini, S.S. and Gurjit Kaur. 1995. Taxonomic studies on the North Indian agarics–the genus *Termitomyces* Heim. *Mushroom Res.* **4**: 7-10.

Atri, N.S., S.S. Saini and Gurjit Kaur 1996. Three species of Agarics from Patiala. *Mushroom Res.* **5**: 77-80.

Atri, N. S., S. S. Saini and Gurjit Kaur 1996. Taxonomic studies on the North Indian Agarics–The genus *Lepiota* (Pers. ex Fr.) Gray. *Mushroom Res.* **5**: 67–76.

Atri, N.S., Saini, S.S. and Kaur, A. 2000b. Taxonomical studies on Agarics from Punjab The genus *Lepiota* (Pers. ex.Fr.), Gray. *Mushroom Research* **9**: 71-77.

Atri, N.S. Saini, S.S. and Mann, D.K. 1991. Genus Russula Pers. in Dalhousie. In "Botanical Researches in India" (eds. N.C. Hery and B.L. Chaudhary) Himanshu Publications, Udaipur, India. pp 92-99.

Atri N.S., S.S. Saini and M.K. Saini 1993. *Gobios New Rports.* **12**(2): 137-140.

Atri, N.S., S.S. Saini and M.K. Saini 1997. Sudies on genus *Russula* Pers. from North West Himalayas. *Mushroom Res.* **6**(1): 1-6.

Atri, N.S. and.T.N Lakhanpal 2002. Conservation of mshroom biodiversity. *Indian J. Mush.* **20** (1 and 2): 45-54.

Azad, K.C., M.P. Srivastava, R.C. Singh and P.C. Sharma 1987. Commercial presentation of mushrooms–I. A technical profile of canning and its economics. *Indian J. Mush.* **12-13**: 21-29.

Aziawa, K. 1998. Antitumour-effective mushroom, meshimakobu, Phellinus linteus. Gendal-shorin, Tokyo.

Benedict, R. G. and Brady, L. R. 1972. Antimicrobial activity of mushroom metabolites. *J. Pharm. Sci.* **61**: 1820–1922.

Beneka, E. A. 1963. *Calvatia*, Calvacin and Cancer, *Mycologia* **55**: 257–270.

Bagchee, K.D. 1953, The fungal diseases of Sal '*Shorea robusta* (Gaertn.) Part I. *Indian For. Rec.*–N.S. **1** (2): 11–23. 199–348.

Bagchee, K.D. 1953. New and noteworthy diseases of trees in India. The sap and heart not disease of *Eucalyptus maculata* Hook. Var. *citriodora* Bailey due to attack of *Tramates cubeuni* (Mont.) Sacc. *Indian For.* **79**: 341-343.

Bagchee, K.D. and Bakshi, B.K. 1951. *Porja monticola* Murr. on Chir (*Pinus longifolia* Roxb.) in India. *Nature* (London) **167**: 4255.

Bagchee, K.D. Puri, Y.N. and Bakshi, B.K. 1954. Principal diseases and decay of Oaks and other hardwood in India-II, *Indian Phytopath.* **7**: 18-42.

Bagchee, K.D. and B. Bakshi, 1954. Studies on Indian Thelephoraceae I. Some species of *Sierium Peniophore and Corticium.* Ind. For. Bull. No. 166: 1–14.

Bagchee, K.D. and V. Singh, 1960. Indian Forest Records, New Series **1** (10): 199.348.

Baker, R., Parton, A. H., Rao, V.B. and Rao, V.J. 1982. The isolation, identification and synthesis of 3,6–dimethylheptan -2. 4-dione, a pheromone of the mushroom fly, *Megaselia halterata* (Diptera: Phoridae). *Tet. Let.* **23**: 3103-04.

Baker, R., Parton, A. H., Rao, V.B. and Rao, V.J. 1982. The isolation, identification and synthesis of 3,6–dimethylheptan -2. 4-dione, a pheromone of the mushroom fly, *Megaselia halterata* (Diptera: Phoridae). *Exp. Appl. Acarol.* **8**: 13-25.

Bakshi, B.K. 1956. Occurrence of *Polyporus squamosus* (Hud.) Fr. in India. *Indian Phytopath.* **9**: 191–194.

Bakshi, B.K. 1971. "Indian Polyporaceae (on trees and timber)".246pp. 1.C.A.R., New Delhi.

Bakshi, B.K. and Y.N. Puri 1978. Edible Fungi their survey and cultivation. *Indian Mushroom. Sci.* **1**: 374–381.

Bakshi, B.K., Y.N. Puri and B. Singh 1955. Two decay fungi on conifers in the Himalayas. *Ind.,J. Mycol. Res.* **1**: 75-79.

Bakshi, B.K. 1974. Mycorrhiza and its role in forestry. Forest Research Institute, Dehra Dun.

Banerjee, S.N. 1946. Some higher fungi of Sikkim Himalayas. *Sci and Cult.* **11**: 444-445.

Banerjee, S.N. 1947. Fungus flora of Calcutta and suburbs. I. *Bull. Bot. Soc. Bengal* **1**: 37- 54.

Banerjee, S.N. and T. Ghosh 1942. Preliminary report on the occurrence of higher fungi on bamboos in and around Calcutta. *Sci. and Cult.* **8**: 194.

Bano, Z. and Singh, N.S. 1972. Steeping preservation of an edible mushroom (*Agaricus bisporus*). *J. Food Sci. Technol.*, **9** (1): 13-15.

Barwal, V.S. 1992. Biochemical changes in relation to enzymatic browning in *Agaricus bisporus*. M.Sc. thesis Dept. of Post-harvest Technology. UHF, Solan, India.

Basu, H. 1955. Monsoon day forays. *Ind. J. Mycol. Res.* **1**: 81-82.

Batra, R. and S.W.T. Batra 1963. Indian Discomycetes, *Univ. Kansas Sci. Bull*. **44**: 109-256.

Behl Nita 1987. Protein and Lipid supplementation for increased yield of *Agaricus Lisporus* In: Development in Crop Sciences **10**, cultivation of Edible Fungi edited by P.J. Wuest, D.J. Roy and R.B. Beelanon. The Pensylvania State University Park, Pennsylvania 1680 U.S.A.–pp 242–244.

Behl, N. and Prasad, D. 1985. Nematodes associated with white button Mushroom. *Indian Bot. Reptr.*, **4**: 74.

Bensaude, M. 1918. Recherches sur le cycle evolutif *et al.*, sexualite chez. les. Basidiomycetes. Ph.D Thesis Univ. of Paris Published Nemours. France.

Berkeley, M.J. 1816. Three fungi from Kashmir, *Grevillea* **4**: 137-138.

Berkeley, M.J. 1850. Decades XXV to XXX. Sikkim Himalayan fungi collected by Hooker. J. D. Hook. *J. Bot.* **2:** 42–51, 76–88, 106–112.

Berkeley, M.J. 1851. Decades XXXII to XXXIII. Sikkim Himalayan fungi collected by Hooker, J.D. Hook. J. Bot **3**: 39–49.

Berkeley, M.J. 1852. Decades of Fungi XXXIX, XL Sikkim and Khassya Fungi. Hooker Jour. Bot. and Kew Gard. Misc. **4**: 130-142.

Berkeley, M.J. 1854. Decades of Fungi XLI-XLIII. Indian Fungi. Hooker Jour. Bot. and Kew Gard Misc. **6**: 129-143.

Berkeley, M.J. 1856. Decayes of Fungi. Decas 1-62. Nos. 1-620. In Hooker's London, *J. Bot.***3-8:** 1844-1856.

Bessinger, S.R. 1990. *Bioscience* **40**: 456-457.

Bhandal, M. S. and Mehta, K.B. 1989. Evaluation and improvement of strains in *Agaricus bisporus. Much. Sci.* **12** (1): 25–35.

Bhargava K.S. and Sehgal A.C. 1954. Additions to fungi of Nainital-I, wood rotting fungi. Proc. 41st Ind. Sci. Congr. Part IV. 25.

Bhaskar. R., Suriachandrasel, M. and Ramchandran. T.K. 1990. Ganoderma with disease of coconut a threat to coconut cultivation in India. *Planter*. **66** (774): 467–471.

Bhatt, R.P., Bhatt V. K. and Gour, R.D. 1995. Fleshy fungi of Gorhwal Himalayas: The genus *Russula* Indian *Phytopathology*. **48** (4): 402–411.

Bhatt, V. K. and Bhatt, R. P. 1999. Studies on the section Lactifluns of the Genus *Lactarius* in India. *Indian Phytopath*. **52** (3): 236–244.

Bhatt, V.K., R.P. Bhatt, R.D. Gour and M.P. Singh 1999. Mushrooms of Garhwal Himalaya: the genus *Amanita* Pers. ex. Hooker. *Mushroom Res.* **8**(2): 1-8.

Bhatt, V.K., Bhatt, R.P. and Gaur, R.D. 2000. Mushrooms of the Garhwal Himalaya: the genus *Lactarius* Pers. ex S.F. Gray. *Mushroom Res.* **9**(1): 11-18.

Bhatt, R.P. and Lakhanpal, T.N. 1988a. *Amanita fulva* (Schaeff. ex) Pers.–An edible mushroom new to India. *Curr. Sci.* **57**(20): 1126-1127.

Bhatt, R. P. and Lakhanpal, T. N. 1988b. New record of fleshy fungi for India. *Indian J. Mycol. Pl. Pathol.* **18** (2): 140–142.

Bhatt, R.P. and Lakhanpal, T.N. 1988c. A new record of edible *Russula* from India. *Curr. Sci.* **57**(22): 1257-1258.

Bhatt, R.P. and Lakhanpal, T.N. 1988d. *Lactarius hygrophoroides* Berk. and Curt-An edible wild milky mushroom new to India. *Curr.Sci.* **57** (1): 38-39.

Bhatt, R. P. and Lakhanpal, T. N. 1990. Fleshy fungi of North Western Himalayas–V. *Indian Phytopath.* **43** (2): 156–164.

Bhatt, R. P. and Lakhanpal, T. N. 1994. The mushroom families Cantharellaceae and Russulaceae in India (In Press).

Bhattacharya, B. and H.K. Baruah 1953. Fungi of Assam. *J. Univ. Gauhati* **4**: 287.312.

Bhavani Devi, S. 1982. Studies on edible mushrooms of kerala with special reference to paddy straw mushroom, *Volvariella* sp. Ph.D. Thesis, Kerala Agriculture University Vellanikkara. pp.221.

Bhavani Devi, S. 1995. Mushroom Flora of Kerala. In: Advances in Horticulture **13**: 276-316–Mushroom (K.L. Chadha and S.R. Sharma, Eds.). Malhotra Publishing House, New Delhi.

Bhavani Devi, M. C. Nair and M. R. Menon 1980. *Termitomyces robustus*- An addition to Indian edible mushrooms, *Kavak* **8**: 53-54.

Bhide, B.P., Pandey Alka, Sathe A.V., Rao V.G. and Patwardhan, P.G. 1987. Fungi of Maharastra(Supp. 1.) MACS Pub.

Bilgrami, K.S., Jamaluddin and M.A. Rizwi. 1979. "Fungi of India" 467 pp. Today and Tomorrow's Printers and Publishers, New Delhi.

Binninger, D. M., Skrzynia, C., Pukkila, P. J. and Casselton, L. A. 1987. EMBO J **6**: 835-840.

Binns, E.S. 1973a. Predatory mites–neglected allies? *Mush. J.* **12**: 540-544.

Binns, E. S. 1973b. *Digamasellus* fallax Leitner (Mesotigmata: Digamasellidae) phoretic on mushroom sciarid flies. *Acarol* **15**: 10-17.

Binns, E.S. 1976. Mushroom pests. Rep. Glasshouse Crop Res. Inst. for 1975. pp 92-93.

Bisht, N.S. and N.S.K. Harsh 2001. conservation strategies for the Fungal diversity of Arunachal Pradesh, *Arunachal Forest News.* **12** (142): 102–109.

Boa, E.R. 2004.Wild edible fungi: A global overview of their use and importance to people. Food and Agriculture Organization, Rome.

Bohus, G., Galz, C. and Scheiber, E. 1961. The antibiotic action of higher fungi on resistant bacteria and fungi. *Acta. Biol. Acad. Sci.* Hung. **12**: 1–12.

Bose, S.R. 1918. Description of fungi in Bengal. *Proc. Indian Assoc. Cult. Sci.*: 109-114.

Bose, S.R. 1920. Records of Agaricaceae from Bengal. *J. Asiatic Soc. Bengal*, N.S. **16**: 354-374.

Bose, S.R. 1921. Possibilities of mushroom industry in India by cultivation. *Agile. J. India* **16**: 643-647.

Bose, S.R. 1923. The fungi cultivated by the termites of Barkuda. *Rec. Indian Mus.* **25**: 253-258.

Bose, S.R. 1919-1928. I. Description of fungi in Bengal. Proc. Indian Association. Cult. Sci. **4**: 109. II. *Ibid*, Proc. Sci. Convention. Indian Assoc. Cult. Sci. for the year 1918. 136-143. III. Fungi of Bengal, Polyporaceae of Bengal, Part III Bull. Cermichad. Med. Coll. Belgachia **1**: 1-5. IV. Polyporaceae of Bengal Part IV *ibid* **2**: 1-5. V Polyporaceae of Bengal Part V *ibid* **3**: 20-25. VI. Polyporaceae of Bengal Part VI Proc. Sci. Convention Indian Assoc. Cult. Sci. for the year 1919. VII. *Ibid* for the year 1920-21, 27-36. VIII. *Ibid* Part VIII. J. Dept. Sci. Calcutta Univ. **9**: 27-34. IX. *Ibid* Part IX. *Ibid* **9**: 35-44.

Bose, S.R. 1934. Polyporaceae of Bengal X. J. Dept. Sci. Calcutta Univ. **11**: 1-18.

Bose, S.R. 1946. Polyporaceae of Bengal XI. J. Dep. Sci. Calcutta Univ. **2**: 53-87.

Bose, S. R. 1953. Antibacterial substances from some higher fungi in India. *Indian J. Pharm.* **15**: 279–291.

Bose. S.R. and A.B. Bose. 1940. An account of edible mushrooms of India. *Sci. and Cult.* **6**: 141-149.

Braun, A., Wolter, M., Zadrazil, F., Flachowsky, G., Mba, C. C. 2000 Bioconversion by *Lentinus tuber regium* and its potentials, utilization as food, medicine and animal feed. *Mushroom Sci.* **15**: 549-558.

Bressa, G., Cima, L. and Costa, P. 1988. *Ecotoxicoi Environ Safety* **16** (2): 85-89.

Brian, P.W. 1951. Antibiotic produced by fungi. *Bot. Rev.* **17**: 357–430.

Brumswik, H. 1924. Untersuchungen uber die Geschlechts und Kernverhaltnisse bei des Hymenomyzetengatung, *Copinus* Bot. Abh. K. Goebels, **5**: 1–152.

Brunnet, H. and Zadrazil, F. 1986. *Eur J Appl Microbiol Biotechnol* **17**: 145-154.

Burrage, K. J. 1981. Ecological biochemistry of the mushroom phorid *Megaselia halterata* (Wood) (Dipt.), BSC. thesis, Univ. Southampton, 24pp.

Butler, E.J. and G.R. Bisby 1931. "The Fungi of India", Imp. Coun. of Age. Res. India. Sci. Mono. 1, XVIII, 237 pp., Calcutta.

Butler, E. J. and Bisby, C. R. 1960. The fungi of India. Indian Agric. Res. Inst., New Delhi. pp 552 (Revised by Vasudeva, R.S.).

Butt. B. White, P.F. and Jarrett, P. 1994. Further development of a dipteran -active strain of *Bacillus thuringiensis* Proc. XXVIIth Ann. Meeting Soc. Imv. path., August, 1994, Montpellier, France.

Callac, P., Billette, C., Imbernon, M. and Kerrigan, R.W. 1993. *Mycologia* **85**: 835-851.

Cantwell, G.E. and Cantelo, W.W. 1984. Effectiveness of *Bacillus thuringiensis* var. *israelensis* in controlling a sciarid fly, *Lycoriella mali*, in mushroom compost. *J. Econ. Ent.* **77**: 473-75.

Castle, A.J., Horgen, P.A. and Anderson, J.B. 1987. *Appl Environ Microbiol* **53**: 816-822.

Cayrol, J.C. 1962. Important de maladies vermicularies dan les champignonniers. *Mushroom Sci.* **5**: 480-96.

Cayrol JC. 1967. Etude du cycle evolutif D' *Aphclenchoides compositicola*. *Nematologica* **13**: 23-52.

Cayrol, J.C. 1967. Etude du cycle volatif D' *Aphclenchoides compositicola*. *Nematologica* **13**: 480-96.

Chahal, D.S. 1963. *Lycoperdon pusillum* Batsch in the Punjab Plains–A new record in Indian Gastromycetes. *Curr. Sci.*, **321**: 282-283.

Chakraborty, Nibha and R. P. Purkaystha 1955. *Pleurotus-pulmonarius*–a new addition to Indian mushrooms. *Kavak* **4**: 51-53.

Chakravarty, D.K. and B.B. Sarkar, 1982. *Tricholoma lobayense,* a new edible mushroom from India. *Curr Sci.* **51**(10): 531-532.

Chakravarty, D.K. and D.C. Khatua 1979. *Termitomyces microcarpus,* a new Indian edible mushroom. *Curr. Sci.* **48** (8): 364-365.

Chakravarty, N. and R.P. Purkayastha 1976. *Pleurotus pulmonarius*–a new addition to Indian mushroom. *Kavaka* **4**: 51-53.

Challen, M.P. and Elliott, T.J. 1994. *Cultivated Mush Newsletter* **2**: 13-20.

Chandra, Aindrila 1974. "Physiological studies on some Indian species of edible mushrooms", Ph.D. Thesis. University of Calcutta, Calcutta.

Chandra, Aindrila and Roy Watling. 1982 Studies in Indian *Armillaria* (Fries per Fries) taude (Basidiomycotina). *Kavaka* **10**: 63.64.

Chang, S.T. 1964. Pure culture spawn for *Volvariella volvacea*. *The Chung Chi. J.* **2**: 222-225.

Chang, S.T. 1980. *Bioscience* **30**: 299,.

Chang, R.Y. 1995. Effective dose of *Ganoderma* in humans. In *Ganoderma*: Systematics, Phytopathology and Pharmacology. 5 th International Congress, Vancouver.

Chang, R.Y. 1999. Role of *Ganoderma* supplementation in cancer management. In Advanced Research in PSP, 1999. Ed. Yang Qing-yao. The Hong Kong Association for Health Care Ltd., Hong Kong, pp. 346-350.

Chang, S. T. 1999. World production of cultivated edible mushroom in 1997 with emphasis on *Lentinus edodes* (Berk.) Sing. in China. *Int. J. Med Mushrooms* **1**: 291-300.

Chang, S.T. and K.Y. Chan. 1973. Qualitative and quantitative changes in protei during morphogenesis of the basidiocarp of *Valvariella volvacea*. *Mycologia* **63** 355.

Chavan, P.B. and Barge, S. N. 1978. Some Fleshy Fungi from Maharastra. *The Botanique* **8**: 122-128.

Chen A.W. 1999. A practical guide for synthetic log cultivation of the medicina mushroom *Grifola frondosa* (Dick.: Fr.) S. Fr. Gray (Maitake). *International Journa of Medicinal Mushrooms* **1**: 153-168.

Chen A.W. and Miles, P.G. 1996. Biomedical research and the application o mushroom nutriceuticals from *Ganoderma lucidum*. Proceedings of 2nd International Conference on Mushroom Biology and Mushroom Products (ed Roycse, D.J.), Penn State University, PA.

Chihara, G. 1992. Immunopharmacology of lentinan, a polysaccharide isolated fron *Lentinus edodes*: –its application as a host defense potentiator. *International Journa of Oriental Medicine*, **17**: 57-77.

Choleva, B. 1969. Studies of the sources of nematode infection in mushroom houses *Rosten Nauki*. **7**(1): 145-52.

Chopra, R.N. and I.C. Chopra 1955. "A Review of Work on Indian Medicinal Plants" 263 58 pp. I.C.M.R., New Delhi (Special Report Series No. 30).

Chopra, S.K., Naquash, G. S. and Chadha, T. R. 1985. Effect of pre-harvest sprays o honey, citric acid and *Euphorbia royleana* latex on the shelf- life of butto mushrooms (*A. bisporus*) (Abstr.). National Symp. on Production an Conservation Forestry. 12-13 April, Solan (H.P.).

Coatney, G. R., Cooper, W.C., Eddy, N.B. and Greenberg, J. 1953. Survey of Antimalaria agents. Public Health Monogr. No.-9. Fed Security Agency Washington, D.C.

Cochran, K. W., Nlshikawa, T. and Beneke, E.S. 1967. Botanical sources of influenz inhibitor. *Antimicrobial agents and Chemotherapy* **1966**: 515–520 pp.

Collins, R.A. and Ng, T.B. 1987. Polysaccharopeptide from *Coriolus versicolor* ha potential for use against human immunodeficiency virus type I infections. *Li Science* **60**: 383-387.

Cooke, M.C. 1870. Kashmir Morels. *Trans. Bot. Soc. Edi*. **10**: 439-443.

Cooke, M.C. 1876. Some Indian fungi. *Grev*. **5**: 14-17.

Cooke, M.C. 1879. Some exotic fungi. *Grev*. **7**: 94-96.

Cooke, M.C. 1880. Fungi of India. *Grev*. **8**: 93-96.

Cooke, M.C. 1892. Himalayan truffles. *Grev*. **20**: 67 (with G. Massee.).

Corner, E. J. 1950. A monograph of *Clavaria* and allied genera. Ann. Bot. Memo. No. London Oxford Univ. Press.

Corner, E.J. and Anand, G.P.S. 1956. The Clavariaceae of the Mussoorie hills-II. *Tran Brit. Mycol. Soc*. **39**: 475-484.

Corner, E.J. and Anand, G.P.S. and Sukhdev S. 1958. The Clavariaceae of the Mussoorie hills (India) -IX. Trans. *Brit. Mycol. Soc.* **41**: 203-206.

Cunningham, G.H. 1942. The Gasteromycetes of Australia and New Zealand, Donedin, N.Z. 236 pp.

Currey, F. 1874. On a collection of fungi made by Mr. Sulpiz Kurz, Curator of the Botanic Garden, Calcutta. Trans. Linn. Soc. London, II. *Ser. Bot.* **1**: 119-131.

Curry, F. 1876. On a collection of fungi made by Mr. Sulpiz Kurz Curator of the Botanic Garden, Calcutta. Trans. Linn. Soc. London Ser **21**: 119–131.

Dang, R. L., Singh, A.P. and Gupta, A. K. 1978. Influence of pre-treatment on the yield and quality of canned mushrooms. *Technologists*, CFTRI, Mysore. 23-25 June.

Dang, R.L. and Singh, R.P. 1978. Preservation of mushrooms. *Indian Mushroom Sci.*, **1**: 215-23.

Das, M.K. and Sinha, M.P. 1990. Agaricales of Estern India–I, *Indian Phytropthology*. **43** (2): 150–155.

Das, Nilanjana, S.C. Mahapatra and R.N. Chattopadhyay 2002. Some edible fungal flora of tribal areas of Midnapur district, West Bengal, Journal of non–timber Forest Products, **9** (3/A): 154–155.

Dastur, J.F. 1946. Report of the Head of the Division of Mycology and Plant Pathology. *Sci. Rept. Agric. Res. Inst.* New Delhi 1945-46. 79-88.

Declaire, J. R. 1978. Economics of cultivated mushrooms. In: The Biology and Cultivation of edible mushrooms. pp. 778–792 (Eds.) Chang, S.T. and Hayes, W.A. Academic Press, New York.

Delmas, J. 1978. The potential cultivation of various edible fungi. in "The Biology and Cultivation of edible mushrooms". p. 251 Eds. S.T. Chang and WA. Hayes, (Academic Press, London).

Deshpande, A.G. and Tamhane, D.V. 1981. Studies on dehydration of mushroom (*Volvariella volvacea*). *J. Food Sci. Technol*, **18**(3): 96-101.

Dey, A. B. 1999. *Podaxis pistillaris* (L. Pers) Fr. Addition to the list of fungi of Rajasthan (India) Indian Forester **125** (6): 634–636.

Devi, S. Bhavani, M.C. Nair and M.R. Menon 1980. *Termitoryees robustus*–An addition to Indian Edible Mushrooms. Kavak **8**: 53–54.

Dhancholia, S. and M.P. Sinha 1988 Aditional studies on Agaricus of Orrisa. *Geobios New Reports* **7**: 169-172.

Dhancholia, S. and Sinha, M.P. 1988a. Noteworthy agarics from Orissa *Indian Phytopath*. **41:** : 474-477.

Dhancholia, S., Bhatt and Pant, S.K. 1991. Studies on some Himalayan agarics. *Acta Botanica Indica* **19**: 104-109.

Dhar, B. L. 1976. Japanese method of cultivation of wood inhabiting mushrooms. *Ind, J. Mushrooms* **2**: 26–32.

Dhar, B.L. 1978. Mushroom recipes. The exotic mushrooms. *Femina*, 23 July, pp. 55-67.

Dhar, B. L. 1997. Mushroom Industry in India–a view. In: Advances in Mushroom Biology and Production by R. D. Rai, B. L. Dhar and R. N. Verma (eds) PP 369-378. *Mushroom Society of India*, Solan.

Diaz-Godinez, G. and Sanchez, C 2002. In situ digestibility and nutritive value of the maize straw generated after *Pleurotus ostreatus* cultivation. *Can. J. Anim. Sci.* **82** 617-619.

Doshi, A. and Sharma, S.S. 1990. A new mushroom *Termitomyces eurhizus* from Rajasthan. India. *Indian J. Mycol. Pl. Pathol.* **20**(3): 279.

Doshi, A. and S. S. Sharma 1997a. Wild Mushrooms of Rajasthan. Advances in Mushroom Biology and Production. Proc. Ind. Mushroom Conf. 1997. Rai, R. D. B. L. Dhar and R. N. Verma(eds) Mushroom Society of India. N.R.C.M., Solan (H.P.) PP. 105-127.

Doshi, A. and S.S. Sharma 1997b. Wild mushrooms of Rajasthan, Advances in Mushroom Biology and Production(eds.) R.D. Rai, B.L. Dhar and R.N. Verma Mushroom Society of India, Solan. PP. 193-203.

Drbal Karel, Pavel Kalac, Alena Sefolva and Jiri Sell. 1975. Content of the trace elements iron and manganese in some edible mushrooms. *Ceska Mykol.* **29** (2): 110-114.

Eaton, T.K., Chang, H. M., Joyce, T.W., Jeffries, T.W. and Kirk, T.K.1982. Method obtains fungal reduction of the color of extraction stage Kraft bleach effluents. *TAPPI Journal* **65**: 89-92.

Editorial, 1997. The magic of mushroom. *The Nature* **388**: 340.

Egli, S., Ayar, F. and Chatelain, F. 1990. Die Einfluss des pilzsammelns auf. die Pilzflora. *Mycol. Helv.* **3**: 417-428.

Elliott TJ 1972. *Mush. Sci.* **8**: 11-18.

Falanghe, H., A.K. Smith and J.J. Rackis 1964. Production of fungal mycelial protein in submerged culture of soyabean whey. *Appl. Microbiol.* **12**: 330-334.

Falok, R. 1923. UBer ein Krystallisiertes Stoft wechsel product. Von Sparusis ramose shiff. *Chem. Ber.* 56 B: 2555–2556.

Farkas, K. and Balazs, S. 1975. The nematode species of horse manure used for cultivation of the field mushroom and possibility of control. *Phytopathologica Acadamii Scienitarum Hungariae* **10**: 153-63.

Farkas, K. and Karenczy, I. 1974. Nematological studies of casing soils used for mushroom beds and possibilities of control. Acta Agronomica Acadamiae Sceinitarum Hungariiae. **23**: 341-58.

Fellner, R. 1989. Mycorrhiza foming fungi as bioindicator of air pollution. *Agric Ecosyst. Environ.* **28**: 115-120.

Fletcher, J.T., White, P.F. and Gaxe, R. H. 1986. Mushrooms: Pest and Disease Control Andover, UK; Intercept Ltd.

Food and Agricultural Organisation (FAO). 1972. Food composition table for use in East Asia. Food Policy and Nutr. Div., Food Agric. Organ. U.N., Rome.

Franklin, M.T. 1957. Aphelenchoides composticcla n.sp. and *A. scprophilus* n.sp. from mushroom cempost and rotting plant tissues. *Nematologica* **1**: 306-13.

Franks, F. l981. Effect of low temperatures on biological membrane (eds.) G.J. Morris and A. Clarkes. Academic Press London, PP 3-19.

Fritsche, G. 1981. Some remarks on the breeding, maintenance of strains and spawns and *Agaricus bisporus* and A. *bitorquis*. Mush. Sci., **11**: 367–385.

Gadd, G.M. 1986. In Microbes in Extreme Environment (RA Herbert and GA Codd eds) p.329. Academic Press, London.

Garcha, H.S. 1980. Mushroom Grows. Punjab Agricultural University, Ludhiana.

Garcha, H. S. 1984 A mannual of Mushroom Grows Punjab Agri. Univ., Ludhiana.

Garcha, H. S. and K. L. Kalra 1977 Weed Mushroom. Symposium on Recent Researches in Plant Sciences 20-21 January, Botany Department, Punjabi University, Patiala P. 8 (Abstr.).

Garcia Mendoza C. 1992. *World J Microbiol Biotechnol* **8** (supplt 1): 36-38.

Gardes, M. Fortin, A. Mueller, G. M. and Kropp, B. R. 1990. *Phytopathology* **80**: 1312-1317.

Gardezi, Syed Riaz Ali and Najma Ayub 2003. Mushrooms of Kashmir VII. *Asian Journal of Plant Sciences.* **2** *(8): 644–652.*

Ghosh, R.N. and N.C. Pathak. 1962. Fungi of India I. *Morchella, Verpa* and *Helvella*. Bull. Nat. Bot. Gardens (Lucknow) No. **71**: 1-19.

Ghosh, R.N. and N.C. Pathak 1965. The genus *Macrolepiota* in India. *Indian Phytopath.* **IX**: 360-362.

Ghosh, RN., N.C. Pathak and B.P, Singh 1974. Studies on Indian Agaricales II. *Proc. Nat. Acad. Sci. India* **44** (13): 125-128.

Ghosh, R.N., N.C. Pathak and B.P Singh 1976. The genus *Chlorophylium* in India (*Chlorophyllurn molybdites*). *Indian Phytopath.* **29**: 50-53.

Ghosh, R.N., N.C. Pathak and Tewari, I. 1967. Studies on Indian Agaricales. *Indian Phytopath.* **20**: 237-242.

Ghosh R.N., Pathak, N. C. and Tiwari, I. 1967. Studies on Indian Agaricales. *Indian Phytopath* **18**: 360-362.

Gillespie, D. R. and Quiring, D.M.J. 1990. Biological control of the fungus gnats, *Bradysia* spp. (Diptera: Sciaridae) and wester flower Thrips, *Frankiniella occidenlailis* (Pergande) (Thysanoptera: Thripidae), in green house using a soil-dwelling predatory mite, Geolaclaps sp. nr. aculeifer (Canestrini) (Acari: Laelapidae). *Can. ent.* **122**: 975-83.

Ginai, M. A. 1936. Further contribution to the knowledge to the Indian Copriphilous fungi. *J. Indian Bot. Soc.*, **15**: 269-84.

Girard, J.E., Hendry. J.B. and Snestsinger, R. 1974. Sex pheromones in a mushroom-infesting sciarid, *Lycoriella mali*. *Mush J.* **13**: 29-31.

Gogoi, Robin, Dipali Majumdar and K. C. Puzari 2000 New addition to mushroom flora of Assam. *Mushroom Reasherch* **9** (1): 55.

Goodey, J.B. 1958. *Paraphelenchus myceliophthorus* n.sp. (Nematoda: Aphlenchoidae). *Nematologica* **3**: 1-5.

Goodey, J.B. 1960. Observation on the effects of the parasitic nematodes *Ditylenchus myceliophagous. Aphelenchoides composticola* and *Paraphelenchus myceliophthorus* on the growth and cropping of mushrooms. *Ann. Appl. Bio.* **48** (3): 655-63.

Graham, R.J.D. 1915. Report of the Economic Botanist, in Report of Agricultural College at Nagpur, of Botanical and Chemical Research, etc., Dept of Agric., Central Provinces and Berar, for the year 1914-15. Govt. Press, Nagpur: 11–17.

Gregory, F. J.; Healy, E. M., Agersborg, Jr. H.P.K. and Warren, G.H. 1966. Studies on antitumour substances produced by Basidiomycetes, *Mycologia* **58**: 80–90.

Grewal, P.S. and Sohi, H.S. 1988. A new and cheaper technique for rapid multiplication of *Arthrobotrys conoides* and its potential as a bio-nematicide. in mushroom culture. *Curr. Sci.* **57**: 44-46.

Grewal, P. S. and C. Smith 1995. Insect parasitic nematodes for mushroom pest control. *Mush. News* **43**: 15-25.

Griffiths R., Richards W., Johnson M., Mc. Cann U., Jesse R 2008. "Mystical type experiences associated by psilocybin mediate the attribution of personal meaning and spiritual significance 14 months later." Journal of Psychopharmacology (Oxford, England) **22**(6): 621-632.

Gulden, G., Hoiland, K., Bendiksen, E., Brandrud, T.E., Foss, B.S., Jennsen, H.B. and Laber, D. 1992 Mycomycetes and air pollution. Bibl. Mycol. Vol. 144.J. Cramer, Berlin, Stuttgrat.

Gunde-Cimerman, N. 1999. Medicinal value of the genus *Pleurotus* (Fr.) P. Karst. (Agaricales SI., " Basidiomycetes*). International Journal of Medicinal Mushrooms* **1**: 69-80.

Gupta, A. K. 1994. Systematic studies on the genus *Agaricus* L. Fr. (Agaricaceae) from NW. India. Ph.D. thesis submitted to Punjabi University, Patiala.

Gupta, D. 1984. Some tropical edible mushrooms of West Bengal. *Indian.J. Mycol. Res.* **22**(2): 149- 150.

Gupta, G. K., B.S. Bajaj and D. Suryanarayana. 1970. Studies on the cultivation of paddy straw mushroom (*Volvariella volvacea* and *V. diplasia*) in India. *Indian Phytopath.* **23**: 615-621.

Gupta, K.K., R.K. Agarwala, S. Kumar and P.K. Seth 1974. Gasteromycetes of Himachal Pradesh. *Indian Phytopath.* **27**: 45-48.

Halpern, G.M. 1999. Cordyceps: China's healing mushroom. Avery Publishing Group, New York, USA, pp. 116.

Harsh, N. S. K., Soni, K.K. and Tiwari, C.K. 1993. *Ganoderma* root rot in *Acacia arboretum*. *European Journal of Pathology*. **23**: 252–254.

Harvey, A., Bistis, G.N. and Leong, I. 1978. Cultural studies of single ascopore isolates of *Morchella esculenta*. *Mycologia*. **70**: ,1269-1274.

Hawksworth, D.L. 1991. *Mycol. Res.* **95**: 641-655.

Hennings, P. 1900. Fungi Indiae Orientalis. *Hedw.* **39**: 150-153.

Hennings, P. 1901. Fungi Indiae Orientales II, ci. W. Gollana 1900 Colecti. *Hedw.* **40**: 323- 324.

Herrman, H. 1966, Cortinellian, eine antibiotisck wirksam substanz ansd *Cortinellus shitake*, *Naturwissenshatten* **40**: 542.

Hesling, J.J. 1978. The role of eelworms. part 3. Methods of controlling eel-worms. *Mush. J.* **84**: 531-36.

Hibbett, D.S. 1992. *Trans Mycol Soc Japan* **33**: 533- 556.

Hishida, I., Nanba, H. and Kuroda, H. 1988. Antitumour activity exhibited by orally administered extracts from fruit-body of *Grifola frondosa* (maitake). *Chemical and Pharmaceutical Bulletin* **36** (5): 1819-1827.

Ho, B. and Chen, Q. 1991. Antifertility action of *Auricularia auricula* polysaccharide. Zhongguo Yacke Daxus Xuebao **22**: 48-49.

Hobbs, C. 1995. Medicinal Mushrooms: An Exploration of Tradition, Healing and Culture. Botanica Presska Cruz, CA.

Hole, R.S. 1927. Mortality of spruce in the jaunsar forests, United Provinces. *Indian For.* **53**: 434–443.

Hooper, D.J. 1958. *Apthelcnchoides dactylocercus* n.sp. and *A. sacchari* n.sp. (Nematoda: Aphlenchoidea). *Nematologica* **3**: 229-35.

Hooper, D.J. 1962. Effects of a nematode on the growth of mushroom mycelium. *Nature* **193**: 496-97.

Hu, B. and But, P. 1987. Chinese material medicines for radiation protection. Abstracts of Chinese Medicines **1**: 475-490.

Hughes, D.H. 1962. *Mushroom Sci.*, **5**: 540.

Humfeld, H. and T.F. Sugihara 1949. Mushroom mycelium production by submerged propagation. *Food Technol.* **3**: 255-356.

Hussey, N.W., Read, W. H. and Hesling, J.J. 1969. The Pests of protected Cultivation. London, Arnold.

Ikekawa, T. 1995. *Enokitake, Flammulina* velutipes: antitumour activity of extracts and polysaccharides. *Food Review International* **11**: 203-206.

Ikekawa, T. 2001. Beneficial effects of mushrooms, edible and medicinal, on health care. *International Journal of Medicinal Mushrooms* abstracts, p. 79.

Iracabal, B. and Labarere, J. 1993. *J Exp Mycol.* **17**: 90-102.

Iracabal, B. and Labarere, J. 1994. *Theor Appl Genet* **86**: 715-718.

Iracabal, B., Zervakis, G. and Labarere, J. 1995. *J Microbiology* (Reading) **141**: 1479-1490.

Itami, H. and Yahagi, N. 1996. Japanese Tochukaso, Challenges for Terminal Cancer. Metamoru shyupan Tokyo, 140 pp.

Ito, H., Shimura, H., Itoh, M. and Kaurode, M. 1997. Antitumour effects of a new polysaccharide protein complex (ATOM) prepared from *Agaricus blazei* (Iwade Strain 101) Himematsuke and its mechanisms in tumour-beairng mice. Anticancer Research Jan-Feb. **17**: (IA), 277-284.

Itvaara, M. 1988. *Trans British Mycol. Soc.* **91**: 295-304.

Jagdale, S. V. and M.S. Patil 1983. Studies in Pyrenomycetes of Maharastra Family Clavicipitaceae. *Indian J. Mycol. Pl. Pathol.* **13**: 150–153.

Jain, S.K., Gujral, G.S., Jha, N.K. and Vasudevan, P. 1988. *Biol. Wastes* **24**: 275-282.

Jalali Sarojini, Meera Pandey, L. N. Doreyappa Gowda and R. P. Tewari 2003. Development of value added products from milky mushroom *Calocybe indica*. In: Current Vistas in Mushroom Biology and Production. R. C. Upadhyay, S. K. Singh, and R. D. Rai (Eds) Mushroom Society of India, Solan PP 238-240.

Jana, K.K. and R.P. Purkayastha 1983. Production of sporocarps and rise in nitrogen level of substrate of two *Pleurotus* species. *Taiwan Mushrooms* **7** (1): 39-45.

Jandaik, C.L. 1976a. Comparative free amino acid composition of sporophores and mycelia of some edible fungi. *Ind. J. Mushrooms* **2** (1): 15-18.

Jandaik, C.L. 1976b. Commercial cultivation of *Pleurotus sajor-caju*. *Ind, J. Mushrooms* **2**: 19-24.

Jandaik, C.L. 1978. Problems and prospects of *Pleurotus* cultivation. *Indian Mushroom Sci.,* **1**: 411-14.

Jandaik, C.L. and I.N. Kapoor 1975. *Pleuotus sajor–caju* (Fr.) Singer from India. *Ind. J. Mushrooms* **1**: 1-2.

Jandaik, C.L. and Sharma, A.D. 1987. Effect of different storage conditions and drying methods on the shelf-life of *Pleurotus* spp. *Indian Mushroom Sci.,* **2**: 212-14.

Jandaik, C.L., Sharma, R.C., Joshi, V.K. and Bhardwal, S.R. 1989. Potential of dehydrated mushroom soup powders. National Seminar on Recent Avances in Post-harvest Management of Temperate Fruits, Vegetables and Ornamental Plants. Abstr., pp. 29.

Janoweiz, K. 1978. Feeding and destructiveness of *Aphelenchoides sacchari* Hooper for mycelium of mushrooms. Rocznik Nauk Rolniczych E **7**(1): 201-220.

Jansen, E. and Van Dobben, HE 1987. Is decline of *Cantharellus cibarius* in The Netherlands due to air pollution? Ambio. **16**: 211-213.

Jennison, M.W., C.G. Richberg and A.E. Krikszens. 1957. Physiology of wood-rotting Basidiomycetes. II. Nutritive composition of mycelium grown in submerged culture. *Appl. Microbiol.* **5**: 87.

Jennison. MW., M.D. Newcomb and R. Henderson. 1955. Physiology of the wood-rotting Basidiomycetes. I. Growth and nutrition in submerged culture in synthetic media. *Mycologinst* **47**: 275.

Jiu, J. 1966, A survey of some medicinal plants of Mexico for selected biological activities. *Loydia* **20**: 250–259.

Jonge, S.C. and Birmingham, J.M. 1992. Medicinal benefits of the mushroom *Ganoderma*. *Advances in Applied Microbiology*. **37**, 101-134.

Joshi, M.C,. N.S. Bishi and N.S.K. Harsh. 1982. Three new records of *Helvella* from India. *Curr. Sci.* **51** (9): 474–475.

Joshi, V.K., Seth, P.K., Sharma, R.C. and Sharma, R. 1991. Standardization of a method for the preparation of sweet chutney from edible mushrooms. *Indian Food Packer*, **45**: 39-43.

Kalitha, M. K., Bhagavati, K. N. and Rathaih, Y. 1997. Some edible fungal flora of Assam-new records, *Mushroom Res* **6**: 51-52.

Kalra, R. and Phutela, R.P. 1991. *Indian Mush* 145-148.

Kamatsu, N., Terakawa, H., Nakanishi, K. and Watanabe, Y. 1963. Flammulin, a basic protein of *Flammulina velutipes* with antitumour activities, *Journal of Antibiotics. Ser. A.*, **16**: 139-143.61.

Kaneda, T., Arat, K. and Tokuda, S. 1964. The effect of dried mushroom, *Cortinellus shitake* on Cholestrol metabolism in rats. *J. Jpn. Soc. Food. Nutri. 16*: 106–108.

Kaneda, T. and Oku, S. 1966. Effect of various mushroom preparations on cholesterol levels in rats. *J. Nutr.* **90**: 371–376.

Kannaiyan, S. and K. Ramasamy 1980. "A Hand Book of Edible Mushrooms", 104 pp. Today and Tomorrow's Printers and Publishers, New Delhi.

Kapoor, J. N. 2010. Mushroom Cultivation. I.C.A.R. Publication, New Delhi. pp.1–89.

Kar, A.K. and B.C. Dewan 1975. Some Discomycetes of Eastern Himalayas. *Indian Phytopath.* **28**: 296-297.

Kar, A.K. and Chakrabarti, H.S. 1977. Some wood Inhabiting Discomycetes from West Bengal. *Indian Phytopath.*, **30**: 103-105.

Kar, A.K. and Gupta, S.K. 1978. Records of *Cordyceps* from India. *Indian Phytopath.* **31**: 331-333.

Kar, A.K. and M.K. Maity. 1970. Some operculate Discomycetes from West Bengal (India). *Mycolgia* **62**: 690-698.

Kar, A.K. and Pal K.P. 1968. *Scutellinia samoensis* from West Bengal. *Indian Phytopath.* **21**: 453-455.

Kar, A.K. and Pal, K.P. 1970. Some operculate Discomycetes from West Bengal. *Mycologia* **62**: 690-698.

Katiyar, R.C. 1985. Evaluation of drying characteristics and storage behaviour of cultivated mushroom (*Agaricus bisporus* (Lange) Sing). M.Sc. thesis. Dept. of Pomology and Fruit Technology, HPKV, College of Agriculture, Solan, India.

Kaul, T.N. 1971. Mushroom rsearch at regional research laboratory, Jammu. 2nd Int. Symp. Pl. Path. New Delhi, p. 136 (Abstr.).

Kaul, T.N. 1978a. Nutritive value of some edible Morchellaceae. *Ind. J. Mushroom* **4** (1): 26-34.

Kaul, T.N. 1978b. Mushroom cultivation and rural development. *Indian Mushroom Sci.* **1**: 1-13.

Kaul, T.N. 1979. Mushroom Cultivation in the Northern India. Paper presented at the Symposium of Plant Disease Problems of North India, 5-7 October, Srinagar J and K, 5 pp.

Kaul, T.N. 1983, Cultivated Edible Mushrooms. Publication and Information Directorate, CSIR, New Delhi. pp.52.

Kaul, T.N. 1992. Conservation of Mushroom Resources of India- Introduction and Action Plan, World Wide Fund for Nature-India, New Delhi.

Kaul, T.N. 1993 -Conservation of mushroom resources in India. *Mush. Res.* **2**: 11-18.

Kaul, T.N. 2002 Biology and Conservations of Mushrooms. Oxford and IBH Publishing Co. Pvt. Ltd., New Delhi.

Kaul, T. N. and B. M. Kapur 1983. Indian Mushroom science II R. R. I. Jammu Tawi PP. 372-374.

Kaul, T. N. and B. M. Kapur 1988. Indian Mushroom Science II RR Jammu Tawai pp. 372–374.

Kaul, T.N. and J.L. Kachroo 1974. Common edible mushrooms of Jammu and Kashmir.J. Bombay *Nat. Hist. Soc.* **71** (1): 26-31.

Kaul, T.N., J. L. Kachroo and A. Raina 1978. Common edible mushrooms of Jammu and Kashmir. *Indian Mushroom Sci.* **1**: 517-529.

Kaur, Amananjeet 2000. Mushroom Flora of Patiala and its surrounding Areas. Ph.D thesis submitted to Punjabi University, Patiala.

Kaur, Amanjeet and Atri, N. S. 2002 Some interesting mushrooms from Punjab plains. *Mushroom Res.* **11**(1): 1-5.

Kaur, Munruchi, Havinder Kaur and N. S. Atri 2008. Study of Punjab Mushrooms: Three new records of Family Tricholomataceae from India. *Indian Jour. of Mushrooms* XXVI (1 and 2): 14–17.

Kaushal, S. C. and Grewal, Kamaljit 1992. Coprophilous from Chattbir. National Symaposium on Botanical Research: Trends and Achievements. 30-31 March 1992. Botany Department, Punjab University, Chandigarh. PP. 19-20 (Abstr.).

Kaviyarasan, V.M., Kumar, R. Siva and K. Natarajan 2006. *Morchella esculenta* a new record from South India. *Mushroom research* **15 (1):** 87–88.

Keil, C.B. 1991. Field and laboratory evaluation of *Bacillus thuringiensis* var. *israclensis* formulation for the control of fly pests of mushroom. *J. Econ. Nt.* **84**: 1180-88.

Kerrigan, R.W., Billette, C., Callac, P. and Velcko, A.J. 1996. In Mush Biol Mush Products (DJ Royse ed.) pp. 25–35, Penn State Univ, USA.

Kerrigan, R.W., Carvlho, D.D., Horgen, P.A. and Anderson, J.B. 1993. 1st Intern Confer on Mush Biol Mush Products p. 128 (Abst), Hong Kong.

Kerrigan, R.W. and Rose, L.K. 1989. *Mycologia* **81**: 433-434.

Khader, V. and Pandya, B.N. 1981. Acceptability studies on weaning foods and a pickle prepared out of paddy straw mushrooms and the keeping quality of the same. *Indian* J. *Mushrooms*, **7 (1 and 2):** 31-36.

Khare, K.B. 1976. Some Gasteromycetes from Uttar Pradesh. *Indian Phytopath.* **29**: 34-38.

Khanna, A.S. 1991. Nematode problems in mushrooms: Management. In Indian Mushroom 'Proceedings of National Symposium on Mushroom pp 242-44. Kerala Agric. Univ. pp. 314.

Khanna, A.S. 1993. Estimation of yield losses in white button mushrooms due to nematode infestation. *Curr. Nematol.* **4**(2): 173-76.

Khanna, A.S. and Sharma, N.K. 1988a. Pathogenicity of various *Aphelenchoides species* on *Agaricus bisporus* (Lange) Singer. *Indian Phytopath.* **41**. (3): 472-73.

Khanna, A.S. and Sharma, N.K. 1988b. Effect of population levels and time of infestation of *Aphelenchoides agarici* on the mycelial growth of *Agaricus bisporus. Nematol Mediterr.* **16** (1) 125-27.

Khurana, I.P.S.1977. Studies on the Clavarioid fungi of India. Ph. D. Thesis, Punjab University, Chandigarh (India).

Khush, R.S., Beckar, E. and Markwach 1992. *Appl. Environ Microbiol* **58** (9): 2971-2977.

Kier, L. B. 1961. Triterpenes of Poria *obligua*. *J. Pharm. Sci.* **50**: 471–474.

Kim, H.S., Kacew, S. and Lee, B.M. 1999. In vitro chemopreventative effects of plant polysaccharides (*Aloe barbardensis* Millar, *Lentinus edodes, Ganoderma lucidum,* and *Coriolus versicolor*). *Carcinogenesis* **20**: 1637-1640.

Kingstad, K.P. and Lindstrom, P.K. 1984. Spent liquors from pulp bleaching. Environ. Sci and Techno. **18**: 236A-248A.

Kirm, Y. (ed.) 1991. Impact experiences for 50 people come back from the depths of death with vegetable wasps and Tochukaso. From the last stages of cancer. Nishinhodo, Tokyo, 207 pp.

Kohlil, M.S. 1984. Judging when to pick an mushroom: An aid to picker, *Mushroom J.*, **138**: 207-09.

Kostelc, J.G. 1977. The chemical ecology of a sciarid fly, *Lcoriella mali* (Fitch). Ph.D. thesis. Penn. State Univ., 196pp.

Kriep, H. 1920. Uber morphologische and Physiologiche: Gesehlehstifferenzierung. Verhandlung. Physiologia Medizin Gesseischafft. Wurzburg. **46**: 1–18.

Krishnamohan, G. 1975. Study on paddy straw mushroom *Volvariella esculenta* (Massee) Sing. M. Sc. (Ag.) Thesis, Tamil Nadu Agric Univ., Coimbatore, Tamil Nadu, India.

Krishnamoorthy, A. S. T., Marimuthu, S. Mohan and R. Jeyarajan 1997. *Plenrotus salmones-stramineus*, an edible Oystermushroom from coconut stump. *Mushroom Res.* **6** (2): 79–80.

Krogagaard, L., 1981. Mushroom probing the gift of GABA. *Science News* **119**: 932.

Kumar, A. 1987. Studies on some mushroom families in NW. Himalayas. Ph.D. thesis, HPU Shimla.

Kumar, A. and Lakhanpal, T. N. 1993. New records of Agaricaeae from India. *Indian Phytoathology*. **46** (4): 418.

Kumar Ashok, Bhatt, R. P. And Lakhanpal, T. N. 1990. The Amanitaceae of India. Bishen Singh Mahendra Pal Singh, Dehra Dun, India. 159 pp.

Kumar Ashwani 1992. Studies on storage and dehydration of white button mushrooms, *Agaricus bisporus* (Lange) Sing. M.Sc. thesis. Dept. of Post-harvest Technology, UHF, Solan, India.

Kumar, K. R., Bano, Z., Subramaniam, L. and Anandaswami, B. 1980. Moisture sorption, storage and sensory quality studies on dehydrated mushrooms. *Indian Food Packer*, **34** (5): 3-8.

Kumar, M. and Nagia, N. 1989. Recipes for button mushrooms. *Indian J. Mushroom*, **15**: 33-35.

Kumar, S. M. and Shukla, C. S. 1995. Mushroom cultivaton in Madhya Pradesh In: Advances in Horticulture Vol. 13- Mushroom eds. K. L. Chadha and S.R. Sharma. Malhotra Publish. House, New Delhi, PP 277-316.

Kumar, S. M., Shukla, C. S. and Agarwal, K. C. 1991. In: Indian Mushroom 1991, Kerala Agricultural University, Vellanikkara, PP 6-7.

Kumar, T. K., Arun and P. Manimohan 2009. Rediscovery of *Trogia cyanea* and a record of T. infundibuliformis (Marasmiaceae, Agaricales) from Kerala, State, India. *Mycotaxon*. **109**: 429-436.

Kundalkar, B.D., M.S. Patil and S.V. Jagdale 1983. New edible *Tremella* from India. Jubilee symposium on "Science and cultivation Technology of edible Fungi" p80 (Abstr.).

Kundalkar, B.D., S.V. Jagdale and M.S. Patil 1983. New edible *Tremella* from India. In: Indian Mushroom science II eds. T. N. Kaul and B. M. Kapur, R. R. Jammatauri P.P 399–401.

Kurashiga, S., Akuzawa, Y. and Eudo, F. 1997. Effects of *Lentinus edodes, Grifola frondosa* and *Pleurotus ostreatus* administration on cancer outbreaks and activities of macrophages and lymphocytes in mice treated with a carcinogen N-butyl-N 1 - butamolinitreso-amine. Immunopharmacology and Immunotoxicology **19**: 175-183.

Labarere, J., Iracabal, B. and Maleville, H. 1993. Report Tottori Mycol Institute **31**: 168-187.

Lakhanpal, T. N. 1986. Survey and studies on mushrooms and toadstools on North West Himalayas. Final progress report DST project H. P. University, Shimla.

Lakhanpal, T.N. 1988. Morphology and anatomy of mycorrhizal fungi from India. In: mycorrhiza Round Table-Being the proceedings of National workshop on mycorrhizae held at JNU, New Delhi March 13-15, 1987. (A.K. Verma et. al. Eds.), IDRC, Canada, pp. 53-83.

Lakhanpal, T. N. 1990. Advances and their cultivation technology. Summer Institute. Recent development in cultivation technology. Edible Mushrooms, NCMRT, Solan, 277-281pp.

Lakhanpal, T.N. 1992. In: Conservation of Mushroom Resources of India-Introduction and Action Plan. World Wide Fund for Nature-India, New Delhi.

Lakhanpal, T.N. 1993. The Himalayan Agaricales Status of Systematics. Mushroom Research **2**: 1-10. A.P.H. Publishing Corporation, New Delhi.

Lakhanpal,T.N.1994. The family Boletaceae in India (in press) Published in 1996.

Lakhanpal TN. 1995. Mushroom flora of North West Himalayas. In: Advances in Horticulture. **13**: 350-373-Mushroom (K.L. Chadha and SR. Sharma, Eds.). Malhotra Publishing House, New Delhi.

Lakhanpal, T.N. 1996. Studies in Cryptogamic Botany: Vol. I: Mushrooms of India: Boletaceae. 170 pp. A.P.H. Publishng Corporation, New Delhi.

Lakkhanpal, T.N. 1997. Diversity of Mushroom Mycoflora in the North Western Himalayas. In: Recent Researches in Ecology Environment and Pollution. Vol. 10. (S.C. Sati et al, Eds.) pp. 35-68. Today and Tomorrow Printers and Publishers, New Delhi.

Lakkhanpal, T.N. and Atri, N.S. 2000. Conservation of Mushroom Biodiversity. 22nd Annual Conference of Indian Society of Mycology and Plant Pathology and Symposium on Mushroom Research in 21st Century May 3- 4. National Research Centre Mushrooms, Chambaghat, Solan. Abstract in *J. Mycol. P1., Pathol.* **30**(2): 279.

Lakhanpal, TN., Bhatt, R.P. and Kamaraja Kaisth. 1987. *Lactarius sanguifluus* Fr.: an edible mushroom new to India. *Curr. Sci.* **56**(3): 148-149.

Lakhanpal, T.N., Kumar, A., and Kaisth, K. 1986. Fleshy fungi of North Western Himalayas-I. A temperate white form of *Volvariella bombycina. Indian J. Mush.* **12**: 1-4.

Lakhanpal, T.N. and Shad, O.S. 1986a. Studies on wild edible mushrooms of Himachal Pradesh (NW. Himalayas)-I. Ethnomycology, production and trade of *Morchella* species (Gucchi). *Indian J. Mush.* **12-13**: 5-14.

Lakhanpal, T.N. and Shad, O. S. 1986b. Studies on wild edible mushrooms of Himachal Pradesh (NW. Himalayas)-II. Ecological relationships of *Morchella* species. *Indian J. Mush.* 12-13.

Lakhanpal, T.N. and Shad, O.S. 1999. A re-appraisal of the genus *Morchella* in India. In Advances in microbial Biotechnology (J.P. Tiwari *et al.*, Eds.) A.P.H. Publishing Corp. New Delhi.

Lakhanpal, T.N. and Sagar, A. 1989. Fleshy fungi of N.W. Himalayas–XIV Three species of *Boletus* new to India. *Kavaka* **17**(1, 2): 32-37.

Lakhanpal, T.N. and Sharma, R. 1988. Fleshy fungi of N.W. Himalayas–XVI: The genus *Strobilomyces* (Boletaceae) *Kavaka* **16**(1, 2): 27-35.

Lakhanpal, T.N., Sharma, R. and Kumar, A., 1987. *Boletus edulis* Bull. ex. Fr.: An edible mushroom new to India. *Curr. Sci.* **57**: 611- 612.

Lakhanpal, T.N., Sharma. R. And Kumar, A. 1988. *Boletus edulis* Bull. ex Fr.–An edible Mushroom new to India. *Curr. Sci.* **57** (11): 611-612.

Lakhanpal, T. N., Sharma, R. and Sharma, J. R. 1985. Fleshy fungi of North Western Himalaya III- the genus *Gyroporus* (Boletaceae) *Kavaka* **13** (2): 91–93.

Lal, B.B., Katyar, R.C. and Jandaik, C.L. 1990. Some studies on mechanical drying of *Agaricus bisporus*, XXIII Int. Hart. Congress. Abstracts of contributed paper I. Poster Frenze (Italy). 27 August-i, September, 3383.

Lange, M. and Smith, A.H. 1953. The *Coprinus ephemerus* group. *Mycologia* **57**: 747-780.

Langton, F. A. and Elliot, T. J. 1980. Genetics of secondarily homothallic basidiomycetes, Heredity **45**(1): 99-106.

Law, W. M., Lau, W. N., Lokl. Wai, L. M., Chiu, S. W. 2003. Removal of biocide pentacholorophenol in water system by the spent mushroom compost of *Pleurotus pulmonarius*, *Chemosphere*, **52**: 1531-1537.

Leelavathy, K. M., Little, Flower Sr. and Suja. C.P. 1983 The genus *Termitomyces* in India. In "Indian Mushroom Science II" (eds. T.N. Kaul and B.M. Kapur) RRI, Jammu Tawi pp. 401–407.

Leelavathy, K. M. and Sr. Little Flower 1986. The genus *Micropsalliota* in India. *Kavaka*, **14**: 17–23.

Li, A. and Hargen, P.A. 1993. Cultivated Mush Newsletter **1**: 11-16.

Lind, R. 1993. Control of mushroom flies with the predatory mite. *Hypoaspis miles.* Contract Rep. Hort. Dev. Council 28pp.

Lintzel, W. 1941. Uber den Nährwert des Eiweisses der speisepilze. Biochem. Z **308**: 413-419. 158-162.

Litchfield, J.H. 1964. Nutrient content of morel mushroom mycelium. B-vitamin composition. *J. Food Sci.*, **29** (1): 690-691.

Liu, B. and Bau, Y.S. 1980. Fungi Pharmacopoeia (Sinicas) Kinoko Co., Oakland, California.

Liu, G.T. 1999. Recent advances in research of pharmacology and clinical applications of *Ganoderma* (P. Korst.) species (Aphyllophoromycetideae) in china. *International Journal of Medicinal Mushrooms* **1**: 63-67.

Lloyd, C.G. 1904-1919. Mycological Letters, 1- 69 (each separately paged), Cincinnati, Ohio, USA.

Lloyd, C.G. 1906. The Nidulariaceae or birds nest fungi. pp. 1-32, 10 pl, 20 figs.

Lloyd, C. G. 1912. Synopsis of the stipitate Polyporoids pp–95–208, 206 figs.

Lloyd, C.G. 1915. Synopsis of the genus *Fomes* pp. 211-288, 41 figs.

Low, Y. B. 1951. A morphological basis for classifying the species of *Auricularia. Mycologia* **43**: 351-368.

Maeta, Kazuhiko, Tomoya ehi, Keisuka Tokimoto, Norihiro Shimomura, Nitaro Maekawa, Nobuhisa Kawaguchi, Makoto Nakaya, Yutaka Kitamoto and Tadanori Aimi 2008. Rapid species identification of cooked Poisonous Mushrooms by using Real Time PCR, *Appl. Environ. Microbiol.* **74** (10): 3306–3309.

Magae, Y., Haga, K., Taniguchi, H. and Sasaki, T., 1990. *Gen Appl Microbiol* **36**: 69-80.

Maini, S.B., Sethi, V., Diwan, B. and Munjal, R.L. 1983. Post-harvestwashing of mushrooms to enhance their shelf life and marketability. 3rd Convention of Association of Food Scientists and Technologists(India), Mysore, 2-4 June (abstr.).

Maini, S.B., Sethi, V., Diwan, B. and Munjal, R.L. 1987. Pre-treating mushrooms to enhance their shelf-life and marketability. *Indian Mushroom Sci.*, **2**: 215-16.

Mane Asha, Nilima Wankhade, Archana Agarkar and G. K. Mane 2003. Oyster mushroom powder: an effective supplement for salty bread of Indians. In: Current Vistas in Mushroom Biology and Production. R. C. Upadhyay, S. K. Singh, and R. D. Rai (Eds) *Mushroom Society of India*, Solan PP 257-260.

Manimohan, P., Vrinda, K. B. and Leelavathy. K. M. 1988. Rare Agarics from Southern India. *Kavaka* **16**(1,2): 50-56.

Manju, N.A. 1933. A contribution to our knowledge of Indian coprophilous fungi. *J. Indian Bot. Soc.* **12**: 153-164.

Manjula, B. 1983. A revised list of agaricoid and boletoid Basidiomycetes from India and Nepal. Roc, *Indian Acad. Sci* (Plant Sci.) **92**(2): 81-213.

Manoharachary, C. and Vijaya Gopal, 1991. Mycofloristics of Agaricales from Andhra Pradesh. In: Indian Mushrooms (MC. Nair Ed. in Chief) Kerala Agricultural University, Vellanikkara, Thiruvanathapurum, pp. 3-5.

Massee, G. 1912. *Fungi Exotici. Kew Bull.* 253-255.

Mathur, R.S. 1936. Notes on the fungus flora of Lansdowne (Garhwal District). *Proc. Ind. Sci.Congr.* **23**: 289.

Matti, Kreula, Maija Saarivirta and Sirkka Lüsa Karanko 1976. On the composition of nutrients in wild and cultivated mushrooms. *Karstenia* **16**: 10-14.

Mc Connell, J.E.W. and W.B. Esselen, 1947. Carbohydrates in cultivated mushrooms. *Food Res.* **12**: 118-121.

Mc. Leod, R. W. 1967. The effect of *Aphelenchoides bicaudatus* a parasite of cultivated mushroom. *Nature* **214**: 1164-65.

Mc. Leod, R. W. 1968. The effects of *Aphelenchoides saprophilus, A. coffeae, Aphlenchus avenae* and *Panagrolaimus* sp. on the cropping of cultivated mushroom. *Nematologica* **14**: 573-76.

Mehta, K. B. 1991. Annual Report of N CMRT, Solan pp. 11-13.

Mehta, K.B. and Jandaik, C.L. 1989. Storage and dehydration studies of fresh fruit bodies of dhingri mushroom-*Pleurotus sapidus*. *Indian J. Mushroom*, **15** 17-22.

Mendel, J.B. 1898. The chemical composition and nutritive value of some edible American fungi. *Am. J. Physiol.* **1**: 225-238.

Miles, P.G and S.T.Chang 2004 Mushroom cultivation, nutritional value, medicinal effect and environmental impact. Boca Raton, F1: CRC Press ISBN 0-8493-1.

Miller, F. C. 1993. Biological control of bacterial blotch. *Mush. News* **41**(8) 18-21.

Miller, R. E. 1971 Evidence of sexuality in the cultivated mushroom *Agaricus bisporus*. *Mycologia* **63**: 630-634.

Mishra, Rishikesh 1999. Studies on some edible mushrooms of Arunachal Pradesh. Ph. D. Thesis submitted to B.R.A. Bihar University, Muzaffarpur.

Mitra, A.K. 1994. Studies on the uptake of heavy metal pollutants by edible mushrooms and its effect on their growth, productivity and mammalian system. Ph.D Thesis, University of Calcutta.

Mitter, J.H. and Tandon, R.N. 1932. Fungus flora of Nainital-I, J. *Ind. Bot. Soc.* **11**: 178-180.

Mitter, J.H. and Tandon, R.N. 1938. Fungi of Nainital Part III, *J. Ind. Bot. Soc.* **17**: 177-182.

Mitra, A.K. and R. P. Purkaystha 1995. Heavy metals and mushrooms–A review. *Mushroom Res.* **4**: 43-48.

Mizuno, T. 1995. Yamabushitake, *Hericium erinaceus*: bioactive substances and medicinal utilisation. *Food Review International* **11**: 173-178.

Mizuno, T. 1999. Bioactive substances in *Hericium erinaceus* (Bull.: Fr.) Pers. (Yamabushitake), and its medicinal utilisation. *International Journal of Medicinal Mushrooms* **1**: 105-119. (contains list of numerous Japanese Patents).

Mizuno, T. 2000. Development of an antitumour biological response modifier from *Phellinus linteus* (Berk. Et Curt.) Teng (Aphyllophoromycetideae) (Review). *International Journal of Medicinal Mushrooms* **2**: 2 1-33.

Mizuno, T., Wasa, T., Ito, H., Suzuki, T. and Ukai, N. 1992. Antitumour-active polysaccharides isolated from the fruiting body of *Hericium erinaceus*: an edible and medicinal mushroom called Yamabushitake or Houton. *Bioscience Biotechnology and Biochemistry* **56**: 347-348.62.

Mohan, M., Meyer, R.J., Anderson, J.B. and Horgen, P.A. 1984. *Curr Genet* **8**: 615-619.

Moncalvo, J.M. 1995. *Mycol. Res* **99**: 1479-1482.

Moncalvo, Jean-Marc, Timothy, J. Baroni, Rajendra P. Bhatt and Steven, L. Stefenson 2004. *Rhodocybe paurii*, a new species from the Indian Himalayas. *Mycologia* **96**(4): 859-865.

Mondal, T. and R.P. Purkayastha. 1983. New addition to Indian edible fungi. *Indian Phytopath.* **36** (4): 736-738.

Mooibroek, H., Vande Rhee, M.D., Rivas, C.S., Mendes, O., Werten, M., Huizing, H. and Wichers, H. 1996. In Mushroom Biol. Mush. Products (DJ Royse ed.) pp. 37-46, Penn State Univ, USA.

Moore, A.J., Challen, M.P. and Elliott, T.J. 1995. *Mush Sci* **14** (part 1): 63-70.

Moser, M. 1949. Untersuchungen uberden Einfluss von Waldbranden auf die Pilzvegetation *J. Sydowia* **3**: 336-383.

Moses, S.T. 1948. A preliminary report on the mushrooms of Baroda. Department of Fisheries. Borada State, Bulletin No. **XIV**: 1-3.

Motskus, A.V. 1977. Biochemical analysis of mushrooms in the order Agaricales: 5. Concentration of protein substances in the fruit-bodies of *Paxillus involutus* and amino acids in their hydrolysates. Liet Tsr Mokslu Akad. *Drab Ser C Biol. Mokslai* (**3**): 123-128.

Mudahar, G.S. and Bains, G.S. 1982. Pre-treatment effect on quality of dehydrated *Agaricus bisporus* mushrooms. *Indian Food Packer*, **28** (5): 19-27.

Mukerji, K.G. and R.C. Juneja 1974. "Fungi of India" supplement to the list of Indian fungi 1962-1972, 221 pp. Emkay Publications, Delhi.

Mundkur, B.B. 1938. Fungi of India. Supplement-I, I.C.A.R. Sci. Monogr. **12**: 54 pp.

Munjal, R.L. and S.C. Chatterjee 1971. Some observations on cultivation of paddy straw mushroom (*Volvariella volvacea*). 2nd Int. Symp. P1. Path. New Delhi: 137 (Abstr.).

Munjal. R.L. and A.D. Sharma 1975. Some observations on the cultivation of *Morchella*. First workshop on mushroom research, Himachal Pradesh University (Agricultural Complex). Solan sponsored by I.C.A.R. (Abstr.).

Munjal, R.L., Shandilya and Seth, P.K. 1974. Mycoflora in mushroom compost. *Ind. J. Mycol. and Pl. Pathol.* **4**: 204-205.

Munoz-Rivas, A., Specht, C.A., Drummond, B.J., Froeliger, V. E. and Novotony, C.P. 1986. *Mol Gen Genet* **250**: 103- 106.

Murrill, W.A. 1924. Kashmir fungi. *Mycologia* **16**: 133.

Musilek, V., Cerna, V.J., Sasek, V. and Semerdzieva and Vondracek, M. 1969. Antifungal antibiotic of the basidiomycete *Oudeman siella mucida*. I. Isolation and cultivation of a producing strain. *Folia Microbiol.* (Prague) **14**: 377–387.

Nag, T. K., K. S. Chauhan and B. L. Jain 1991. Studies on mushroom mycoflora of E. Rajasthan -I Some Agaricales from Jaipur district. *Adv. Mushroom Sci.* P. 7 (Abstr.).

Nair, L.N. and D. S. Patil 1978. Fleshy fungi from Western India. Gasteromycetes. *Kavaka* **5**: 19-23.

Natarajan, K. 1975. South Indian Agaricales I. *Kavaka* **3**: 63-66.

Natrajan, K. 1977a. A new species of *Termitomyces* from India. *Curr. Sci.* **46**(19): 679-680.

Natarajan. K. 1977b. South Indian Agricales-III. *Kavaka* **5**: 35-39.

Natarajan, K. 1978. South Indian Agaricales VI. *Kavaka* **6**: 65-70.

Natrajan, K. 1979. South Indian Agaricales-V *Termitomyces heimii*. *Mycol*. **71**: 653-655.

Natrajan, K. 1995. Mushroom flora of South India (except Kerala). In: Advances in Horticulture **13**: 387-397- Mushroom (K.L. Chadha and S.R. Sharma, Eds.). Malhotra Publishing House, New Delhi.

Natarajan, K. and B. Manjula 1978. Studies on *Lentinus* polychorus Lev. *Squarrosulus* Mont. *Indian Mushroom Sci*. **1**: 451–53.

Natarajan, K. and B. Manjula 1982. South Indian Agaricales XV. *Indian Phytopath*. **35** (1): 57-64.

Natrajan, K., Mohan, V. and Kaviyaresan, V. 1988. On some ectomycorrhizal fungi occurring in southern India. *Kavak* **16** (1 and 2): 1–7.

Natarajan, K. and Purshothama, K. B. 1986. South Indian Agaricales XXI, *Kavaka* **14** (142): 47-60.

Natarajan, K. and Raman, N. 1980. South Indian Agaricales IX *Sydowia* **33**: 225-235.

Natarajan, K. and Raman, N. 1981. *Lentinus cladopus*–a potential edible mushroom. 3rd Int. Symp. Pl. Path., New Delhi. (Abstr.).

Natrajan, K. and Raman. N. 1983a. South Indian Agaricales–A preliminary study on some dark spored species: *Bibliotheca Mycologica* Vol **89**. *J. Cramer Vaduz*. 203 pp.

Natrajan, K. and Raman, N. 1983b. South Indian Agaricales XX–some mycohizal species. *Kavaka* **11**: 59–66.

Natrajan, K. and Raman. N. 1984. Occurrence of *Pleurotus cystidiosus* in India. *Curr. Sci*. **53**(12): 658-659.

Neda, H. and Nakal, T. 1995. *Mush. Sci* **14**: 161-168.

Newstead, R. 1906. A large crop of mushroom destroyed. Q Fi. Inst. *Comm. Res. Trop.* **1**: 23.

Newton, 1986. In Rangaswami, G. 1956 Studies on *Volvariella displasia* (Berk and Br.) *Madras Agric. J.* **43**: 182-191.

Ng., T.B. 1998. A review of research on the protein-bound polysaccharide (polysaccharopeptide PSP) from the mushroom *Coriolus versicolor* (Basidiomycetes: Polyporaceae). *General Pharmacology* **30**: 1-4.

Nita, Behl 1987. Protein and lipid supplementation for increased yield of *Agaricus bisporus*. Development in Crop Sciences **10**, Cultivation of Edible Fungi edited by P. J. Wuest, D. J. Roys and R. B. Beelman. The Pennsylvania State University Park. Pennsylvania 1680 U. S. A. PP. 242-244.

Nita, Behl 1988. Hand Book of Mushrooms. Oxford and IBH Publish. House Co. Pvt. Ltd. PP. 105-114.

Ohenoja, E. 1988. Effect of forest management procedures on fungal fruit body production in Finland, Acta, Bot. Fenn. **136**: 81-84.

Ohmasa, M. and Furukawa, H. 1986. *Trans Mycol Soc Japan* **27:** 79-90.

Padwick, G.W. and Merh, J.L. 1943. Notes on Indian fungi *J. Mycol. Pap.* I. M. I. No. **7**: 7 pp.

Paesler, F. 1957. Beschreibung Einiger Nematoden Aus Champgnon Beetwen. *Nematologica* **2**: 314.

Pandey, M.C. and Aich, J.C. 1989. Equilibrium moisture content of dehydrated mushrooms (*Pleurotus sajor-caju*). *J. Food Sci. Technol.*, **26**(2): 108-10.

Pandotra, V.R. 1966. Notes on fungi of Jarnmu and Kashmir *J. Proc.Acad. Sci.* **54**: 68-73.

Paracer, C.S. and D.S. Chahal. 1962. A new edible species of genus *Agaricus* in Punjab (India). *Myocopath. Myol. appl.* **18**(4): 267-270.

Parndekar, S.A. 1964. A contribution to the fungi of Maharashtra. *J. Univ. Poona*. **26**: 56-64.

Pathak, N.C., Ghosh, R. N. and Singh, M.S. 1976. The genus *Volvariella* in India In: Indian Mushroom Science (eds. Atal, C.K., Bhat, B.K. and Kaul, T. N.) pp. 295–301 Indo American Literature House, New Delhi.

Pathak, N.C. and Gupta, Savita 1982. Agaric flora of Lucknow. In "Advances in Mycology and Plant Pathology" p. 76. Eds. S. B. Chattopadhyay, N. Sarnajpati and. Sri Raju Primlani of Oxford and IBH Publishing Co.

Pathak, N.C., R.N. Ghosh and M.S. Singh 1978. The genus *Volvariella* Speg. in India. *Indian Mushroom Sci*. **1**: 295-303.

Patil, B.D., Jadhav, S.W. and Sathe, A.V. 1995. Mushroom flora of Maharashtra. In: Advances in Horticulture. **13**: Mushrooms (K.L. Chadha and S.R. Sharma, Eds.) Malhotra Publishing House, New Delhi. pp 317-328.

Patil, S.D., L.N. Nair and B.P. Kapdnis 1979. Studies on fleshy fungi of Western India. I. Univ. Poona. Sci. Tech. **52**: 349-354.

Pegler, D.N. 1975. The classification of the genus *Lentinus* Fr. (Basidiomycota). *Kavaka* **3**: 11-20.

Pegler, D.N. 2002. Useful fungi of the world: The poor man's truffle and Arabia' and Manna' of the Israellites. *Mycologist* **16**: 8-9.

Pegler, D. N., B. M. Spooner and Young, T. W. K. 1993. British Truffles. A revision of British hypogeous fungi pp. 216 Kew Royal Botanic Gardens.

Pegler, D. N., D. J. Lodge and K. K. Nakasone 1998. The pantropical genus *Macrocybe.* gen. nov., *Mycologia*, **90** (3): 494–504.

Pegler, D. N. Rayner, W. 1969. *Kew Bulletin*. **23**: 347–412.

Pegler, D. N. and Vanhaecke, M. 1994 *Termitomyces* of Southeast Asia. *Kew Bulletin* **49** (4): 717-7 36.

Petch, T. 1916. Ceylong Lentini. *Ann. R. Bot. Gard. Peradeniya* **6**: 145-152.

Pethybridge, N. J. 1991. The systematic use of bioassay methods to determine the effectiveness of *Bacillus thuringiensis* isolates for the control of the mushroom sciarid *Lycoriella auripila* (Diptera: Sciaridae). *Mush. Sci.* **XIII**: 453-56.

Pighi, L., Puempel, T. and Schinner, F. 1989. *Biotechnol Lett.* **11**: 275-280.

Poinar, G. O. 1979. Nematodes for Biological Control Insects. Florida: CRC Press Inc.

Poitou, N. and Olivier, J.M. 1990. *Agric Ecosys Environ.* **28** (1-4): 403-408.

Pomerleau, R. 1951, "Mushrooms of Eastern Canada and the United States", 300 pp. Les Editions Chantecler. Ltee, Montreal.

Pradeep, C.K., Vrjnda, K.B., Simimathew and Abraham, T.K. 1998. The genus *Volvariella* on Kerala State, India. *Mushroom Res.* **7(2)**: 53-62.

Prasad, N.N. 1971. Nematode infection on *Volvariella diplasia*. 2nd Int,. Symp. *Pl. Path.*: 137 (Abstr.).

Pruthi, J.S., Gopalkrishnan, M. and Bhat, A.V. 1978. Studies on the dehydration of tropical paddy straw mushroom (*Volvariella volvacea*). *Indian Food Packer*, **32**(2): 7-15.

Pruthi, J.S., Manan, J.K., Raina, B.L. and Teotia, M.S. 1984. Improvement in whiteness and extension of shelf-life of fresh and processed mushrooms (*Agaricus bisporus* and *Volvariella volvacea*). *Indian Food Packer*, **38** (2): 55-63.

Puri, Y.N. 1955. Rusts and wood rotting fungi on some of the important Indian conifers. For, Bull. Dehra Dun (N.S.) *Mycology* **179**: 10 pp.

Puri, Y.N. 1956. Studies on Indian *Poria*. *J. Indian Bot. Soc* **35**: 277-283.

Puri, Y.N., P.S. Rehill and B. Singh. 1981. Cultivation trials of *Pleurotus fossulatus*. *The Mushroom Journal* **102**: 209-214.

Purkayastha, R.P. and Aindrila Chandra 1974. New species of edible mushroom from India. *Trans. Br. Mycol. Soc.* **62** (2): 415-418.

Purkayastha. R,P. and Aindrila Chandra 1975a. Recent studies of some Common edible mushrooms of West Bengal with special reference to growth and improvement of *Calocybe indica* P. and C. First workshop on mushroom research, Himachal Pradesh University (Agicultural complex), Solan; sponsored by I.C.A.R. (abstr.).

Purkayastha, R.P. and Aindrila Chandra 1975b. *Termitomyces eurhizus*, a new Indian edible mushroom. *Trans. Br. Mycol. Soc.*: **64** (1): 168-170.

Purkayastha, R.P. and Aindrila Chandra 1976. A new technique for in vitro production of *Calocybe indica*–an edible mushroom of India. *The Mushroom Journal* **40**: 1-2.

Purkayastha, R. P. and Aindrilla Chandra 1985. International Bioscience Monograph–16: Manual of Indian Edible Mushroom. Today and Tomorrow Printers and Publ., New Delhi.

Purkayastha, R.P. and K.K. Jana 1983. A step forward for the commercial cultivation of *Pleurotus eous*. *The mushroom Journal* (In press).

Rahi, Deepak K., Kamlesh K., Shukla, R. C. Rajak and A. K. Pandey 2003. Agaricales of Central India–I Two new species. *Indian J. Mushroom* XXI (1 and 2): 29–31.

Rai, M. K. 1997. Wild edible mushrooms of tribal areas of Seoni district, Madhya Pradesh. *Mushroom Res.* **6**(2): 107-108.

Rai, M.K., Jagdish Singh and Deepak Acharya 1999. *Morchella conica*, a new report from Madhya Pradesh. *Mushroom Res.* **8**(1): 61-62.

Rai, R. D., Chandrashekhar, V. and Arumuganathan, T. 2003. Post harvest technology of mushrooms. In: Current Vistas in Mushroom Biology and Production. R. C. Upadhyay, S. K. Singh and R. D. Rai (Eds) Mushroom Society of India, Solan PP 225-236.

Rai, R. D. and Saxena, S.1988. Effect of storage temperature on vitamin C content of mushrooms (*Agaricus bisporus*). *Curr. Sci.*, **57** (8): 434-35.

Rai. R.D. and Saxena, S. 1989. Biochemical changes during post-harvest storage of button mushroom *Agaricus bisporus*. *Curr. Sci.*, **59**: 508-10.

Rajarathnam, S., Bano, Z. and Patwardhan, M.V. 1983. Post-harvest physiology and storage of the white oyster mushroom (*Pleurotus flabellatus*). *J. Food Sci. Technol.*, **18**(2): 153-62.

Rajarathnam, S. and Zakia Bano 1991. Biological utilization of edible fruiting fungi. In: "Hand Book of Applied Mycology-Foods and Feeds." eds. DilipK. Arora, K. G. Mukerjee and Elmer, H. Marth. Vol. 3, pp. 241-275.

Ramachandra, B.S. and Ramanathan, P.K. 1978. Production technology of dehydrated foods. *Indian Food Packer*, **32**(1): 103-107.

Ramakrishnan, K. and Subramanian, C.V. 1952. The fungi of India–a second supplement. *J. Madras Univ.* **21** B: 303-305.

Ramakrishnan, T.S. and C.V. Subramanian. 1952. The Fungi of India. A second supplement. *J. Madras Univ.* B. **22**: 1-65.

Ramakrishnan, T.S., K.V. Srinivasan and N.Y. Sundaram 1952. Addition to the fungi of Madras–XIII. *Proc Indian. Acad. Sci. Sec.* B. **36**: 85-95.

Ramasamy, K. and Kandaswamy, T.K. 1978. Possible causes for the quick deterioration of quality of paddy straw mushroom in storage. *Indian Mushroom Sci.*, **1**: 329-35.

Rangaswami, G. 1956. Studies on *Volvariella diplasia* (Berk and Br.) The paddy straw mushroom. *Madras Agric. J.* **43**: 182-191.

Rangaswami, G., Seshadri, V.S. and Lucy Channamma K.A. 1970. Fungi of South India. University of Agricultural Sciences, Bangalore pp. 193.

Rangnathan, R. and K. Swaminathan 2004 Biological treatment of a pulp and industry effluent by *Pleurotus* spp. World Jour. of Microbio. and Biotech. **20**: 383-393. Kluwer. Academic Publishers, Netherland.

Rao, M.S. and Pandey, M. 1991. Comparative efficacy of Karanj leaf and carbofuran on *Aphelenchoides composticola*. *Indian J. Nematol.* **21**: 158-59.

Rao, M. S., R. P. Tewari and M. Pandey 1992. Effect of Saprophytic nematode *Rhabditis* sp. on the yield of *Agaricus bisporus. Mush. Res.* **1**: 59-60.

Rao, M.S., Reddy, P.P. and Tawari, R.P. 1991. Comparative efficacy of certain oil cakes on nematode *Aphelenchoides sacchari* and their effect on yield of *Agaricus bisporus*. **21** 101-6.

Rao, T.A. 1964. Some interesting Gasteromycetes in Maharashtra, *India. Botanique* **7** (1): 30- 36.

Raper, C.A/ 1976. The Biology and Cultivation of Edible Mushrooms (ST Chang and WA Hays eds.) pp. 83- 117, Academic Press, New York, USA.

Raper, C.A., Raper, J.R. and Miller, R.E. 1972. *Mycologia* **64**: 1088-1117.

Rath, G.C. 1962. The Genus *Volvariella* Speg. in Lucknow. *J. Indian Bot. Soc.* **41**: 524-530.

Rawla, G. S. and A. M Narula 1983. Some edible Boletes from India. In Indian Mushroom Science II, Eds. T. N. Kaul and B. M. Kapur PP- 408-410.

Rawla, G.S. and S. Arya 1983. Some edible and poisonous *Agaricus* new to India In: Indian Mushroom Science II (eds. T. N. Kanl and B. M. Kapur), RRI, Jammutauri pp. 375–382.

Rawla, G. S., Sarwal, B. M. and Arya, S. 1982. Agarics new to India I. *Nova Hedwigia* **36**: 433-43.

Rawla, G. S. and Sarwal, B. M. 1983. *Bibliotheca Mycologica* **91**: 23-46.

Rawla, G.S., M.P. Sharma and A.M. Narula. 1983. Some edible boletes from India In: Indian Mushroom Science II (eds. T. N. Kanl and B. M. Kapur), RRI, Jammutauri pp. 408–410.

Ray, S. and N. Samajpati 1979. Gasteromycetes of West Bengal. *Ind. J. Mycol. Res.* **17**: 13-20.

Ray, S. and N. Samajpati 1980. Agaricales of West Bengal–V. *Ind. J. Mycol. Res.* **18**: 33-37.

Razak, A.A. 1989. *Biol Trace Elem Res* **22** (3): 277-286.

Rea, C. 1922. British Basidiomycetes XI + 799 pp. Cambridge.

Reed, J. N., Crook, S., He, W. 1995. Harvesting mushroom by robot. *Mushroom Sci.* **15**: 385-391.

Reed, J. N., Miles, S. J., Butler, J., Baldwin, M. 1997. Influence of mushroom strains and population density on the performance of robotic harvester. *J. Agric, Eng. Res.* **68**: 215-222.

Reed, J. N., Miles, S. J., Butler, J., Baldwin, M., Noble R. 2001. Automatic mushroom harvester development. *J. Agric, Eng. Res.* **28**: 15-23.

Reusser, F., J.F.T. Spencer and H.R. Sallans 1958. Protein and fat content of some mushrooms grown in submerged culture. *Appl. Microbiol.* **6** (1): 1-4.

Richardson, P.N. 1987. Susceptibility of mushroom pests to the insect-parasitic nematodes *Steinernema feltiae* and *Heterorhabditis heliothidis*, Ann. Appl. Biol. **111**: 433-38.

Richardson, P. N. and Chanter, D.O. 1979. Phorid fly (Phoridae: *Megaselia halterata*) longevity and the disssemination of nematodes (Allantonematideae: *Howardula hysseyi*) by parasitized females. *Ann. Appl. Biol.* **93**: 1-11.

Richardson, P.N. and Grewal, P.S. 1991. Comparative assessment of biological (Nematoda: *Steinernema feltiae*) and chemical methods of control for the mushroom sciarid *Lycoriella auripila* (Diptera: Sciaridae). *Bio. Sci. Tech.* **1**: 217-28.

Robbins, W. J., Hervey, R. W., Davidson, R. W., Ma, R. and Robbins, W.L. 1945. A survey of some wood destroying and other fungi for antibacterial activity. *Bull. Torrey Bot. Club.* **72**: 165–190.

Roland, J. F., Chinielewlcz, Z. F., Weiner, B.A., Gross, A.M., Boening, O.P., Luck, L.Y., Bardos, T.J., Reilly, H.C., Suguira, K., Stock, C.C., Lucas, E.H., Byerrum, R.U. and J.A. Stevens. 1960. Calvacin a new antitumor agent. *Science* **132**: 1897.

Romaine, C.P., Schlagnhaufer, B. and Goodin, M.M. 1994. *Curr Genet* **25**: 128-134.

Roux, P. and Labarere, J. 1990. *J. Exp Mycol.* **14**: 101-102.

Roy, A. and N. Samajpati 1978. Agaricales of West Bengal 11. *Ind. J. Mushrooms.* **4** (2): 17- 23.

Roy, A. and Samajpati, N. 1979. Gasteromycetes of West Bengal, *Indian J. Mycol. Res.* **17**: 13-20.

Roy, A. and Samajpati, N. 1981. Edible mushrooms of West Bengal IV. *Termitomyces striatus* (Beeli) Heim. *Indian J. Mycol. Res.* **19**(2): 47-50.

Roy, M. K. and Bahl, Nita 1984. Studies on gamma radiation preservation of *Agaricus bisporus. The mushroom Journal* **142**: 411-414.

Royer, J.C., Hintz, W.E., Kerrigan, R.W. and Horgen, P.A. 1992. *Genome* **35**: 694-698.

Royer, J.C. and Horgcn, P.A. 1991. in Genetics and Breeding of *Agaricus* (LJLD Van Griensven ed.) pp. **13**: 5-139, PUDOC, Wageningen, The Netherlands.

Royse, D.J. and May, B. 1993. In Genetics and Breeding of Edible Mushrooms (ST Chang PA Buswell and PG Miles eds.) pp. 225-248, Gordon and Breach Science Publishers.

Ruelius, H. W., Jannssen, F.W., Kerwin, R.M., Goodwin, C.W. and Schilling, R.T. 1968. Poricin, an acidic protein with antitumor activity from a basidiomycete I. Production, isolation and purification. *Arch. Biochem. Biophys.* **125**; 126–135.

Saccardo, P.A. 1925. "Syloge Fungorum". XXIII.

Sagar, A., Champa, G. and A. K. Shehgal 2007. Studies on some medicinal mushrooms of Himachal Pradesh. *Indian Jour, of Mushrooms.* XXV (1 and 2): 8–14.

Sagar, A. and Lakhanpal, T. N. 1989. Edible species of boltetes in N. W. Himalays. *Indian J. Mushroom* **15**: 1-3.

Sagar, A. and Lakhanpal, T. N. 1993. Fleshy fungi of N.W. Himalayas–XV. Six species of *Boletus* new to India. *Indian J. Mycol. P1. Pathol.* **23** (3): 227–231.

Sagawal, T. M. and Nagata,Y. 1992. *J Gen Appl Microbiol* **38**: 597-603.

Saini, S. S. and Atri, N. S. 1981. *Russula foetens* (Pers.) Fr.-A new record to India. *Curr. Sci.* **50**: 460-61.

Saini, S. S. and Atri, N. S. 1982. North Indian Agericales-1 *Indian Phytopath* **35**(2): 265-272.

Saini, S. S. and Atri, N. S. 1984. Studies on North-West Himalayan Russulaceae. Geobios New Reports **3**: 4-6.

Saini, S. S. and Atri, N. S. 1989. North Indian Agaricales- IX Section Ingratae Quel of *Russula* Pers. *Kavak* **17**(1,2): 21-27.

Saini, S. S. and Atri, N. S. 1989a. Some noteworthy taxa of *Agaricus* from North India. National Symposium on Plant Systematics. Advances and Trends. 10-12 March. Punjab University, Chandhgarh. P. 9 (Abstr.).

Saini, S. S. and Atri, N. S. 1990. Two noteworthy texa of *Lactarius* Pers. From India. *J. Indian Bot. Soc.* **69**: 475-476.

Saini, S.S. and Atri, N. S. 1993. Studies on genus *Lactarius* from India. *Indian Phytopath.* **46**(4): 360-364.

Saini, S.S. and Atri, N. S. 1995. Mushroom Flora of Punjab. In: Chadha K.L. and S.R. Sharma (eds) *Advances in Horticulture*, vol 13, Mushroom, Malhotra Publish. House, New Delhi PP. 375-386.

Saini, S. S., Atri, N. S. and Bhupal, M. S. 1988. North Indian Agaricales V. *Indian Phytopath.* **41**: 622-25.

Saini, S.S., N.S. Atri, Anjula and M.K. Saini 1993. 16th All India Botanical Conference, Decembe 3-5. Abstract vol. 72 No. 1136 R.D., University, Jabalpur.

Saini, S. S., N. S. Atri and A. K. Gupta 1991. Additional studies on North West Indian Agaries. *Indian Mushrooms* PP. 7-12.

Saini, S.S., N. S. Atri and A. K. Gupta 1992. Geobios New Reports **11**: 109-112.

Saini, S. S., Atri, N.S. and Gupta, A. K. 1993. Systematic Studies on the genus *Agaricus* L; Fr. Proc. 80th Indian Sci. Cong. Part–III, Goa pp. 26 (abstr.).

Saini, S. S., N. S. Atri and A. K. Gupta 1997. Studies on the genus *Agaricus L. Fr.*–the subgenus *Agaricus* section Sanguinolenti Schaeff. et Moller from North West India. *Mushroom Res.* **6**(2): 53-58.

Saini, S. S., N. S. Atri and Kanwarjit Singh 1983. *Volvariella hypopithys*- a new record for India. *Indian Phytopath.* **36**: 180-182.

Saini, S.S., Atri, N.S. and M.K. Saini 1989 *J. Indian Bot. Soc.* **68**: 205-508.

Sexena, A.S., Mukerji, K.G. and Agarwala, M.K. 1969. Spread of fungal spores causing allergic diseases. Aspects of Allergy. *Appl. Immun.* **2**: 175-180.

Saksena, S.B. and K.M. Vyas 1962-64. The wood decaying fungi of Saugar, Madhya Bharat. *J. Univ. Saugar* **11-31B**: 15-28.

Saksena, R.K. and Mehrotra, B.S. 1952. Fungus flora of an Allahabad soil. *Proc. Nat. Acad. Sci. India* **22**B: 22-43.

Sandhu, G.S. 1995. Management of mushroom insect pests. In: Advances in Horticulture, Vol.13, Mushroom (eds. K.L. Chadha and S. R. Sharma) MPH, New Delhi. pp 239-60.

Sarwal, B.M. 1984. Taxonomic studies on Indian Agarics–II, *Indian Phylopath.*, **37**: 228–233.

Sasek, V. and Musilek, V. 1967. Cultivation and antibiotic activity of mycorrhizal basidomycetes. *Folia Microbiol.* (Prague) **12**: 515–523.

Sastry, C.A. 1986. Color removal from pulp and paper mill wastes. *Indian Jour. of Environ. Protect.* **6**: 105-113.

Sathe, A.V. and J. Daniel 1980. Agaricales (Mushrooms) of Kerala State. MACS Monograph **1**: 75-108.

Sathe, A.V. and K. C. Sasangan 1977. Agaricaies from South-West India-Ill. *Biovigyanam* **3**: 119-121.

Sathe, A. V. and K. C. Sarangan 1978 a new species of *Lepista* from South West India. *Curr. Sci.* **47**: 739-740.

Sathe, A. V. and Deshpande, S. 1979 a. *Chlorolepiota* a new genus of Agaricales from India, *Curr. Sci.* **48** (15): 693-695.

Sathe, A. V., Deshpande, S. 1979 b. Agaricales of Maharashtra. in Advances in Maycol. and Pl. Pathol. Chattopadhyay, S.B. and Samajpati, N. Eds. pp. 81–88.

Sathe, A.V. and S. Deshpande 1980. Agaricales (Mushrooms) of Maharashtra State. MACS Monograph No. **1**: 9-42.

Sathe, A.V. and S. Deshpande 1982. Agaricales in Maharashtra in"Advances in,Mycology and Plant Pathology", p. 81, Eds. N. Samajpati and SB, Chattopadhya (Sri Raju Primlani of Oxford and IBH Pubiishing Co.).

Sathe, A.V. and S.M. Kulkarni 1980, Agaricales (Mushrooms) of Karnataka State. MACS Monograph No. **1**: 43-73.

Sathe, A.V. and S.R. Rahalkar 1975. Agaricales from South-West India 1. *Biovigyanam* **I**: 75-78.

Sathe, A.V. and S.R. Rahalkar 1976. Agaricales from South West India-II. *Ind. J. Mushrooms.* **2**: 77-80.

Sathe, A. V. and Rahalkar, S. R. 1978. Agaricales from South-West India–I *Biovigyanam* **1**: 75-78.

Saxena, M.C. 1960. Some fleshy fungi of Raipur District. Proc. 47th Indian Sd. Congr. Part **III**: 322-323.

Saxena, M. and Kapoor, J.N. 1988. New records of Agarics from India. *Indian Phytopathology*. **41** (1): 30–33.

Saxena, S. and Rai, R.D. 1988. Storage of button mushrooms (*Agaricus bisporus*). The effect of temperature, perforation of packs and pre-treatment with potassium metabisulphite. *Mushroom J. Tropics*, **8**: 15-22.

Schillimgs, R. J. and Ruelius, H. W. 1968. Poricin, an acidic protein with antitumor activity from a basidioinycete. II. Crystallization, Composition and properties. *Arch. Biochem. Biophys.* **127**: 672–679.

Seth, P.K. 1971. Evaluation and selection of *Agaricus bisporus* strain's for commercial cultivation of mushroom. 2nd Int. Symp. Pl. Path. New Delhi: 135 (Abstr.).

Seth, P.K. 1980. Harvesting, packing, preservation and recipes. *Indian J. Mushroom*, **6**: (*1 and 2*): 64-67.

Seth, A. and Sharma N.K. 1986. Five new species of genus *Aphelenchoides* (Nematoda Aphlenchoida) infesting mushroom in Northern India. *Indian J. Nematol.* **16**: 205-15.

Sethi, V. and Anand, J.C. 1978. Processing of mushrooms. *Indian Mushroom Sci.*, **1**: 233-38.

Sethi, V. and Anand, J.C. 1982. Enjoy mushroom products. *Indian Farming*, **32** (6): 1-7.

Sethi, V. and Anand, J.C. 1983. Processing of mushrooms. Symposium on Science and Cultivation Technology of Edible Fungi. 9-11, September. R.R.L., Srinagar (J and K). pp. 56. (abstr.).

Sethi, V., Bahi, N. and Bhagwan, J. 1989. Steeping preservation of mushrooms. Symp. Impact of Pollution in and from Food Industry and its Management. CFTRI, Mysore. p. 50 (abstr.).

Sethi, V., Bhagwan, J., Bahl, N. and Lal, S. 1991. Low cost technology for preserving mushrooms (*Agaricus bisporus*) *Indian Food Packer* **45**(6): 22-26.

Shad, O.S. and Lakhanpal, T.N. 1991. Sociobiology of morels in the N.W. Himalayas. In: Indian Mushrooms (MC. Nair, Ed. in Chief), Kerala Agricultural University, Vellanikara, Thiruvananthapurum Pp. 1-3.

Shajahan, M. D., Roy Choudhary, N., Saha A. K. and Samajpati, N. 1988. Mushroom flora of Khasi Hills (Meghlalaya) India. *Indian J. Mycol. Res.* **26** (2): 75-85.

Shankar, M. 1994. Agaricaceae and Bolbitiaceae from India. Indian Phytopathology. **47 (1):** 113.

Shankar, N. and Ram Dayal 1983. Fresh water Chytrids from Varansai two new records. *Indian Phytopath.* **36**: 376–377.

Sharda, R.M. 1983. Studies on the Clavarioid fungi of Eastern Himalayas and adjoining hills. Ph. D. Thesis, Punjab university, Chandigarh (India).

Sharda, R. M., Kaushal, S. C. and Negi, G. S. 1997. Edible fungi of Garhwal Himalayas, *Mushroom Res.* **6**: 11-14.

Sharma, A.D. and C.L. Jandaik 1978. The genus *Ramaria* Holmsk. ex S.F.Gray in Himachal Pradesh. *Ind. J. Mushroom* **4** (1): 5-7.

Sharma, A.D. and Munjal, R.L. 1977. Some fleshy fungi from Himachal Pradesh. *Ind. J. Musshroom*. **3**: 18-21.

Sharma, A.D., Jandaik, C.L. and Munjal, R.L. 1977. Some fleshy fungi from Himachal Pradesh, *Ind. J. Mushroom*. **3**: 12-15.

Sharma. A.D., C.L. Jandaik, R.L. Munjal and P.K. Seth. 1978. Some fleshy fungi from Himachal Pradesh-L *Ind. J. Mushrooms*. **4**: 1-4.

Sharma, A.D., R.L. Munzal and P.K. Seth 1978. Some fleshy fungi from Himachal Pradesh-III. *Ind. J. Mushroom*. **4**(2): 27-29.

Sharma, A.D. and Vineeta Thakur 1978. Some fleshy fungi from Himachal Pradesh-III. Ind, J. Mushrooms **4**: 27–29.

Sharma, R., Sharma, J. R. and Lakhanpal, T. N. 1986. Fleshy fungi of N. W. Himalayas–IV. *Austroboletus*–A new genus record from India. *Kavaka*. **14**: 41-43.

Sharma, J.R. and Lakhanpal, T.N. 1981. Mycorrhiza forming species of Boletaceae. In. Proc. Symposium on Improvement of Forest Biomass(P.K. Khosla, Ed.). Indian Society of Tree Scientists, Solan, H.P. pp. 455-457.

Sharma, N.K. 1995. Management of mushroom nematodes. In Advances in Horticulture **13**: 261-75 (KL Chadha and SR Sharma Eds.), Malhotra Publishing House, New Delhi, India.

Sharma, N. K. and Khanna, A. S. 1992. Nematode pests of mushroom. pp. 267-74. In Netnatode pests of crops (DS Bhatti and R.K. Walia Eds.) CBS Publishers and Distributors, Delhi pp 381.

Sharma, N. K. Thapa, C. D. Nath, A. 1981. Pathogenicity and identity of myceliophagous nematodes infesting *Agaricus bisporus* (Lang) Sing. in India. *Indian J. Nematol*. **11**: 230–231.

Sharma, N.K., Thapa, C.D. and Kaur, D.J. 1985. Pathogenicity and identity of *Ditylenchus myceliophagous* in India. *Indian J. Nematol*. **15**: 233-34.

Sharma, R.C., Jandaik, C.L. and Bhardwaj, S.R. 1991 a. Effect of blanching media on weight loss and appearance of button mushroom. *Advances in Mushrooms Sci*., pp. 23 (abstr.).

Sharma, R.C., Jandaik, C.L. and Bhardwaj, S.R. 1991 b. Enrichment of cookies/biscuits with mushroom prepared from unmarketable mushroom portions. *Advances in Mushroom Sci*., pp. 102 (abstr.).

Sharma, R. K. and B.L. Dhar 2009. Mushroom cultivation–a highly remunerative crop for Indian Farmers. Indian Farming November 2009 Fifth International Conference of Indian Phytopathol. Soc. pp. 31–36.

Sharma, S.R. and Vijay, B. 1996. Prevalence and interaction of competitor and parasitic moulds in *Agaricus bisporus*. Mush. Res. **5**: 13-18.

Sharma, S. S., A. Doshi and A. Trivedi 1992. Edible fleshy fungi fom Rajasthan. *Mush. Res*. **1**: 141.

Sharma, V.P. 1994. Potential of *Pleurotus sajor-caju* for biocontrol of *Aphekhchoides composticola* in *Agaricus bisporus* cultivation. *Mushroom Res*. **3**: 15-20.

Sharma, Y.K. and A. Doshi 1996. Some studies on an edible wild fungus *Phellorinia inquinans* Berk. in Rajasthan, India. Mush. Res. **5**: 51-54.

Shu-Ting Chang 1991. Cultivated Mashrooms. In: "Hand Book of Applied Mycology-Foods and Feeds." eds. DilipK. Arora, K. G. Mukerjee and Elmer, H. Marth. Vol. 3, pp. 221-240.

Singer, R. 1961. "Mushrooms and Truffles", 272 pp. London, Leonard Hill (Books) Limited. Interscience Publishers, INC. New York.

Singh, B., Mohankumar, B.L. and Jayaraman, K.S. 1984. Studies on dehydration of mushroom (*Agaricus bisporus*). In Proceedings of the Fourth Indian Convention of Food Scientists and Technologists. pp. 51.

Singh, A., Rai N., Basu, M. and Bihari Lal 2001. Some wood inhabiting fungi of Allahabad. In: Intenational Symposium on "Frontiers of fungal diversity and diseases in South East Asia "held in department of Botany DDU. Gorakhpur University, Gorakhpur, U.P.pp. 42.

Singh, Harminder 1977. Mushroom marketing–a problem. *Indian Journal of Mushroom* Vol. III: 1-4.

Singh, N.S. and Bano, Z. 1977. Standardization of mushroom pickle (*Pleurotus* spp.) in oil. Indian Food Packer, **31**(5): 18-19.

Singh, N. S. and Bano, Zakia 1977. *India Food Packer* **3**: 3-8.

Singh, N.S. and S. Rajarathnam 1977. *Pleurotus eous* (Berk.) Sacc. A new cultivated mushroom. *Curr. Sci.* **46** (17): 617-618.

Singh, P, U.K. Pandey, K. S. Suhag and Ajay Singh 2004. Factors constraining mushroom production in Haryana. *Mushroom Res.* **13**: 39–44.

Singh, R.D. 1994. Edible mushrooms of arid zone of Rajasthan. National Symposium on Mushrooms. Abstract: p. 8. April. 8-10, NR CM, Solan.

Singh, R.N. and Tewari, V.P. 1976. A new record of Agaricales from India. *Indian Phytopath.*, **29**: 68-70.

Singh, R.P., Dang, R. L. Bhatia, A.K. and Gupta, A.K. 1982. Waterbinding additives and canned mushroom yield. *Indian Food Packer*, **36** (1): 39-43.

Singh, S. M., R. N. Verma and K. S. Bilgrami 1992. *Lentinus conatus* Berk the Bamboo mushroom of Mizorum Bioved **3** (1): 117-118.

Singh, S. Mukta, Th. Chittaranjan Singh, N. Irabanta Singh and R. N. Verma 2001. *Pleurotus* species in Manipur and their artificial cultivation. *Mushroom Research* **10 (1):** 5–8.

Sinha, M.P. and B. Padhi 1978. The genus *Pleurotus* in Orissa. *Ind. J. Mycol. Res.* **16**(2) 294-300.

Sivaprakasham, K. and T.K. Kandaswamy 1980. Effect of cultivation methods on sporophore production of *Pleurotus sajor-caju*. *Ind. J. Mushrooms* **6**: 13–15.

Sivaprakasham, K., F.S. Rajan and R. Jeyarajan 1986. *Pleurotus citrinopilatus* a new Indian edible mushroom. *Curr. Sci.* **55**: 1203.

Smith, A. N. 1949. Mushroom in their natural habitats. Hafner Press, New York PP. 626.

Sohi, H.S., P.K. Seth and S. Kumar 1965a. Some interesting fleshy fungi from Himachal Pradesh-1. *J. Indian Bot. Soc.*, **44**: 69-74.

Sohi, H.S., P.K. Seth and S. Kumar 1965b. Grows of the Common European mushroom *Psalliota bispora* (Lange) Moller and J. Schaeffer in Himachal Pradesh. Indian J. Hort. **22**: 365-369,.

Sohi, H. S. and R. C. Upadhyay 1990. Natural occurrence of different *Auricularia* species in Himachal Pradesh. *Mush. J. Tropics* **10**: 47-51.

Sohi, H.S., S. Kumar and P.K. Seth 1964. Some interesting fleshy fungi from Himachal Pradesh. *Indian Phytopath.*, **17**: 317-322.

Srivastava, H.C. and Bano, Z. 1970. Nutrition requirements for *Pleurotus flabellatus. Applied Microbiology*. **19**(1): 166- 69.

Srivastava, R.K., Shukla, K.G., Tripathi, M.P., Joshi, M. and Verma, R.A. 1987. Influence of soaking and additives on canning quality of mushrooms (*Agaricus bisporus*). *Prog. Hot.*, **19**: 319-24.

Stamets, P. 1999 and 2000. (3rd edition). Myco-Medicinals, an informative booklet on medicinal mushrooms. Mycomedia, Olympia, Washington.

Starling, A.P. and Ross, I.S. 1991. *Mycol Res* **95**: 712- 714.

Subramanian, C.V. 1952. Fungi isolated and recorded from Indian soils. *J. Madras Univ.* **22**B: 206-222.

Subramanian, CV. 1973. Facetes of life and strategy of moulds and mushrooms in soil. *J. Indian Bot. Soc.* **52**: 17–28.

Sundararaman, S. and Marudara, Jan, D. 1925. Some Polyporaceae of the Madras Presidency. Madras Agric. Dept. Year Book 1924, 69-75.

Sunagawa, M. 1992. Research bulletins College Experiment Forests **49**(2): 219-259.

Sunagawa, M., and Mura, K. 1992. *Trans Mycol Soc Japan* **33**: 375-383.

Sunagawa, M., Mura, K., Ohmasa, M., Shino, Y., Nobuo, Y. and Toshinaga, I. 1992. *Mukuzai Gakkaishi* **38**(4): 386-392.

Sunagawa, M., Ohmasa, M., Shinso, Y., Nobuo, Y. and Toshinaga, I. 1989. *Mukuzai Gakkaishi* **35**(12): 1131-1138.

Sunagawa, M., Ohmasa, M., Shinso. Y., Nobuo, Y. and Toshinaga, I. 1991. *Mokuzai Gakkaishi* **37**(11): 1069- 1077.

Swaminathan, M.W. 1982. Indian Agariculture at the cross roads, *Current Science*. **51**: 13–24.

Sydow, H. and E.J. Butler, 1911. Fungi lndae Orientalis Part 111. *Ann. Mycol.* **9**: 372-421.

Tanga, A.Q. 1974. Weight loss in mushroom during canning. *Indian Mushroom Science*, **1**: 225-32.

Teng, S.C. 1996. In Fungi of China, Ko, R.P. ed. Mycotaxon Ltd., Ithaca, NY.

Termorshuizen, A. and Schaffers, A. 1991. The decline of carpophores of ectomycorrhizal fungi in stands of *Pinus sylvestris* L. in the Netherlands: Possible causes. *Nova Hedwigia* **53**: 267-289.

Tewari, R.P. 2005. Mushrooms, their role in nature and society. In: Frontiers in Mushroom Biotechnology(eds.) R.D. Rai, S.R. Sharma and R.P. Tewari, PP. 1-8, NR CM, Solan.

Tewari, R. P. and H. S. Sohi 1978. Studies on the effect of depth of casing soil on mushroom production in *Agaricus bitroquis* (Quel). *Indian Mushroom Sci.* **1**: 263-265.

Thakur, N. (Mrs. Chakraborty) 1980. Studies on the physiology of some tropical edible fungi- Ph.D. Thesis, Calcutta University.

Thapa, C.D., Kumar, S., Jandaik, C.C., and Seth, P.K. 1977. Some weed fungi occurring in mushroom (*Agaricus bisporus*) beds- I. *Ind. J. Mushroom.* **3** (2): 27-28.

Theissen, F. 1911. Fungi aliquot Bornbayensis, a Rev. Ed. Blatter Collecti *Annls. Mycol.* **9**: 153-159.

Thind, KS. 1961. "The Clavariaceae of India", 197 pp. I.C.A.R., New Delhi.

Thind, K.S. 1973. The Aphyllophorales in India. *Indian Phytopath.* **26**: 2–23.

Thind, K.S. and Chatrath, M.S. 1957. The Polyporaceae of the Mussoorie hills-II. *Res. Bull. Punjab Univ.* **125**: 431-442.

Thind, K.S. and Dhanda, R.S. 1979. The Polyparceae of India. XII. The genus *Albatrellus. Indian Phytopath.* **32**: 55-60.

Thind, K.S. and Khera, H.S. 1975. The Hydnaceae of North Western Himalayas-II. *Indian Photopath.* **28**: 57-65.

Thind, K.S. and Sukhdev 1957. The Clavariaceae of the Mussoorie Hills *J. Ind. Bot. Soc.* **36**: 92-103.

Thind, K.S. and Rattan, S.S. 1967. The Clavariaceae of India XI. Proc. Ind. Acad. Sci. **66**B: 143-156.

Thind, K.S. and Rattan, S.S. 1971. The Polyporaceae of India VIII. *Res. Bull. Punjab Univ.* (N.S.) **22**: 27-34.

Thind, K.S. and Raswan, G.S. 1958. The Clavariaceae of the Mussoorie hills X. *J. Ind. Bot. Soc.* **37**: 453-469.

Thind, K.S., Bindra, P.S., and Chatrath, M.S. 1957. The Polyporaceae of the Mussoorie Hills-III. *Res. Bull. Punjab Univ.* **129**: 471-483.

Thind, K.S., E.K. Cash and J.S. Sethi 1957. The Pezizaceae of Mussoorie Hills-V. *Mycologia* **49**: 831–836.

Thind, K.S. and G.P.S. Anand 1956. The Clavariaceae of Mussoorie Hills. V. *J. Indian Bot. Soc.* **35**: 92-102.

Thind, K.S. and I.P.S. Thind 1982. The Gasteromycetes of the Himalayas-I. *Kavaka* **10**: 35-45.

Thind, K.S. and J.S. Sethi 1957. The Peizaceae of Mussoorie Hills-III. Indian Phytopath. **10**: 26-37.

Thind, K.S. and K.S. Waraitch 1964. The Pezizales of India-VII. *J. Indian Bot. Soc.* **43**: 459–475.

Thind, K. S. and L.R. Batra 1957. The Pezizaceae of Mussoorie Hills-I. *J. Indian Bot. Soc.* **36**: 51-60.

Thind, K.S. and M.S. Chatrath 1960. The Polyporaceae of Mussoorie Hills. *J. Indian Phytopath.* **13**: 76-89.

Thind, K.S. and Sukh Dev 1956. The Clavariaceae of the Mussoorie Hills V. *J. Indian Bot. Soc.* **35**: 512-521.

Thind, K.S. and Sukh Dev 1957. The Clavariaceae of Mussoorie Hills. VIII. *J. Ind. Bot. Soc.* **36**: 475-485.

Thite, A. N. and Kulkarni, A. K. 1975. fungal flora of Panhala. *J. Bombay Nat. Hist. Soc.*, **63**: 456-463.

Thite, A. N., Patil, M. S. and More, T. N. 1976. Some fleshy fungi from Maharastra. *Botanique*. **7**: 77-78.

Thomas, K. M., Ramkrishnan, T. O., Narasimhan, T. T. 1943. Paddy Straw Mushroom. *Madras Agric. J.* **31**: 51.

Tillet, R. D. and Batchelor, B. G. 1991. An algorithm for locating mushrooms in a Grows bed. *Comput. Electr. Agric.* **6**: 191-200.

Toyomasu, T. 1991. Overseas Edible Mush **1**: 42-43.

Toyomasu, T., Takazawa, H. and Zennyonji, A. 1992. *Biosci Biotech Biochein* **56**: 359.

Trevors, J.T., Stratton, G.W. and Gadd, G.M. 1986. *Can J Microbiol* **32**: 447-464.

Trivedi, T.K. 1972. Agaricales of Nagpur. I. The Botanique (Nagpur). **3** (1): 53-59.

Tsukagoshi, S. 1984. Krestin (PSK). Cancer Treatment Reviews **11**: 131-155.

Upadhayay, R.C. and Amanjeet Kaur 2003. New additions to the Indian fleshy fungi from North Western Himalayas. *Mushroom Res.* **12**(1): 9-14.

Upadhyay, R.C. and Sohi, H.S. 1987. *Strobilurus stephanocystis* (Hora) Singer–a new record from India. *Curr. Sci.* **56** (7): 309–310.

Upadhyay, R. C., Amarjeet Kaur and A. Gulati 2005. Dark Spored Agarics from North Eastern Himalaya. *J. Mycol. Pl. Pathos.* **35** (1): 15–20.

Upadhyay, R.C., Amanject Kaur and K.C. Semwal 2007. New records of fleshy fungi from North Western Himalaya In, Mushroom Biology and Biotechnology Eds. R.D. Rai, S.K. Singh, M. C. Yadav and R.P. Tewari Mushroom Society of India.

Uppla, B.N., Patel, M.K. and Kamat, M.N. 1935. The fungi of Bombay VIII: 1-56.

Van de Rhee, M.D., Graca, P. and Mooibroek, H. 1994. Fifth International Mycological Congress Van couver Canada p. 228 (Abst.).

Vandendries, R. 1923. Recherches sur le determinisme sexuelledes Basidiomycetes Mem. Acad. R. Beig. *Cl. Sci.* **5**: 1–98.

Vandendries, R. 1924. Recherches experimentales sur le bipolarite sexuelle des Basidiomycetes. *Bull. Soc. R. Bot. Belg.* **57**: 75–78.

Venkatakrishnaiya, N.S. 1956. Anabe roga of coconut and areca nut caused by *Ganoderma lucidum* and its irradiation. *Mysore Agric. J.* **31**: 227-231.

Vasudeva, R.S. 1957. Report of the Division of Mycology and Plant Pathology. *Sci. Rept. Agric. Res. Inst.* New Delhi 1954-55, 87-101.

Vasudeva, R,S. 1960. "The Fungi of India" by E.J. Butler and G R. Bisby. Revised. 255 pp. I.C.A.R. New Delhi.

Vasudeva, R.S. 1962. "Fungi of India", Supplement I, 206 pp. I.C.A.R. New Delhi.

Verma, R.K. 1996. Some new records of fungi associated with salfi palm (*Caryota urens*). *Indian Phytopathology*. **49** (1): 22–25.

Verma, A. and Sarbhoy, A. K. 1996. Diversity **12**: 83-84.

Verma, R. N., G. B. Singh and S. M. Singh 1985 Some Gastromycetes from Manipur and Meghalaya, *Meghalaya Sci. Soc. J.*, **7** and **8**: 6-9.

Verma, R. N., G. B. Singh and K. S. Bilgrami 1987. Fleshy fungal flora of N. E. H. India -I, Manipur and Meghalaya. *Indian Mush.Sci* **2**: 414-421.

Verma, R.N., Singh, G.B. and Singh, Mukta 1995. Mushroom Flora of North East-Himalayas. In: Advances in Horticulture **13**: 229-349- Mushroom (K.L. Chadha and SR. Sharma,Eds.). Malhotra Publishing House, New Delhi.

Verma, R. N. and R.C. Upadhyay 2000. Mushroom genetic resources in India In: Characterization, conservation, evolution and utilization of Mushroom Genetic Resources for food and agriculture. Global Network on Mushrooms under the aegis of Labarere F. A. O. eds. Jacques E. and Umberato G. Menini. Pp 153–157.

Verma, R. N., S. M. Singh, G. B. Singh and K.S. Bilgrami 1989. *Gomphus floccocus*- a new record from India. *Curr. Sci.* **58**: 1370-1371.

Verma, R.N. and T.H. Gourbidhu Singh 1981. Investigation on edible fungi in the North Eastern Hills of India. *Mushroom Sci.* **11**: 89-99.

Vijaya, Joseph, A., T.K. Abraham, K.B. Vrinda and C.K. Pradeep 1995. New Agarics from South India. *Mushroom Res.* **4**: 1-6.

Vijayan, A. K. and Rehill, P.S. 1990. *Schizophyllum commune* Fr., First record on seeds of forest trees from India. *Indian Journal of Forestry*. **13** (1): 67–68.

Vilgalys, S. and Sun, B. L. 1994. *Proc Natl Acad Sci* USA **91**: 4599-4603.

Vilgalys, R., Moncalvo, J. M. Liou, S. R. and Volvosek, M. 1996. In Mushroom Biol Mush Products (DJ Royse ed) pp. 91-101, Penn State Univ PA, USA.

Vogel, F.C., Mc Garry, S.J., Kemper, L.A.K. and Graham, D.C. 1974. Bacteriocidal properties of a class of qumoid compounds related to sporulation in the mushroom Agaicus bisporus, *Am. J. Pathol*, **76**: 165–174.

Vogel, F.S., Kemper, L.A.K., Mc Garry, S.J. and Graham, D.G. 1975. Cytostatic, cytocidal and potential antitumor properties of a class of quinoid compounds, intiaater of the dormant state in the spores of *Agaricus bisporu*. Am. J. pathol. **78**: 33–48.

Vok Ton 2001. *"Hypomycos Lectifluorum*. the lobster mushroom" fungus of the month. http://botit.botanywise. edu/tomsfungiaug 2001.html. Retrieved on 2008–10–13.

Vrinda, K. B., Pradeep, C. K. SIBI Mathew and Abraham T. K. 1999a. Agaricales from Western Ghats–VI. Indian Phytopath. **52** (2): 198-200.

Vrinda, K. B., Pradeep, C. K. SIBI Mathew and Abraham T. K. 1999b. Agaricales from Western Ghats–VII. *Mushroom Res.* **8** (2): 9-12.

Vrinda, K. B. and C. K. Pradeep 2006. *Microcybe lobayensis* an edible mushroom from Western Ghats of Kerala. *Mushroom Research* **15** (2): 157-158.

Vrinda, K.B., C.K. Pradeep, Sibi Mathew and T.K. Abraham 1995. Agaricales from Western Ghats-VII. *Mushroom Res.* **8**(2): 9-12.

Vrinda, K.B., C.K. Pradeep, Sibi Mathew and T.K. Abraham 2000. Some pleurotoid agarics from Western Ghats, Kerala. J. *Mycopathol Res.* **38**(1): 45-47.

Vrinda, K. B., C. K. Pradeep and S. S. Kumar 2005. Occurrence of lesser known edible *Amanita* in the western Ghats of Kerala. *Mush. Res.* **14** (1): 5--8.

Vrinda, K.B., C.K. Pradeep and Abraham, T.K. 2000. A large bolete from the Western Ghats. *Mushroom Res.* **9**(2): 113-115.

Vrinda, K.B., C. K. Pradeep and T.K. Abraham 2001. Additions to Indian Mushroom flora. *Mushroom Research.* **10**(1): 1-4.

Vrinda, K.B., C.K. Pradeep and T.K. Abraham 2002. *Termitomyces umkowaani* (Cook and Mass.) Reid.–an edible mushroom from the Western Ghats. *Mushroom Res.* **11**(1): 7–8.

Vrinda, K.B., C.K. Pradeep and T.K.Abraham 2003. *Psathyrella velutina* (Pers. ex. Fr.) Sing–An edible mushroom from Western Ghats. *Mushroom Res.* **12**(1): 17-18.

Wakatsuki T, Michiko I and Imahara H 1988. J Ferment. *Technol* **66**(3): 257-266.

Wang, S.Y., Hsu, M.L., Hsu, H.C., Tzeng, S.S., Lee, S.S., Shiao, M.S. and HO, C.R. 1997. The antitumour effect of *Ganoderma Iucidum* is mediated by cytokines released from activated macrophages and T-lymphocytes. *International Journal of Cancer* **70**(6): 669-705.

Waraitch, K.S. 1976. The genus *Morchella* in India. *Kavaka* **4**: 69-76.

Wargel, R.J., W.L. Aim and J.F. Roland, Jr. 1974, A -and E- hydroxylation of keto acid by mushrooms. *Lipids* **9** (ll): 943-944.

Wassen, G.R 1968. "Divine Mushroom of Immortality". XIII pp. 318 Harcourt Brace and World Inc., New York.

Wassen, G. R. 1969. Divine Mushroom of Immortality XII Harcourt Brace and World Inc., New York, 318 PP.

Wassser, S.P. and Weis, A.L. 1999a. Medicinal properties of substances occurring in higher Basidiomycetes mushrooms: current perspectives (Review). *International Journal of Medicinal Mushrooms* **1**: 31-62.

Wasser, S.P. and Weis, A.L. 1999b. General description of the most important medicinal higher Basidiomycetes mushrooms. 1. International Journal of Medicinal Mushrooms **1**: 351-370.

Watling, R. and N.M. Gregory. 1980. Larger Fungi from Kashmir. *Nova Hedwigia* **XXXII**: 493–564.

Watling, R. and S. P. Abraham 1992. Ectomycorhizal Fungi of Kashmir Forest. *Mycologia*. **2**: 81–87.

Watt, G. 1980. Fungi in the dictionary of the economic products of India **1**: 130-133, 1889, **3:** 259, 455-458.

White, P.F. and Jarrett, P. 1990. Laboratory and field tests with *Bacillus thuringiensis* for the control of the mushroom sciarid *Lycoriella auripila*. Proc. ECPC–Pests and Diseases pp. 373-78.

Williams, J.G.K., A.R. Kubelik, K.J. Livok, J.A. Rafalski and S.V. Tingey 1990. DNA polymorphisms amplified by arbitrary primers are useful as genetic markers. Nucls. Acid Res.**18**: 6531-6535.

Worgan, J.T. 1968. Culture of the Higher Fungi. *Progress in Industrial Microbiology* **8**: 74-139. Ed. D.J.D. Blockenhull (J. and A Churchill Ltd., London).

Wright, E.M. and Chambers, R.J. 1994. The biology of the predatory mite *Hypoaspis miles* (Acari: Laelapidae), a potential biological control agent of *Bradysia paupera* (Dipt: Sciaridae). *Entomophaga* **39**: 225-35.

White, P. P. 1995. Biological control of mushroom pests: an evaluation. *Mush. Sci.* **XIV**: 475-84.

Xawek, V., Bhatt, T., Cajthami, T., Malachova, K. and Lednika, D. 2003. Compost mediated removal of polycyclic aromatic hydrocarbons from contaminated soil. *Arch. Environ. Contam. Texicol*. **44**: 336-342.

Yang, G.Y. 1993. A new biological response modifier. PSP. In Mushroom Biology and Mushroom Products. Chinese University Press, 247-259.

Yasui, A., Chuichi, T., Masonori, T. and Takeshi, M. 1988. *J. Jpn. Soc. Food Sci. Technol.* **35**(3): 160-165.

Ying, J. Z., Mao, X.L., Ma, Q.M., Zong, Y.C. and Wen, H.A. 1987. Icons of Medicinal Fungi from China (Trnsi. Xu, Y.H.), Science Press, Beijing.

Yoshika, P., Ikekawa, T., Moda, M. and Fuknoka, I. 1972. Studies on antitumour activity of some factions from basidiomycetes. I. an antitumour acidic polysaccharide faction from *P. ostrpeatus* (Fr.) Quel. *Chem. Pharm. Bull.* **20**: 1175–1180.

Zang, J.M., Zheng, Y.P. and Chen, M.Y., 1992. *Acta Botanica Yunnanica* **14** (3): 283-288.

Zas, Lulu 2003. Cooking God's own food–the South Indian way. In: Current Vistas in Mushroom Biology and Production. R.C. Upadhyay, S.K.Singh and R.D. Rai (Eds.). Mushroom Society of India, Solan. pp 253-256.

Zervakis, G. and Labarere, J. 1992. *J Gen Microbiol* **138**: 635-645.

Zhang, B.C. and D.W. Minter 1988. *Tuberhimalayensi* sp nov. with notes on Himalayan truffles. *Trans. Br. Mycol. Soc.*, **91**: 593–597.

Zhang, C.K., Chen, Y.L., Qin, H.M. and Liang, Y. 1996. In Mushroom Biol. Mush Products (DJ Royse ed.) pp. 103-111, Penn State Univ. USA.

Zhang, C.K., Liang Y. and Qin, H.M. 1994. *J Microbiol* **5**: 9-14.

Zhang, H., Gong, F., Feng, Y. and Zhang, C. 1999. Flammulin purified from the fruit bodies of *Flammulina velutipes* (Curt.: Fr.) P. Karst. *International Journal of Medicinal Mushrooms* **1**: 89-92,.

Zhuang C. and Mizuno, T. 1999. Biological responses from *Grifola frondosa* (Dick.: Fr. S.F. Gray Maitake Aphylophoromycetideae). *International Journal of Medicinal Mushrooms* **1**: 317-324.

Zhuk, Yu T., I.E., Tsapalova and E.N., Stepanova 1981. Lipids of some edible mushroom grows in Siberia. U.S.S.R. *Rastit Resur* **17** (1): 109-114.

Index

I

J

K

L

N